Lecture Notes in Mathematics 1588

Editors:
A. Dold, Heidelberg
B. Eckmann, Zürich
F. Takens, Groningen

Claus Scheiderer

Real and Étale Cohomology

Springer-Verlag
Berlin Heidelberg New York
London Paris Tokyo
Hong Kong Barcelona
Budapest

Author

Claus Scheiderer
Fakultät für Mathematik
Universität Regensburg
D-93040 Regensburg, Germany

E-Mail: claus.scheiderer@mathematik.uni-regensburg.de

Mathematics Subject Classification (1991): 14F20, 14P, 18F10, 12G05, 14F25, 18B25, 20J

ISBN 3-540-58436-6 Springer-Verlag Berlin Heidelberg New York

CIP-Data applied for

© Springer-Verlag Berlin Heidelberg 1994
Printed in Germany

Typesetting: Camera ready by author
SPIN: 10130132   46/3140-543210 - Printed on acid-free paper

Contents

Introduction

This work wants to reveal some of the intimate connections that exist between the étale site of a scheme X and the orderings of the residue fields of X. The emphasis is laid on cohomological aspects. It is well known that the existence of an ordering on some residue field influences directly the qualitative behavior of étale cohomology. For a basic example take X to be the Zariski spectrum of a number field k. If k is totally imaginary then $H_{et}^n(X, A) = 0$ for $n > 2$ and any torsion coefficients A. But if k is real then $H_{et}^n(X, \mathbb{Z}/2) \neq 0$ for any $n \geq 0$.

More generally, it is often felt that étale cohomology with 2-torsion coefficients has "bad" properties if there exists an ordering on some residue field of X; and so this case has frequently to be excluded or needs special consideration. The existence of an ordering implies $H_{et}^n(X, \mathbb{Z}/2) \neq 0$ for all n, no matter how well-behaved X is otherwise, and no matter what the cohomological 2-dimension $\mathrm{cd}_2(X'_{et})$ of $X' = X[\sqrt{-1}]$ is. The infinity of $\mathrm{cd}_2(X_{et})$ is in some sense accidental, and a more sensible cohomological dimension is exhibited only after killing all real phenomena by adjoining a square root of -1, i.e. passing to X'. But this being said, what then is the significance of the groups $H_{et}^n(X, A)$ for $n \gg 0$ and A 2-torsion?

I propose that there is a very satisfactory answer to this question. The orderings of all residue fields of X form a topological space X_r, the real spectrum of X. It turns out that étale cohomology of X (with 2-primary coefficients) is in high degrees just cohomology of the real spectrum X_r! This is not a very precise formulation, but for the moment it gives the right idea about one of the main results of this treatise.

Before I summarize the contents of this book more systematically, I would like to sketch what was the "point of departure" for this work, and then to highlight some of its main ideas.

I was drawn to the questions treated here when I studied the paper *Real components of algebraic varieties and étale cohomology* by Colliot-Thélène and Parimala [CTP]. Let X be an algebraic variety over the real numbers $\mathbb{R}$, and denote by $\mathcal{H}^n$ the Zariski sheaf on X associated with the presheaf $U \mapsto H_{et}^n(U, \mathbb{Z}/2)$. The main theorem of [CTP] says that if X is smooth, there are canonical isomorphisms $H^0(X, \mathcal{H}^n) \cong H^0(X(\mathbb{R}), \mathbb{Z}/2)$ for $n > \dim X$. More generally, the result is proved over arbitrary real closed fields, where classical topology is replaced by semi-algebraic topology. An essential ingredient of the proof is quadratic form theory, and in particular, Mahé's theorem on the separation of real connected components by such forms.

When I was discussing some simplifications and generalizations with Colliot-Thélène, he pointed out to me Cox's paper on the étale homotopy type of $\mathbb{R}$-varieties [Co]. From Cox's results one could easily deduce the main theorem of [CTP] for arbitrary, not necessarily smooth $\mathbb{R}$-varieties. But what was needed from [Co] is obtained there through the complicated machinery of étale homotopy theory; moreover Cox's construction uses transcendental methods. Therefore it became desirable to find a different approach. Before proceeding further I want to recall Cox's theorem. Based on an idea of M. Artin, it says:

Theorem [Co]. — *Let X be a scheme of finite type over $\mathbb{R}$. Let $G = \mathrm{Gal}(\mathbb{C}/\mathbb{R}) = \mathbb{Z}/2$ act on $X(\mathbb{C})$ by conjugation. Then there is a weak homotopy equivalence*

$$\{X_{et}\}^{\wedge} \;\simeq\; \left(X(\mathbb{C})_G\right)^{\wedge}$$

of the pro-finite completions.

Here $\{X_{et}\}$ denotes the étale homotopy type of X in the sense of Artin-Mazur [AM], and $X(\mathbb{C})_G = EG \times_G X(\mathbb{C})$ is the total space of the fibre bundle over $BG = EG/G$ associated with $X(\mathbb{C})$, where EG is a free contractible G-space.

This can be seen as a real analogue of the comparison theorem of [AM] for the étale homotopy type of complex algebraic varieties. It implies in particular that étale cohomology of X with finite constant coefficients M can be calculated as G-equivariant cohomology on $X(\mathbb{C})$. In this way one derives a long exact sequence [Co, Prop. 1.2]

$$\cdots H^n\big(X(\mathbb{C})/G,\, X(\mathbb{R});\, M\big) \;\longrightarrow\; H_{et}^n(X, M) \;\longrightarrow\; H_G^n(X(\mathbb{R}), M) \cdots \quad (1)$$

to which I will refer as to the "Cox sequence". Here $H_G^n(X(\mathbb{R}), M) = H^n(X(\mathbb{R}) \times BG, M)$, and both this group and the first group in (1) are singular cohomology groups. This sequence shows, in particular, that there are isomorphisms for $n > 2d$ (with $d := \dim X$)

$$H_{et}^n(X, \mathbb{Z}/2) \;\xrightarrow{\sim}\; \bigoplus_{i=0}^{d} H^i(X(\mathbb{R}), \mathbb{Z}/2) = H^*(X(\mathbb{R}), \mathbb{Z}/2). \quad (2)$$

The present work grew out of an attempt to understand (1) and its consequences in a more elementary way. Besides I wanted to see whether (1) was special to $\mathbb{R}$-varieties, or rather whether it could be generalized to other situations. It was clear that in the hoped-for generalization the real spectrum of X would have to replace the space $X(\mathbb{R})$. It was less obvious, however, what could play the role of the quotient space $X(\mathbb{C})/G$. Also it wasn't clear initially how to give a purely algebraic construction of the homomorphisms $H_{et}^n(X, M) \to H_G^n(X_r, M)$; Cox's construction for $X/\mathbb{R}$ uses transcendental methods. The idea that (1) or (2) could possibly be generalized was supported by a theorem of Arason ([Ar], completed in [AEJ]), which says: If k is any field, char $k \neq 2$, then there is a natural isomorphism

$$\lim_{n \to \infty} H^n_{et}(k, \mathbb{Z}/2) \xrightarrow{\sim} H^0(\text{sper } k, \mathbb{Z}/2) = H^*(\text{sper } k, \mathbb{Z}/2), \tag{3}$$

where sper k is the space of all orderings of k (the real spectrum of k) and the transition maps on the left are cup-product with the class of -1 in $k^*/k^{*2} \cong H^1_{et}(k, \mathbb{Z}/2)$. So one meets the same phenomenon here as for $\mathbb{R}$-varieties, namely that in high degrees, étale cohomology stabilizes against cohomology of the real spectrum.

The key to the desired generalizations, and more generally to a better understanding of the relationship between étale site and real spectrum, is to see the situation as an equivariant one. This is best explained by way of analogy with a space with operators. If T is a topological G-space (G being the group of order two, say), there is a well-developed theory which relates equivariant cohomology of G-sheaves on T to cohomology of both the space of fixpoints T^G and the quotient space T/G. For example, the sequence (1) is just an application of this theory to the G-space $X(\mathbb{C})$.

Let now X be a (general) scheme and consider the G-action on $X' := X \otimes_{\mathbb{Z}} \mathbb{Z}[\sqrt{-1}]$ over X. If X has no points of characteristic 2, it follows by descent that étale sheaves on X are the same thing as G-equivariant étale sheaves on X'; and so étale cohomology of X can be identified with G-equivariant étale cohomology of X'. Now the real spectrum X_r of X (or rather, the topos $\widetilde{X}_r$ of sheaves on X_r) must be thought of as the "fixobject" (or "fixtopos") of the G-action on X'_{et}, in a similar way as $X(\mathbb{R}) = X(\mathbb{C})^G$ in the case of an algebraic $\mathbb{R}$-variety! Indeed, there is a natural topos morphism ν from $\widetilde{X}_r$ to $\widetilde{X}'_{et}$; although ν is by no ways an embedding, it plays in a precise sense the role of the inclusion of the fixpoints in a G-space. By means of ν one can pull back any étale sheaf A on X to a G-sheaf $\nu(G)^*A$ on the real spectrum X_r. This purely algebraic construction yields in particular homomorphisms in cohomology which generalize the right arrows in (1); and it works for arbitrary schemes and sheaves of coefficients.

To get something which corresponds to the quotient space one has to form a new Grothendieck topology, namely the intersection of the étale and the real étale topology of X. The study of this site, which is denoted by X_b, occupies a considerable part of this book. The topos of sheaves on X_b contains both $\widetilde{X}_{et}$ and $\widetilde{X}_r$ as full subcategories, in such a way that $\widetilde{X}_{et}$ is an open subtopos and $\widetilde{X}_r$ is its closed complement. In this way one arrives at a long exact sequence (Theorem (6.6)) which exists for arbitrary schemes over $\mathbb{Z}[\frac{1}{2}]$ and arbitrary sheaves of coefficients, and which generalizes the Cox sequence.

The obvious question becomes then, what can one say about the cohomological properties of X_b, in particular about its cohomological dimensions? The answer to this question includes also a comparison of the cohomology of X_r and of X'_{et}. Here it is shown that if $\frac{1}{2} \in \mathcal{O}(X)$, the functor ν_* (on abelian sheaves) is always exact; and so all cohomological dimensions of X_r are bounded above by those of X'_{et}. The main result of this paper (Section 7) says (in a simplified form) that X_b has the same cohomological dimension for 2-primary torsion sheaves as X'_{et}, or possible one higher. (There is a similar result for odd torsion sheaves, but this is easy to prove.) All one needs is that the scheme X is quasi-compact and quasi-separated, and that 2 is invertible on X. As a corollary one gets that the homomorphisms

$$H^n(X_{et}, A) \longrightarrow H^n_G(X_r, \nu(G)^* A) \tag{4}$$

are isomorphisms for $n > \text{cd}_2(X'_{et})$ and any 2-primary sheaf A on X_{et}. In this sense, high-dimensional étale cohomology of X is cohomology of the real spectrum X_r. Actually the theorem makes a non-trivial statement also in the case $\text{cd}_2(X'_{et}) = \infty$, since for A annihilated by 2 it asserts that (4) is an isomorphism "in the limit" $n \to \infty$. For example this shows that the localization of the cohomology ring $H^*(X_{et}, \mathbb{Z}/2) = H^*(X_{et}, \mu_2)$ with respect to the class $(-1) \in H^1(X_{et}, \mu_2)$ is isomorphic to the full cohomology ring $H^*(X_r, \mathbb{Z}/2)$ of the real spectrum!

A different line of investigation is taken up in the second part of the paper. The aim is to reach a better understanding of why the real spectrum X_r behaves so much like a fixobject of the G-action on the étale site of X'. For this one needs the notion of G-toposes, which are toposes with a (pseudo) G-action. The analogue of the space of fixpoints for a G-topos is its inverse limit topos (or "fixtopos"). It is characterized by a 2-categorical universal property. In Section 10 this inverse limit of a G-topos is explicitly constructed (G may be any finite group here), and in particular, its existence is shown. Then it is proved for an arbitrary scheme X with $\frac{1}{2} \in \mathcal{O}(X)$ that the real topos $\widetilde{X}_r$ is the inverse limit of the G-topos $\widetilde{X}'_{et}$.

It becomes therefore interesting to take other examples of G-toposes, try to determine their inverse limit toposes and see whether one can prove similar theorems about cohomology. Apart from the basic example of a topological G-space there is one major example studied here, namely group extensions: Any such extension

$$1 \longrightarrow \Delta \longrightarrow \Gamma \longrightarrow G \longrightarrow 1 \tag{5}$$

makes the category (Δ-sets) into a G-topos. In the case where G is finite and the groups are discrete or profinite, the inverse limit of this G-topos is determined in Section 11.2.

The motivation for the study of this case, and for its inclusion into this paper, is as follows. The proof of the main theorem was reduced in Section 7 to the case of fields; and in Section 9 it was shown that for fields this main result can be deduced from Arason's theorem (3) mentioned above. Now the proof of the latter uses very specific (cohomological) properties of absolute Galois groups of fields. On the other hand, Arason's theorem fits exactly a general pattern which is given by theorems of K.S. Brown for *discrete* groups: If Γ is any discrete group of finite virtual cohomological dimension d, Brown has shown that cohomology of Γ in degrees $n > d$ is Γ-equivariant cohomology of a certain simplicial complex formed by finite subgroups of Γ. It is quite straightforward how a conjectural analogue of Brown's theorem for profinite groups should look like. In the case of the absolute Galois group of a field, this conjectural theorem would become just the main theorem of Section 9! This makes one wonder, of course, whether there is indeed such a profinite analogue of Brown's result. In fact, such a theorem exists. In Section 12, it is proved only under a special hypothesis, namely in the "rank one" case. Nevertheless, this includes absolute Galois groups of fields. In this way an independent and completely different approach to Arason's theorem is given, which does not use any of the special properties of Galois groups. The general profinite version of Brown's theorem will be proved in a forthcoming paper [Sch2].

Now the interesting point is this. Let the group G in (5) be cyclic of *prime* order p, and let $d := \text{cd}_p(\Delta)$ be finite. The groups with which the cohomology groups $H^n(\Gamma)$ for $n > d$ are identified (by Brown's theorem in the discrete case, by Section 12 in the profinite case) are nothing but the G-equivariant cohomology groups of the fixtopos of the G-topos (Δ-sets). This follows from the identification of this fixtopos in 11.2. So from the perspective of G-toposes, the main results of Section 7 (for general schemes) and Brown's (discrete or profinite) theorem (in the case just considered) have *identical formulations*!

It is shown in Section 13 that also the (easier) case of a topological G-space fits the same pattern.

I now give a systematic overview of the contents of this paper. For some more specific information see also the introductory remarks to the respective sections.

In Section 1 it is proved that sheaves on the real spectrum X_r of a scheme are, up to equivalence, the same as sheaves on the real étale site X_{ret}. Although this theorem is well known, at least for affine schemes (proved by Coste-Roy and Coste), the proof given here is new and — as I think — more elementary than the proofs existing before. It uses neither methods from mathematical logic nor the concept of strict real localization. The result is basic for much of what follows, since it allows to switch freely between two quite different descriptions of the real topos of X, each of which has its specific advantages.

In Section 2 the b-topology of a scheme is defined as the intersection of the étale and the real étale topologies. It is shown that $\widetilde{X}_b$ (the category of sheaves on X_b) is the result of glueing $\widetilde{X}_{et}$ and $\widetilde{X}_{ret}$, and some consequences of this fact are obtained. Also the next section presents basic material: Limit theorems for sheaf cohomology, the description of the stalk functors, stalks of (higher) direct images and the like. Here as well as in later sections it happens frequently (but not always) that results for the b-topology are obtained by combining results for the étale and the real (étale) topologies. In most of the cases, the étale part is the harder one, of course.

The next topic is the study of the topos morphism ν from $\widetilde{X}_{ret}$ to $\widetilde{X}'_{et}$ in Section 5. Geometrically ν corresponds to the Weil restriction functor with respect to $X' \to X$. The main result is that ν_* is exact on abelian sheaves, provided that 2 is invertible on X. For this one has to study the real spectrum of the Weil restriction of a strictly henselian local ring, and to show that its sheaf cohomology vanishes. I could not decide whether this remains true when characteristic 2 points are present.

For lack of suitable references I have inserted before this a section in which all facts about Weil restrictions are proved which are used later on (Section 4).

In Section 6 the fundamental long exact sequence is established which relates cohomology of X_{et}, X_r and X_b, and which generalizes the Cox sequence (1). In Section 7 the proof of the main result on cohomology of X_b is taken up. Assuming that the theorem is true for fields, it is first extended to schemes of finite type over spec $\mathbb{Z}$, and then by limit and glueing arguments to general quasi-compact, quasi-separated schemes. The case of fields is treated in Section 9. For this one needs a third description of the real topos of a field k. The absolute Galois group Γ of k acts on the space T of all real closures of k (inside a fixed algebraic closure). If Γ' denotes the subgroup of Γ which fixes $\sqrt{-1}$ then sheaves on the real spectrum of k can be identified with Γ'-equivariant sheaves on T. Here a "continuous" notion of equivariant sheaves is required which takes care of the fact that Γ' carries a topology. Since I do not know any reference for this concept, I have again inserted a preparatory section (Section 8) in which some necessary foundations are laid. As already remarked, the proof of the main theorem is finally reduced to Arason's theorem (3). The proof of the latter is discussed at the end of Section 9, and the question is raised whether the specific cohomological properties of Galois groups which it uses are necessary for the theorem to hold.

In Section 12 it is shown that this is not the case. The approach there is completely different from Arason's. One considers a profinite group Γ which has an open subgroup Δ with $\mathrm{cd}_p(\Delta) < \infty$ (p may be any prime). Let $\mathfrak{T}$ be the (boolean) space of all subgroups of order p in Γ. Under the assumption that Γ contains no subgroup isomorphic to $\mathbb{Z}/p \times \mathbb{Z}/p$, it is proved that the natural homomorphisms

$$H^n(\Gamma, A) \longrightarrow H^n_\Gamma(\mathfrak{T}, A)$$

are isomorphisms for $n > d$ and for any p-primary Γ-module A. As already remarked, this theorem has a counterpart for discrete groups, which is due to K.S. Brown and whose proof is based on Farrell cohomology. If one tries to imitate Brown's proof one runs into difficulties in the profinite case. The proof given here proceeds differently, but is limited to the case where Γ contains no $\mathbb{Z}/p \times \mathbb{Z}/p$ (but see [Sch2] for a proof covering the general case, based on the same ideas). In any case it is essential that one uses *projective* resolutions of modules, instead of injective ones. This cannot be done with discrete Γ-modules alone (there aren't enough projectives in the profinite case), rather one has to consider also profinite Γ-modules. The necessary backgroung material is summarized at the beginning of Section 12.

Before Section 12, however, there are two sections on G-toposes. In Section 10 I give a detailed review of the basic concepts, and then construct the fixtopos of an arbitrary G-topos E in the case where the group G is finite. If F denotes this fixtopos and $\nu: F \to E$ is the corresponding topos morphism, then all composite topos morphisms $g \circ \nu: F \to E$ $(g \in G)$ are "coherently" isomorphic (but not in general equal); and ν is universal for this property. This fixtopos is determined in Section 11 in the two cases which are important for this paper. Namely, if X is a scheme on which 2 is invertible, the fixtopos of $\widetilde{X}'_{et}$ is identified as the real topos $\widetilde{X}_r$. Thus a precise meaning is given here to the feeling that the real spectrum behaves like sort of a fixobject of the G-action on X'_{et}. Second, if (5) is an extension of discrete or profinite groups, and if G is finite, the fixtopos of the G-topos (Δ-sets) is shown to be the category of Δ-equivariant sheaves on the space of splittings of (5). Note how these two examples agree on their common "intersection", namely the spectrum (resp. Galois group) of a field.

The material on G-toposes may appear quite technical and "dry". But there seems no doubt that this perspective is essential if one wants to grasp the right idea about the relations between the sites X'_{et}, X_{et}, X_r and X_b. Moreover, if one assumes this point of view, one is rewarded in several ways: By analogy with topological G-spaces the main results of the first part become very natural, if not to be expected. Besides I found it quite satisfactory to see a common principle at work in so different situations. To describe the basic idea, consider a group G of *prime* order p which acts on a (nice) topological space T. If $Z = T^G$ is the set of fixpoints then the restriction maps $H^n_G(T, A) \to H^n_G(Z, A|_Z)$ in equivariant cohomology are bijective for $n > \dim T$. Alternatively one can read this as a statement on the cohomological dimension of T/G, by a long exact sequence similar to Cox's sequence. Now replace T by a G-topos E and Z by its fixtopos F. The idea is that there should be a good chance for an analogous theorem to hold. Indeed, the results of this paper give three different examples for such a situation: In the case of the G-topos $\widetilde{X}'_{et}$ (X a scheme over $\mathbb{Z}[\frac{1}{2}]$) the above principle corresponds to the main theorem of Section 7. In the case of a group extension (5), with $|G| = p$ and $cd_p(\Delta) < \infty$, it corresponds to particular cases of Brown's theorem (for discrete groups) resp. of the main result of Section 12 (for profinite groups). Finally, if E is the topos of sheaves on a G-space T then F is the category of sheaves on $Z = T^G$, as shown in Section 13. For a summarizing discussion I refer to the end of Section 14.

Also the notion of quotient is considered for G-toposes. In general there does not seem to be an obvious topos-theoretic construction which corresponds to the quotient space T/G of a topological G-space T. In Section 14 I propose for G-

toposes E, in the case where G is of prime order, to form a glued topos by mimicking the construction of $\widetilde{X}_b$, and to regard this as a substitute for the topological quotient. That this yields the right thing in the situation of a topological G-space is proved in Section 13.

In the third part of this book I return to the study of the three topologies et, b, ret on a scheme X. Section 15 contains comparison results. For example it is shown that, for a (separated) scheme X of finite type over $\mathbb{R}$, the cohomology of X_b with torsion coefficients is the cohomology of the quotient space $X(\mathbb{C})/G$. A similar result holds over arbitrary real closed fields. This shows once more that X_b has to be considered as the "topological" (or "non-free") quotient of X'_{et} mod G. Also a very easy deduction of the Cox exact sequence (1) is given which does not need étale homotopy theory. However it uses the Comparison Theorem between classical and étale cohomology of complex varieties, which of course is also a non-trivial tool.

In Section 16 the proper and smooth base change theorems are proved for the real étale topology, and from this one gets corresponding theorems also for the b-topology (using the étale theorems, of course). Proper base change for real spectra has been proved before by H. Delfs; here a new proof is given. The smooth base change theorem seems to be new in the real setup. The reason why it hasn't been considered so far may be that smoothness doesn't make sense in the context of semi-algebraic spaces and maps (the structure sheaves are simply too large), nor in the more general framework of real closed spaces. I explain however how to weaken the smoothness hypothesis in such a way that one can prove a corresponding base change theorem for real closed spaces.

Section 17 contains finiteness theorems for the real and the b-topology, and for finitely presented proper morphisms. So these theorems are saying that the higher direct image functors of such morphisms preserve constructible abelian sheaves (and similarly for set-valued sheaves). As a corollary one gets a theorem on the smooth specialization of the cohomology of proper schemes. By making use of the étale finiteness theorem one can again reduce the study of the b-topology to the real topology. In the latter case the proof reduces to a semi-algebraic situation in which the finiteness theorem is proved without any properness hypothesis. The main technical tool is semi-algebraic triangulation.

The aim of Section 18 is to show, for a d-dimensional affine variety over a real closed field, that b-cohomology vanishes in degrees $> d$. Unfortunately it seems that there is no way of getting this result as a corollary to the étale case (which is well known) plus the real case (which is obvious). Rather one has to mimic the proof of the corresponding étale theorem. In particular this means to consider more generally a relative affine situation between varieties over a field.

Section 19 relates the three topologies on a scheme X to the Zariski topology. In particular it is proved that the direct image functor of the support map $X_r \to X$ is exact on abelian sheaves. This simple fact is quite useful, as is demonstrated by various applications. Among them are very far reaching generalizations of results of Colliot-Thélène and Parimala from [CTP].

The last section (Section 20) is devoted to some explicit computations and applications. First, smooth curves over a real closed field are considered and their étale and b-cohomology (with coefficients $\mu_n^{\otimes i}$) is determined. Also, classical theorems on real curves by Weichold, Witt and Geyer are reproved by means of results of this work. Then it is studied (on general schemes) what one gets from the results of Part One for some specific étale sheaves. Most interesting is here the case of the multiplicative sheaf $\mathbb{G}_m$; its n-th étale cohomology group, for $n \gg 0$, is the part of $H^*(X_r, \mathbb{Z}/2)$ of the same parity as n. This subsection contains also a variety of side remarks and other complements. Then fields are considered again: Some remarks are made on the b-cohomology of a field, and on the fundamental group of the b-topology. At the end of Section 20 a few historical remarks are made on the relations of this work to work of other authors.

There are two appendices. The first assembles results on spectral spaces which are used in several places throughout the paper and which I could not find in the literature. The second is an application of results of this paper to Artin-Schreier structures. The notion of an Artin-Schreier structure was invented by D. Haran and M. Jarden in the course of their study of absolute Galois groups of PRC fields, where it plays a key role [HJ]. For example, every field k gives rise to an Artin-Schreier structure $\mathfrak{A}(k)$, which consists essentially of the absolute Galois group $\mathrm{Gal}(k_s/k)$ together with its distinguished subgroup $\mathrm{Gal}(k_s/k(\sqrt{-1}))$ and its action on the space of real closures of k. In [Ha3] Haran proposes a cohomology theory for these structures. Among its features is that it yields reasonable (finite) cohomological 2-dimensions for real fields, and that it allows a cohomological characterization of real projective groups. (The class of these groups comprises exactly the class of absolute Galois groups of PRC fields [HJ].) For non-real fields this cohomology coincides with Galois cohomology. Haran's cohomology is defined by *ad hoc* methods, and the definition is somewhat mysterious, as the author remarks by himself. In Appendix B a natural explanation of this cohomology theory is given in terms of the b-topology. For example, in the case of the Artin-Schreier structure $\mathfrak{A}(k)$ of a field, Haran's cohomology coincides with sheaf cohomology on the site $(\mathrm{spec}\, k)_b$; and a similar characterization is true in general. By the results of Section 9 (in the field case) or Section 12 (in general) it is an easy corollary to determine the cohomological dimensions of arbitrary Artin-Schreier structures. A particular corollary is that a field k is real projective if and only if $k(\sqrt{-1})$ is projective. These latter results were proved independently by Haran, who used different methods [Ha4].

After this introduction the reader can find a summary of the general notations, definitions and conventions valid in this paper. This part is referred to by labels of the form $(0.xx)$. Otherwise the first entry of a label indicates the number of the section in which it can be found.

I should add the remark that, although this paper is devoted to cohomological studies, non-abelian cohomology has not been considered anywhere in it. This mainly in order to save space and time.

This book is a slightly revised version of my Habilitationsschrift with the same title, which was submitted to the Mathematische Fakultät of Universität Regensburg in November 1992. The idea to think about relations between étale cohomology and the real spectrum owes a lot to Louis Mahé, whom I want to thank warmly for stimulating questions and discussions through which he whetted my appetite. He always stressed the similarity between étale cohomology classes and quadratic forms. Although later this work turned into somewhat different directions, it was Louis who had shown me in his friendly way that here is something to think about. I am also grateful to Dan Haran for sending me his papers and preprints; in particular I profited from his exposition of profinite group modules in [Ha2].

Moreover, I am indebted to several people who critically read the whole manuscript or parts of it, for their constructive criticism. Above all, Jean-Louis Colliot-Thélène contributed a great number of detailed suggestions and comments which were incorporated into this revision and improved the presentation. I also received valuable advice from Claudio Casanova, Michel Coste, Manfred Knebusch, Louis Mahé, Manuel Ojanguren and Jean-Pierre Serre. It is a pleasure to thank all of them here.

Finally I would like to express my special gratitude to Manfred Knebusch, who has actively supported and encouraged me for many years now.

Leitfaden

The following diagram indicates the main relations between the sections of the book. An arrow from X to Y signalizes that results or techniques of Section X are used in Section Y. A dotted such arrow indicates a weaker sort of dependency.

The reader should start with Sections 1-3, and then proceed with Sections 5, 6, 7 and 9 in order, using Sections 4 resp. 8 as technical references for Sections 5 resp. 9. The lecture of most of Section 5 may be skipped at a first reading, if one is willing to accept in Section 7 that Theorem (5.9) is true.

After the study of Part One, the reader may proceed with either Part Two or Part Three (if he or she hasn't lost any interest at all), or may even pass directly to Sections 19 and 20, where applications of Part One can be found. Generally, Part Two emphasizes topos theoretic techniques; but Section 12 doesn't use such techniques, and can practically be read independently of the rest of the book, perhaps with a glance into Section 8 at some points. (See the Introduction for why this section is placed here.) Sections 15-18 are devoted to fundamental theorems for the real and the b-topology, and do neither depend seriously on Part Two nor on the main results from Section 7.

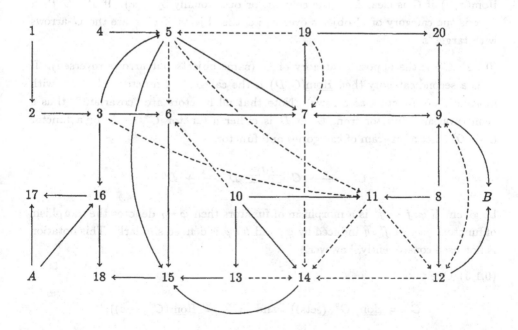

General notations and conventions

The following serves the purpose of fixing notations and conventions which are used throughout the paper. It also gives the precise sense in which some general concepts are used later on, either by explicitly recalling definitions or by giving references to the literature.

(0.1) Categories and functors

In general I have tried to maintain definitions and notations from [SGA4 I]. Throughout all set-theoretic questions are ignored. In particular, no attempt has been made to keep track of a hierarchy of universes, as is done in [SGA4].

(0.1.1) Let C be a category. Examples are

$$
\begin{array}{rcl}
\text{(sets)} & = & \text{the category of sets,} \\
\text{(Ab)} & = & \text{the category of abelian groups,} \\
\text{(Top)} & = & \text{the category of topological spaces,}
\end{array}
$$

the morphisms being the obvious ones. I often write "$x \in C$" to indicate that x is an object of C. The set of C-arrows from x to y is denoted $\mathrm{Hom}_C(x,y)$, or simply $\mathrm{Hom}(x,y)$ if C is clear from the context; or occasionally by $y(x)$. If $x \in C$ then C/x is the category of C-objects over x, i.e. the objects of C/x are the C-arrows with target x.

(0.1.2) C° is the opposite category of C (same objects but arrows reversed). If D is a second category then $\underline{\mathrm{Hom}}(C, D)$ is the category of functors $C \to D$ (with morphisms of functors as arrows). Note that all functors are "covariant"; thus a "contravariant" functor from C to D is either a functor $C^\circ \to D$ or a functor $C \to D^\circ$. Let a diagram of categories and functors

$$
C' \xrightarrow{\;g\;} C \xrightarrow[\;\;]{\;f,f'\;} D \xrightarrow{\;h\;} D'
$$

be given. If $\varphi \colon f \to f'$ is a morphism of functors then $\varphi * g$ denotes the morphism of functors $f \circ g \to f' \circ g$ induced by φ; and $h * \varphi$ is defined similarly. This notation is not used consequently, however.

(0.1.3) One puts

$$
\widehat{C} := \underline{\mathrm{Hom}}(C^\circ, \text{(sets)}) \quad \text{and} \quad C^\vee := \underline{\mathrm{Hom}}(C, \text{(sets)});
$$

the objects of $\widehat{C}$ are called *presheaves* (of sets) on C. The *Yoneda embedding* of C is the fully faithful functor

$$h\colon C \longrightarrow \widehat{C}, \quad x \longmapsto h_x := \mathrm{Hom}_C(-, x).$$

(0.1.4) A category I is *right filtering* if $I \neq \emptyset$ and
    (1) for $x, y \in I$ there exists a diagram in I

$$x \longrightarrow \bullet \longleftarrow y;$$

    (2) if $f, f'\colon x \rightrightarrows y$ are two arrows in I, there is an arrow $g\colon y \to \bullet$ in I such
        that $g \circ f = g \circ f'$.

I is *left filtering* if I° is right filtering. (Note that in [SGA4] the terms "filtrante" resp. "cofiltrante" are used for "right filtering" resp. "left filtering".)

(0.1.5) A *pro-object* in C is a functor $x\colon I \to C$, $i \mapsto x_i$ from some (small) left filtering category I to C. Symbolically x is often written as $\{x_i\}_{i \in I}$. Such x determines an object $\bar{x}$ of $C^{\vee}$ by setting

$$\bar{x}(z) := \varinjlim_I \mathrm{Hom}_C(x_i, z) \quad (z \in C).$$

The category $\mathrm{pro}(C)$ of pro-objects in C is defined in such a way that $x \mapsto \bar{x}$ becomes a fully faithful functor $\mathrm{pro}(C)^\circ \to C^{\vee}$. See [SGA4 I.8.10].

(0.1.6) If A is an abelian category then $\mathrm{D}^+(A)$ is the derived category of bounded below complexes in A. If A has sufficiently many injectives, if B is a second abelian category and $f\colon A \to B$ is an additive functor then $\mathrm{R}f\colon \mathrm{D}^+(A) \to \mathrm{D}^+(B)$ is the right derived functor of f, and $\mathrm{R}^n f\colon A \to B$ are its components ($n \geq 0$).

(0.1.7) If M is an abelian group (or more generally, an object of an abelian category) and n is an integer, the kernel (resp. the cokernel) of multiplication $M \xrightarrow{n} M$ by n will be denoted by $_nM$ (resp. by M/nM, or frequently just M/n).

(0.2) Sites, sheaves and toposes

The concepts of site and topos are always used in the sense of Grothendieck [SGA4].

(0.2.1) A *site* is a pair (C, τ) where C is a category and τ is a topology on C [SGA4 II.1]. Frequently only the notation C is used, when the topology is clear from the context. A sheaf on (C, τ) without further specification means a sheaf of sets. The category of sheaves on (C, τ), i.e. the *topos* associated with the site C, is written $\widetilde{C}$ or $\widetilde{C}_\tau$ or $(C, \tau)^{\sim}$. The initial object of $\widetilde{C}_\tau$ is $\emptyset = \emptyset_\tau$, the final object of $\widetilde{C}_\tau$ is $*$. If M is a set, the corresponding constant sheaf on C is again denoted by M, or sometimes by $\underline{M}$ for emphasis. The (abelian) category of sheaves of abelian groups on C is $\mathrm{Ab}(C)$ or $\mathrm{Ab}(C, \tau)$.

(0.2.2) $a_\tau\colon \widehat{C} \to \widetilde{C}_\tau$ denotes the functor "associated τ-sheaf". Thus $a_\tau = L_\tau \circ L_\tau$ where $L_\tau\colon \widehat{C} \to \widehat{C}$ is the functor of [SGA4 II.3]. The composite functor $C \xrightarrow{h} \widehat{C} \xrightarrow{a_\tau} \widetilde{C}_\tau$ is written ϵ_τ, so $\epsilon_\tau(x)$ is the τ-sheaf associated with the presheaf h_x represented by $x \in C$.

(0.2.3) A *morphism of sites* from C to D is a continuous functor from D to C, usually denoted by a symbol like f^{-1}, such that its prolongation $f^*\colon \widetilde{D} \to \widetilde{C}$ is left exact [SGA4 IV.4.9]. The induced topos morphism from $\widetilde{C}$ to $\widetilde{D}$ is written $f = (f^*, f_*)$. Occasionally the *presheaf* inverse image functor $f^\cdot$ induced by f^{-1} has to be considered; this is the functor $f^\cdot\colon \widehat{D} \to \widehat{C}$ which is left adjoint to the presheaf direct image functor $f_\cdot\colon \widehat{C} \to \widehat{D}$, $f_\cdot = (-) \circ f^{-1}$ [SGA4 I.5].

If E and E' are toposes then $\underline{\mathrm{Homtop}}(E', E)$ is the category of topos morphisms from E' to E. Its set of objects is the set $\mathrm{Homtop}(E', E)$ of topos morphisms $f = (f^*, f_*)$ from E' to E. Via the direct (resp. the inverse) image component, $\underline{\mathrm{Homtop}}(E', E)$ is a full subcategory of the category of all functors $\underline{\mathrm{Hom}}(E', E)$ (resp. of the opposite category $\underline{\mathrm{Hom}}(E, E')^\circ$). See [SGA4 IV.3.2].

(0.2.4) A *point* of a topos E is a topos morphism $p = (p^*, p_*)$ from (sets) to E. The category of points of E will be written $\underline{\mathrm{pt}}(E) := \underline{\mathrm{Homtop}}((\mathrm{sets}), E)$. Note that the functor $p \mapsto p^*$ from $\underline{\mathrm{pt}}(E)^\circ$ to $\underline{\mathrm{Hom}}(E, (\mathrm{sets}))$ is fully faithful and has as its image those left exact functors $E \to (\mathrm{sets})$ which preserve arbitrary direct limits. Such a functor is also called a *fibre functor* of E. See [SGA4 IV.6].

(0.2.5) If E is a topos then $H^0(E, -) := \Gamma(E, -) := \mathrm{Hom}_E(*, -)$ is the *global sections* functor $E \to (\mathrm{sets})$. Given $x \in E$ one also writes $H^0(x, -)$ or $\Gamma(x, -)$ for the functor $\mathrm{Hom}_E(x, -)\colon E \to (\mathrm{sets})$. The (abelian) category of abelian group objects in E is denoted by $\mathrm{Ab}(E)$. The left derived functors of $H^0(E, -)\colon \mathrm{Ab}(E) \to (\mathrm{Ab})$ (resp. of $H^0(x, -)\colon \mathrm{Ab}(E) \to (\mathrm{Ab})$) are written $H^n(E, -)$ (resp. $H^n(x, -)$), $n \geq 0$. If $E = \widetilde{C}$ is the topos of sheaves on a site C one of course writes also $H^n(C, -)$ instead of $H^n(\widetilde{C}, -)$.

(0.3) Rings and schemes

(0.3.1) Let A be a ring (this always means that A is commutative and has a unit). The Zariski spectrum of A is $\mathrm{spec}\, A$. The residue field of a prime ideal $\mathfrak{p} \in \mathrm{spec}\, A$ is $\kappa(\mathfrak{p}) = A_\mathfrak{p}/\mathfrak{p}A_\mathfrak{p}$. If A is a local ring then $\mathfrak{m}_A$ denotes the maximal ideal of A and $\kappa_A = A/\mathfrak{m}_A$ its residue field.

(0.3.2) Let X be a scheme. The structure sheaf of X is written $\mathcal{O}_X$ or simply $\mathcal{O}$. Its stalk at a point $x \in X$ is the local ring $\mathcal{O}_{X,x}$ with maximal ideal $\mathfrak{m}_{X,x}$ and residue field $\kappa(x)$. If X is a scheme over $\mathrm{spec}\, A$ for some ring A and B is an A-algebra then $X(B) := \mathrm{Hom}_{\mathrm{spec}\, A}(\mathrm{spec}\, B, X)$ denotes the set of B-valued points of X. By an algebraic variety (or just a variety) over a field k I mean a scheme of finite type over $\mathrm{spec}\, k$.

(0.3.3) (Sch) is the category of schemes. If X is a scheme, (Sch/X) denotes the category of all X-schemes. The full subcategory of (Sch/X) consisting of all étale X-schemes $U \to X$ is written Et/X. The surjective families form a pretopology on Et/X, and the associated topology is called the *étale topology* on X, abbreviated by "*et*". I use the notation $X_{et} := (\mathrm{Et}/X, et)$ for this site (the "small" étale site of X). Accordingly, $\tilde{X}_{et}$ is the étale topos of X, $\mathrm{Ab}(X_{et})$ is the category of abelian étale sheaves on X, etc. Recall the common notations for some frequently used sheaves: $\mathbb{G}_a$ (the structure sheaf), $\mathbb{G}_m$ (the sheaf of its multiplicative units), μ_n (the sheaf of n-th roots of unity), ...

(0.3.4) As usual, $\mathbb{Z}$, $\mathbb{Q}$, $\mathbb{R}$, $\mathbb{C}$ denote the rings of integers, rational numbers, real numbers, complex numbers, respectively. The ring of ℓ-adic integers (ℓ any prime) is written $\mathbb{Z}_\ell$; its quotient field is $\mathbb{Q}_\ell$.

(0.4) Real spectrum

(0.4.1) Here are some reminders on the real spectrum, cf. [CR] or [BCR]. Let A be a ring. The *real spectrum* of A, denoted sper A, is the topological space of all pairs $\xi = (\mathfrak{p}, \alpha)$, where $\mathfrak{p} \in \mathrm{spec}\, A$ and α is an ordering of the residue field $\kappa(\mathfrak{p})$. The real closure of the ordered field $(\kappa(\mathfrak{p}), \alpha)$ is written $k(\xi)$. For $a \in A$ one writes $a(\xi) > 0$ iff the image of a in $k(\xi)$ is positive; the meaning of $a(\xi) \geq 0$, $a(\xi) = 0$ etc. is defined similarly. The topology of sper A is generated by the domains of positivity $D(a) := \{\xi \in \mathrm{sper}\, A : a(\xi) > 0\}$, $a \in A$, which form a subbasis of open sets. The space sper A is spectral (0.6.3). In addition it has the property that the closure $\overline{\{\xi\}}$ of any point ξ is totally ordered by the specialization relation. The *support map* supp: sper $A \to \mathrm{spec}\, A$, $\xi \mapsto \mathfrak{p}$ is a spectral map. For brevity the notation supp will often be replaced by φ.

(0.4.2) Let X be a scheme. If $\{U_\lambda\}$ is a covering of X by open affine subschemes then the real spectra sper $\mathcal{O}_X(U_\lambda)$ can be glued together. The resulting topological space, the real spectrum of X, is denoted by X_r. It does not depend on the particular covering $\{U_\lambda\}$. The support map $\varphi: X_r \to X$ is a locally spectral map of locally spectral spaces (0.6.3). Again $k(\xi)$ denotes the real closed field associated with $\xi \in X_r$.

(0.4.3) Frequently the following description of X_r (as a set, at least) is convenient. The elements of X_r are equivalence classes of morphisms of schemes $f: \mathrm{spec}\, k \to X$ where k is a real closed field. Here two morphisms $f': \mathrm{spec}\, k' \to X$ and $f'': \mathrm{spec}\, k'' \to X$ lie in the same equivalence class if and only if there is a real closed field k and factorizations

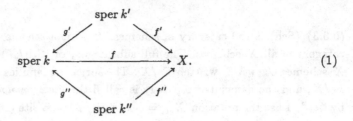

$$(1)$$

(If one reverses the arrows g' and g'' in (1) this gives another equivalent condition; the reason is that real closed fields can be amalgamated.)

(0.4.4) The real spectrum X_r depends functorially on the scheme X. If $f: Y \to X$ is a morphism of schemes then $f_r: Y_r \to X_r$ denotes the induced (locally spectral) map between the real spectra.

(0.5) Groups and group actions

(0.5.1) Group actions are from the left unless stated otherwise. If a group G acts on a topological space T (which may be a discrete set) I write T^G, or also $H^0(G, T)$, for the subspace of fixpoints, and T/G for the orbit space. (These notations are used both for left and right actions.) By G_x I denote the subgroup of G which fixes a point $x \in T$. If H is a subgroup of G then $C_G(H)$ resp. $N_G(H)$ is the centralizer resp. the normalizer of H in G.

(0.5.2) All topological groups are assumed to be Hausdorff. If G is a topological group I write (G-sets) for the category of *discrete* (left) G-spaces, i.e. G-sets all of whose point stabilizers are open subgroups. The category (G-sets) is a topos (compare (8.5)). Similarly (G-mod) is the abelian category of discrete (left) G-modules. The right derived functors of $H^0(G, -)$: (G-mod) $\to$ (Ab) are denoted by $H^n(G, -)$, $n \geq 0$. If G carries a topology, these functors need not coincide with discrete group cohomology.

(0.5.3) Let $f: H \to G$ be a continuous homomorphism between topological groups. Then pullback of actions via f is called the *restriction functor*

$$f^* = \mathrm{res}_H^G \colon (G\text{-sets}) \longrightarrow (H\text{-sets}).$$

The functor f^* preserves all direct and all finite inverse limits. Therefore it admits a right adjoint f_*, and the pair (f^*, f_*) is a topos morphism from (H-sets) to (G-sets) which again is denoted by f. If G is either compact or discrete then the direct image functor

$$f_* = \mathrm{coind}_H^G \colon (H\text{-sets}) \longrightarrow (G\text{-sets})$$

is the *coinduction* (or *multiplicative induction*) *functor*: It sends a discrete H-set N to

$$f_* N = \mathrm{coind}_H^G(N) = \mathrm{Hom}_H(G, N),$$

the set of all H-equivariant locally constant maps $a: G \to N$, on which G acts by the rule

$$(g \cdot a)(x) := a(xg) \quad (g, x \in G).$$

(The hypothesis on G implies that the stabilizers of this action are open subgroups of G.)

If on the other hand the homomorphism f is open then f^* has also a left adjoint

$$f_! = \operatorname{ind}_H^G \colon (H\text{-sets}) \longrightarrow (G\text{-sets}),$$

the *induction functor*. It sends a discrete H-set N to

$$f_! N = \operatorname{ind}_H^G(N) = G \times_H N.$$

By definition, $G \times_H N$ is the orbit space $(G \times N)/H$ where H acts on $G \times N$ by $h \cdot (x, n) = (xf(h)^{-1}, hn)$. Writing $[x, n]$ for the class of (x, n) in $G \times_H N$, the group G acts on $G \times_H N$ by $g \cdot [x, n] = [gx, n]$. Since f is open the space $G \times_H N$ is discrete, and the G-action on this set is continuous. Also the restriction functor $f^*\colon (G\text{-mod}) \to (H\text{-mod})$ for group modules has a left adjoint $f_!$ which is given by $f_! N = \operatorname{ind}_H^G(N) = \mathbb{Z}G \otimes_{\mathbb{Z}H} N$, for N a discrete H-module.

In this paper I will occasionally favor the notations $f_!$, f^*, f_* above the more classical ones ind_H^G, res_H^G, coind_H^G.

(0.6) Topological spaces

(0.6.1) Let T be a topological space. If $x, y \in T$ are such that $x \in \overline{\{y\}}$ one writes $y \succ x$ and says that x is a *specialization* of y, or that y is a *generalization* of x. If S is a second topological space then $C(T, S)$ sometimes denotes the set of continuous maps $T \to S$.

(0.6.2) The category of sheaves (resp. abelian sheaves) on T is written $\widetilde{T}$ (resp. $\operatorname{Ab}(T)$). Sheaves may be regarded both as espaces étalés over T or as contravariant functors on the ordered set of open subsets of T, without further comment. If A is a sheaf on T and $x \in T$ then A_x is the stalk of A at x. If s is a local section of A around x then s_x denotes the image of s in A_x. Otherwise standard notations are used. Unless specified differently all cohomology on a topological space is sheaf cohomology.

(0.6.3) A *spectral space* is a topological space T which is T_0 and quasi-compact, in which every non-empty closed irreducible subset has a generic point, and which has a basis of quasi-compact open sets which is stable under finite intersections. The open quasi-compact subsets of T and their complements form a subbasis for a second (finer) topology on T, the *constructible topology*. The constructible topology is compact and totally disconnected, i.e. *boolean*. The clopen (:= closed and open) subsets in this topology are called the *constructible sets* in T; arbitrary intersections of such sets are called *pro-constructible sets*. A map between spectral spaces is called *spectral* if it is continuous with respect to both the original and the constructible topologies. A space is *locally spectral* if it has an open covering by spectral spaces. The constructible topology on such a space is obtained by glueing the constructible topologies of such a covering. A map between locally spectral spaces is *locally spectral* if it is continuous with respect to the original and the constructible topologies.

Part One

1. Real spectrum and real étale site

In this section the real étale site of a scheme X is considered, and a new proof is given for the fact that the category of sheaves on this site is equivalent to the category of sheaves on X_r, the real spectrum of X. This theorem is originally due to Coste-Roy and Coste, at least in the affine case. It will be of central importance for what follows since it allows to switch freely between the real étale topology and the real spectrum. The former is much better suited for a comparison with the étale topology — after all, such a comparison is the central issue of this work! — since both are topologies on the same underlying category. On the other hand, any study of finer properties of the real étale topos is greatly facilitated (if made possible at all) by the fact that one is studying sheaves on a topological space, a space which in addition is known to have quite specific and remarkable properties.

Recall (0.4) that X_r denotes the real spectrum of a scheme X. Most of the time X_r will be regarded as a topological space, without consideration of a structure sheaf.

(1.1) Definition. A family $\{f_\lambda : U_\lambda \to U\}_{\lambda \in I}$ of morphisms of schemes is said to be *real surjective* if it is surjective on real spectra, i.e. if U_r is the union of the images $(f_\lambda)_r((U_\lambda)_r)$, $\lambda \in I$.

(1.2) For any scheme X the real surjective families in (Sch/X) satisfy the axioms for the coverings of a pretopology [SGA4 II.1.3]. The reason is the amalgamation property for real closed fields. Hence these families define a topology on (Sch/X). I am interested in the induced topology on Et/X, the category of étale X-schemes:

(1.2.1) Definition. Let X be a scheme. The topology on Et/X defined by the real surjective families is called the *real étale topology* of X, abbreviated *ret*. The site $(\mathrm{Et}/X, \mathrm{ret})$ is called the *real étale site* of X and is denoted by X_{ret}. Accordingly, the category $\widetilde{X}_{\mathrm{ret}}$ of sheaves on X_{ret} is called the *real étale topos* of X.

(1.2.2) Observe that the real étale topology is definitely *not* coarser than the canonical topology on Et/X (as long as $X \neq \emptyset$). In other words, representable presheaves are not in general sheaves on X_{ret}.

The main object of this section is to give a proof of

(1.3) Theorem. — *For any scheme X the toposes $\widetilde{X}_{ret}$ and $\widetilde{X}_r$ are naturally equivalent.*

(1.4) This theorem, in the case of an affine scheme X, is due to M.-F. Coste-Roy and M. Coste [CRC]. Historically it was in this way that the real spectrum was found, since the initial object of interest for Coste-Roy and Coste was the real étale topos $(\operatorname{spec} A)\widetilde{}_{ret}$ of a ring A. They wondered if this topos is "spatial", i.e. equivalent to the topos of sheaves on some topological space, and observed that (the sobrification of) this topological space would necessarily have to be the real spectrum sper A [CC, pp. 77-78]. Using a general criterion for coherent toposes to be spatial, itself proved by methods of mathematical logic [CC, Appendix 2] they reduced the original question to a rigidity problem between the strict real henselizations of A. This problem was subsequently solved ([CRC], compare also [R], [AR] and Corollary (3.7.4) below).

The proof given here seems to be new. Apart from the Tarski-Seidenberg principle it does not use ideas or methods from mathematical logic, and it works for arbitrary schemes. It follows a pattern which is quite common in comparison problems like the given one: One constructs an auxiliary site X_{aux} which comes along with natural site morphisms to both X_{ret} and X_r, and one then proceeds to show that both these site morphisms induce equivalences of the associated toposes. Thus, while it may be difficult to compare $\widetilde{X}_{ret}$ and $\widetilde{X}_r$ directly, it is easy to do so via an intermediate third.

While the equivalence of $\widetilde{X}_{aux}$ with $\widetilde{X}_{ret}$ will be trivial, the proof of the equivalence with $\widetilde{X}_r$ relies on the fact that étale morphisms of schemes induce local homeomorphisms between the real spectra. This fact is mentioned in [CRC] as a *consequence* of the theorem, but has since been given at least two independent proofs ([R] and [AR]). Here I give a third one which perhaps may be considered to be simpler than the others. It does not use the concept of strict real henselization. Instead it emphasizes real closed valuation rings to some extent, and uses only elementary and well-known properties of real spectra and étale morphisms.

(1.5) By definition, a *real closed valuation ring* is a valuation ring B whose field of fractions and whose residue field are both real closed. Equivalently, the real closed valuation rings are the convex subrings of real closed fields. The support mapping sper $B \to \operatorname{spec} B$ is a homeomorphism, and in particular, sper B is totally ordered by the specialization relation.

It is well known that specializations in the real spectrum can be described using these rings. Recall how this is done:

(1.5.1) **Definition.** Let $v: V \to X$ be a morphism of schemes, where V is the spectrum of a real closed valuation ring. The *specialization in* X_r *determined by* v is $v_r(\zeta') \succ v_r(\zeta)$, where ζ' resp. ζ are the generic resp. the closed point of V_r.

(1.5.2) Any specialization $\xi' \succ \xi$ in X_r is determined in this way. In fact there is a unique minimal choice of v; i.e. there is a morphism $v: V \to X$, with V the spectrum of a real closed valuation ring, such that v determines $\xi' \succ \xi$, and such that any other such $v': V' \to X$ which determines $\xi' \succ \xi$ factors uniquely as

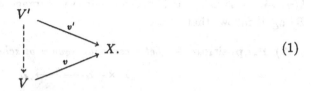

$$\tag{1}$$

(Take $V = \operatorname{spec} B$ with B the convex hull of the image of $\mathcal{O}_{X, \operatorname{supp} \xi}$ in $k(\xi')$, the real closed field associated with ξ'.) Note however that the factorization $V' \to V$ in (1) need not be a *local* morphism, i.e. the closed point of V' is not necessarily mapped to the closed point of V.

(1.6) **Lemma.** — *Let $f: Y \to X$ be a morphism of schemes, and suppose that for every $y \in Y$ the field extension $\kappa(y) \supset \kappa(f(y))$ is algebraic. (For example, f could be étale, or quasi-finite.)*

a) *If $\xi \in X_r$ is represented by a morphism of schemes $\alpha: \operatorname{spec} R \to X$, with R a real closed field (cf. (0.4.9)), then the natural map*

$$\operatorname{Hom}_X(\operatorname{spec} R, Y) \longrightarrow Y_r$$

is a bijection from the first set to $f_r^{-1}(\xi)$, the real spectrum fibre of ξ in Y_r.

b) *Let $v: V \to X$ be a morphism of schemes, with V the spectrum of a real closed valuation ring. Suppose that a specialization $\eta' \succ \eta$ in Y_r is given such that $f_r(\eta') \succ f_r(\eta)$ is the specialization determined by v. Then there is a unique X-morphism $V \to Y$ which determines $\eta' \succ \eta$.*

Proof. a) is immediately reduced to the case where X and Y are Zariski spectra of fields, in which case the assertion is well known to be true, compare e.g. [BCR 1.3]. (The point is that a real closure of a field has no automorphisms over the base field.) In b) the uniqueness assertion is clear, since the restriction of $V \to Y$ to the generic point of V is unique by a). For the existence one reduces to the case where $X = \operatorname{spec} M$, $Y = \operatorname{spec} N$ with local domains M and N, where f is induced by a local inclusion $M \subset N$, and where the support of η' resp. η is the generic resp. the closed point of Y. Write $\xi' := f_r(\eta')$ and $\xi := f_r(\eta)$. Let K resp. L be the quotient fields of M resp. N, and let $L \hookrightarrow R$ be the real closure of L with respect to η'. Hence $K \subset L \hookrightarrow R$ is the real closure of K with respect to ξ'. It suffices to assume that v is minimal for $\xi' \succ \xi$ in the sense of (1.5.2), i.e. that $V = \operatorname{spec} B$ with B the convex hull of M in R (and v induced by $M \subset B$). Hence what one has to show is $N \subset B$. Let $C \supset B$ be the convex hull of N in R. Consider the sequence of homomorphisms

$$M \longrightarrow k(\xi) \longrightarrow k(\eta) \longrightarrow \kappa_C.$$

Here $k(\xi)$, being a real closure of κ_M, is archimedean over the image of M. Since $\kappa_M \subset \kappa_N$ is an algebraic field extension, $k(\xi) \to k(\eta)$ is an isomorphism; and $k(\eta) \to \kappa_C$ is again archimedean. Since the image of M in κ_C is contained in $B/\mathfrak{m}_C$ it follows that $B = C$. $\qquad\square$

(1.7) Proposition. — *Let a cartesian square of schemes*

$$
\begin{array}{ccc}
Y \times_X Z & \xrightarrow{\ q\ } & Z \\
{\scriptstyle p}\downarrow & & \downarrow{\scriptstyle g} \\
Y & \xrightarrow{\ f\ } & X
\end{array}
\qquad\qquad (2)
$$

be given, and assume that $\kappa(y)$ is algebraic over $\kappa(f(y))$ for every $y \in Y$. Then the natural map

$$\gamma \colon (Y \times_X Z)_r \longrightarrow Y_r \times_{X_r} Z_r$$

is a homeomorphism. (The right hand side is the fibre product in the sense of topological spaces.)

(1.7.1) Corollary. — *For any scheme X the real spectrum functor $U \mapsto U_r$ from* Et/X *to* $(\mathrm{Top})/X_r$ *preserves all finite inverse limits.* $\qquad\square$

Proof of (1.7). Write $W := Y \times_X Z$. Note that q has the same property as f. From (1.6a) it follows that γ is bijective: If $\zeta \in Z_r$ is represented by $\alpha \colon z \to Z$, with z the spectrum of a real closed field, then the diagram

$$
\begin{array}{ccc}
\operatorname{Hom}_Z(z, W) & \xrightarrow{\ \approx\ } & \operatorname{Hom}_X(z, Y) \\
{\scriptstyle \approx}\downarrow & & \downarrow{\scriptstyle \approx} \\
q_r^{-1}(\zeta) & \xrightarrow{\ \gamma\ } & f_r^{-1}\big(g_r(\zeta)\big)
\end{array}
$$

commutes, in which the top horizontal arrow is induced by p and the vertical arrows are the maps of (1.6a). To show that γ^{-1} is continuous one may assume that the schemes in (2) are affine. Then γ is a spectral map. Let $\xi', \xi \in W_r$ with $\gamma(\xi') \succ \gamma(\xi)$, i.e. $p_r(\xi') \succ p_r(\xi)$ and $q_r(\xi') \succ q_r(\xi)$. Choose a morphism $v \colon V \to Z$, with V the spectrum of a real closed valuation ring, such that v determines the specialization $q_r(\xi') \succ q_r(\xi)$. By (1.6b), $g \circ v$ can be lifted (uniquely) to a morphism $u \colon V \to Y$ which determines $p_r(\xi') \succ p_r(\xi)$. Since γ is bijective (proved above) it follows that the specialization in W_r determined by the morphism $(u, v) \colon V \to W$ is $\xi' \succ \xi$. Hence the inverse map γ^{-1} preserves specializations, and so it is continuous. $\qquad\square$

(1.8) Proposition. — *For any étale morphism* $f: Y \to X$ *the map* $f_r: Y_r \to X_r$ *is a local homeomorphism.*

Proof. That is, Y_r has a covering by open subsets W_i such that the restrictions $W_i \to X_r$ of f_r are open embeddings. First, to prove openness of f_r one can assume that X and Y are affine. Since in this case f is finitely presented, it is a standard consequence of the Tarski-Seidenberg theorem that f_r maps constructible sets to constructible ones [CR Prop. 2.3]. Thus it suffices to show that the map f_r is generalizing, which means, given $\eta \in Y_r$ and $\xi' \in X_r$ with $\xi' \succ f_r(\eta)$, one has to find $\eta' \in Y_r$ with $\eta' \succ \eta$ and $f_r(\eta') = \xi'$.

Represent $\xi' \succ f_r(\eta)$ by a morphism $v: V \to X$, with V the spectrum of a real closed valuation ring (1.5.2). Let $z \hookrightarrow V$ be the inclusion of the closed point. Using (1.6a) one gets a commutative diagram (solid arrows)

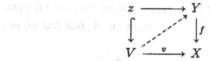

in which $z \to Y$ represents η; one has to show that the dotted lift exists, leaving the diagram commutative. After base-changing to V one can assume $X = V$ and $v = $ identity, and has to show that f has a section. But this follows from the henselian property of a real closed valuation ring.

So f_r is open. Since the diagonal map $Y \to Y \times_X Y$ is an open embedding, Corollary (1.7.1) shows that also the diagonal $Y_r \to Y_r \times_{X_r} Y_r$ is an open embedding. This proves the proposition, since a map $h: T \to S$ of topological spaces is a local homeomorphism if and only if both h and the diagonal $T \to T \times_S T$ are open maps. $\square$

(1.9) The proof of Theorem (1.3) proceeds now as follows. Let X be a scheme. Let $\mathcal{A}$ be the following category. An object of $\mathcal{A}$ is a pair (U, W) with $U \in \mathrm{Et}/X$ and W an open subset of U_r. An arrow $(U', W') \to (U, W)$ in $\mathcal{A}$ is a morphism $f: U' \to U$ of schemes over X with $f_r(W') \subset W$. A family $\{f_\alpha: (U_\alpha, W_\alpha) \to (U, W)\}_{\alpha \in I}$ of arrows in $\mathcal{A}$ is called a covering of (U, W) iff $W = \bigcup_\alpha (f_\alpha)_r(W_\alpha)$. All finite inverse limits exist in $\mathcal{A}$, and the notion of coverings defines a pretopology on $\mathcal{A}$. The associated site will be denoted by X_{aux}.

Regard the space X_r as a site in the usual way. Both Et/X and $O(X_r)$ (:= the category of open subsets of X_r) are full subcategories of $\mathcal{A}$ in natural ways. These inclusions define morphisms of sites

$$X_{ret} \xleftarrow{\varphi} X_{aux} \xrightarrow{\psi} X_r$$

since they transform coverings into coverings and preserve fibre products. The Comparison Lemma [SGA4 III.4.1] shows that the topos morphism $\varphi: \widetilde{X}_{aux} \to \widetilde{X}_{ret}$

is an equivalence. Indeed, one only has to check that each object (U, W) of $\mathcal{A}$ can be covered by objects (V, V_r) with $V \in \mathrm{Et}/U$. In other words, given $\xi \in U_r$ and an open neighborhood W of ξ in U_r, one has to find an étale morphism $f: V \to U$ with $\xi \in f_r(V_r) \subset W$. One may assume U affine, in which case this is elementary.

So it remains to see that also the topos morphism induced by ψ is an equivalence.

Let S be a sheaf on X_{auz}. For every $U \in \mathrm{Et}/X$ the assignment $W \mapsto S(U, W)$ (for $W \subset U_r$ open) is a sheaf on U_r which will be denoted by S_U. If $f: V \to U$ is a map in Et/X, the restriction maps of S define an f_r-morphism from S_U to S_V, i.e. a sheaf map $f_r^* S_U \to S_V$ on V_r.

(1.10) Lemma. — $f_r^* S_U \to S_V$ *is a sheaf isomorphism.*

Proof. It suffices to show that, for $f: U \to X$ étale, $f_r^* S_X \to S_U$ is an isomorphism. Fix an open subset W of U_r such that $f_r|_W$ is injective, i.e. an open embedding (1.8). Since $f: (U, W) \to (X, f_r(W))$ is a covering in $\mathcal{A}$, one has an exact sequence (of sets)

$$(f_r^* S_X)(W) = S_X(f_r(W)) \longrightarrow S_U(W) \xrightarrow{\mathrm{pr}_1^*, \mathrm{pr}_2^*} S_{U \times_X U}(W \times_{X_r} W), \quad (3)$$

with $W \times_{X_r} W \subset U_r \times_{X_r} U_r = (U \times_X U)_r$ (1.7.1). Since $f_r|_W$ was assumed to be injective, the diagonal map $W \to W \times_{X_r} W$ is bijective. So the diagonal morphism $(U, W) \to (U \times_X U, W \times_{X_r} W)$ is a covering in $\mathcal{A}$. But this implies that the two maps pr_1^*, pr_2^* in (3) coincide, and hence that $(f_r^* S_X)(W) \to S_U(W)$ is bijective. This proves the proposition, since by (1.9) U_r has a basis $\{W_i\}$ of open sets such that $f_r|_{W_i}$ is injective for every i. $\qquad\square$

(1.11) Now the proof of the theorem is readily completed: For $F \in \widetilde{X}_r$, the sheaf $\psi^* F$ on X_{auz} is the sheaf associated with the presheaf $(U, W) \mapsto F(f_r(W))$; here $f: U \to X$ denotes the structural morphism of $U \in \mathrm{Et}/X$. Using again (1.8) one sees that

$$(\psi^* F)(U, W) = (f_r^* F)(W), \quad \text{i.e.} \quad (\psi^* F)_U = f_r^* F.$$

On the other hand, ψ_* sends $S \in \widetilde{X}_{auz}$ to S_X. So Lemma (1.10) says that $\psi^* \psi_* S \to S$ is an isomorphism. Since the other adjunction $F \to \psi_* \psi^* F$ is trivially one this completes the proof of the theorem. $\qquad\square$

(1.12) Notation. If there is need for a notation for the equivalences just constructed, I will use the following: If G is a sheaf on X_{ret} put

$$G^\sharp := \psi_* \varphi^* G \in \widetilde{X}_r;$$

if F is a sheaf on X_r put

$$F^\flat := \varphi_* \psi^* F \in \widetilde{X}_{ret}.$$

So the compositions $\flat \circ \sharp$ and $\sharp \circ \flat$ are naturally isomorphic to the identity functors.

(1.13) In order to make $G^\sharp$ more explicit, some other notation has to be introduced. Given any open subset W of X_r, let I_W be the category of all pairs (U, s), where $U \in \mathrm{Et}/X$ and $s: W \to U_r$ is a continuous section of $U_r \to X_r$ over W. By definition, an arrow $(U', s') \to (U, s)$ in I_W is an X-morphism $f: U' \to U$ for which $s = f_r \circ s'$. From (1.7.1) one sees that the category I_W has fibre products. Moreover I_W is left filtering. This again follows easily from Corollary (1.7.1): Given a diagram

$$V \underset{f, g}{\rightrightarrows} U$$

$$q \searrow \quad \swarrow p$$

$$X$$

with p, q étale and $p \circ f = p \circ g = q$, and a section $s: W \to V_r$ of q_r over W with $f_r \circ s = g_r \circ s$, let $h: E \to V$ be the kernel of f and g. Then h is étale, and $h_r: E_r \to V_r$ is the kernel of f_r and g_r by (1.7.1), so s factors uniquely through h_r.

If $W' \subset W$ is an inclusion of open subsets of X_r, there is a natural ("restriction") functor $I_W \to I_{W'}$. Thus if P is a presheaf on Et/X, the functor

$$P^\dagger: \quad W \longmapsto \varinjlim \left\{ \begin{array}{ccc} I_W^\circ & \longrightarrow & (\text{sets}) \\ (U, s) & \longmapsto & P(U) \end{array} \right. \qquad (W \subset X_r \text{ open}) \tag{4}$$

is a presheaf on the topological space X_r. Now we are ready to describe $\sharp$ and $\flat$:

(1.14) Theorem. —

a) *If F is a sheaf on X_r then $F^\flat$ is the sheaf on X_{ret} which sends $(f: U \to X) \in \mathrm{Et}/X$ to $H^0(U_r, f_r^* F)$.*

b) *If P is a presheaf on Et/X and G is the associated X_{ret}-sheaf then $G^\sharp$ is the sheaf on X_r which is associated to the presheaf $P^\dagger$ (4). In short, the diagram of functors*

$$(\mathrm{Et}/X)^\widehat{} \xrightarrow{\quad \dagger \quad} O(X_r)^\widehat{} = \{\text{presheaves on } X_r\}$$

$$a_{ret} \downarrow \qquad\qquad \downarrow a_r$$

$$\widetilde{X}_{ret} \xrightarrow{\quad \sharp \quad} \widetilde{X}_r$$

commutes up to natural isomorphism.

Proof. a) is obvious from (1.9) and (1.11). To prove b) let x be the Zariski spectrum of a real closed field and let $\alpha: x \to X$ be a morphism of schemes, representing a point $\xi \in X_r$. Then the stalk of the presheaf $P^\dagger$ in ξ is

$$P_\xi^\dagger = \varinjlim_{\substack{W \subset X_r \text{ open} \\ \xi \in W}} P^\dagger(W) = \varinjlim_{\substack{x \to U \\ x}} P(U),$$

where the second direct limit is formed over the category of all X-maps from x into étale X-schemes U.

Consider the natural morphism $P \to a_{ret}(P) = G$ of presheaves on Et/X. For any ξ as above the induced map $P_\xi^\dagger \to G_\xi^\dagger$ is bijective. Hence the induced map $P^\dagger \to G^\dagger$ of presheaves on X_r becomes an isomorphism after applying the associated sheaf functor, since it is bijective on stalks.

This shows that it suffices to prove b) when $P = G$ is a *sheaf* on X_{ret}. Since $\sharp$ and $\flat$ are known to be quasi-inverses of each other, it is enough to prove for any sheaf F on X_r that F is isomorphic to the sheaf associated to $(F^\flat)^\dagger$. To this end write $G := F^\flat$. If $W \subset X_r$ is open and $(U, s) \in I_W$, there is a natural map $G(U) \to F(W)$, namely the pullback by s

$$s^*: \quad G(U) \underset{a)}{=} (f_r^* F)(U_r) \quad \longrightarrow \quad (s^* f_r^* F)(W) = F(W).$$

These maps fit together to give a map $G^\dagger \to F$ of presheaves on X_r. It is immediate that the stalk maps $G_\xi^\dagger \to F_\xi$ are bijective, and so $a_r(G^\dagger) \xrightarrow{\sim} F$, as required. $\square$

(1.15) Remark. Every étale sheaf on X (i.e. sheaf on X_{et}) may be regarded as a presheaf on Et/X, and as such may be sheafified with respect to the topology X_{ret}. (This is the functor $\rho: \widetilde{X}_{et} \to \widetilde{X}_{ret}$ studied in the next section.) If one starts with the étale structure sheaf $\mathcal{O}_X = \mathbb{G}_a$, the sheaf on $\widetilde{X}_r$ which corresponds to $\rho \mathbb{G}_a$ is the sheaf $\mathcal{N}_X$ of abstract Nash functions, as introduced by Roy [R]. This follows immediately from (1.14b). In particular, if X is a smooth algebraic variety over the real numbers $\mathbb{R}$, the sheaf $\rho \mathbb{G}_a$ corresponds to the classical sheaf of Nash (= algebraic and analytic) functions on $V(\mathbb{R})$. Roy's definition of the abstract Nash sheaf $\mathcal{N}_X$, for X a general scheme (namely as the sheaf associated to $\mathbb{G}_a^\dagger$, in the above notation), was inspired by the Artin-Mazur description of the Nash sheaf in the classical situation, which it copies. The above interpretation $\mathcal{N}_X \cong \rho \mathbb{G}_a$ gives a different point of view from which $\mathcal{N}_X$ is a natural object of study.

(1.16) Remark and Notation. Finally an obvious remark on functorial behavior has to be added: A morphism of schemes $f: Y \to X$ gives rise to the base-change functor $f^{-1}: \mathrm{Et}/X \to \mathrm{Et}/Y$, which is continuous with respect to the real étale topologies and hence is a morphism of sites $Y_{ret} \to X_{ret}$. The associated topos morphism from $\widetilde{Y}_{ret}$ to $\widetilde{X}_{ret}$ is written $f_{ret} = (f_{ret}^*, f_{ret*})$. (The subscript "ret" is appended to the functors since later one has to deal with different topologies at the same time.) The equivalence between $\widetilde{X}_{ret}$ and $\widetilde{X}_r$ constructed above is natural, i.e. it is compatible with morphisms $f: X \to Y$ of schemes.

(1.17) Remark. Without proof I mention that the material in this section can be generalized as follows. If X is a scheme and Z is an arbitrary subset of the real spectrum X_r, there is a mod Z real étale topology on Et/X (which may be denoted ret_Z) in which only orderings lying over Z are relevant for the definition of coverings. One can show that the topos of sheaves on the site $(\mathrm{Et}/X, ret_Z)$ is equivalent to the topos of sheaves on the topological space Z.

2. Glueing étale and real étale site

It is intuitively clear that, for relating the étale topology of a scheme X to the real spectrum X_r, the result from Sect. 1 should be instrumental. It allows one to replace X_r by X_{ret}, a site modelled on the same underlying category Et$/X$ as X_{et}. However, the two topologies X_{et} and X_{ret} cannot be compared directly since none of them is finer than the other (except in the trivial and uninteresting case when the real spectrum of X is empty).

A natural idea is to try a comparison via an intermediate third topology. Since the topologies on any category form a lattice, there are canonical choices. The topology generated by X_{et} and X_{ret} is the discrete topology on Et$/X$, so it cannot be helpful at all. Indeed, every scheme can be covered, in the étale sense, by a family of schemes (or a single scheme) all of whose residue fields are nonreal; and a scheme with empty real spectrum is covered, in the real étale sense, by the empty scheme. A more interesting object is the intersection of the two topologies; this finest common coarsening X_b of X_{et} and X_{ret} is introduced in this section. It will be used extensively in the sequel. From the argument just mentioned it follows easily (2.4.1) that the topos $\widetilde{X}_b$ of sheaves on X_b contains the étale topos $\widetilde{X}_{et}$ as an open subtopos and that the real étale topos $\widetilde{X}_{ret}$ is the closed complement. This fact is basic for what follows.

The reasoning which shows that $\widetilde{X}_b$ is the result of glueing $\widetilde{X}_{ret}$ to $\widetilde{X}_{et}$ is of completely general nature, and will be presented in a general set-up. The construction can be carried out with numerous other examples (some of which may also deserve to be studied?).

(2.1) Let C be a category. See (0.2), in particular (0.2.2), for some notations used below. Fix two topologies τ and τ' on C, and consider their intersection $\tau \cap \tau'$ [SGA4 II.1.1.3], i.e. the finest topology which is coarser than both τ and τ'. Let $j = (j^*, j_*)\colon \widetilde{C}_\tau \to \widetilde{C}_{\tau \cap \tau'}$ and $i = (i^*, i_*)\colon \widetilde{C}_{\tau'} \to \widetilde{C}_{\tau \cap \tau'}$ be the natural topos morphisms. These are embeddings (= *plongements*) in the sense of [SGA4 IV.9], which means that the functors j_* and i_* are fully faithful.

(2.2) **Proposition.** — *The following are equivalent conditions:*
(i) *j makes $\widetilde{C}_\tau$ an open subtopos of $\widetilde{C}_{\tau \cap \tau'}$, and i makes $\widetilde{C}_{\tau'}$ its closed complement;*
(ii) *every $X \in C$ has a covering sieve R with respect to τ such that, for every $U \to X$ in R, the empty sieve is τ'-covering for U.*

Proof. See [SGA4 IV.9] for the concepts of open or closed subtopos of a topos. Let $\emptyset_{\tau'}$ be the initial object of $\tilde{C}_{\tau'}$. The empty sieve is τ'-covering for $U \in C$ if and only if $\epsilon_{\tau'}(U) = \emptyset_{\tau'}$ [SGA4 II.4.6.1]. Write $W := i_*(\emptyset_{\tau'})$, so

$$W(X) = \begin{cases} * & \text{if } \epsilon_{\tau'}(X) = \emptyset_{\tau'}, \\ \emptyset & \text{otherwise}, \end{cases}$$

for $X \in C$. As a subsheaf of the final sheaf $*$, this W defines an open subtopos $\tilde{C}_{\tau \cap \tau'}/W$ of $\tilde{C}_{\tau \cap \tau'}$.

(ii) $\Rightarrow$ (i): First it is shown that j is isomorphic to the embedding corresponding to the open subtopos $\tilde{C}_{\tau \cap \tau'}/W$. For $F \in \tilde{C}_{\tau \cap \tau'}$ the τ-sheaf j^*F is the τ-sheaf associated to the presheaf F. If $X \in C$ satisfies $\epsilon_{\tau'}(X) = \emptyset_{\tau'}$ then $(j^*F)(X) = F(X)$, since $\epsilon_{\tau'}(U) = \emptyset_{\tau'}$ holds also for all $U \to X$, and so every τ-covering sieve of U is also $\tau \cap \tau'$-covering. Hence $W \times j_*j^*F$ is the $\tau \cap \tau'$-sheaf

$$X \longmapsto \begin{cases} F(X) & \text{if } \epsilon_{\tau'}(X) = \emptyset_{\tau'}, \\ \emptyset & \text{otherwise}. \end{cases}$$

So if $F \in \tilde{C}_{\tau \cap \tau'}$ has $\text{Hom}(F, W) \neq \emptyset$, then $F \to W \times j_*j^*F$ is an isomorphism. From (ii) it follows that $j^*W = *$, the final (pre-)sheaf, and hence for every $A \in \tilde{C}_\tau$ one has $j^*(W \times j_*A) = A$. So the restriction of j^* to $\tilde{C}_{\tau \cap \tau'}/W$ is an equivalence from $\tilde{C}_{\tau \cap \tau'}/W$ to $\tilde{C}_\tau$, a quasi-inverse being given by $A \mapsto W \times j_*A$. In other words, j is isomorphic to the open embedding corresponding to the open subtopos $\tilde{C}_{\tau \cap \tau'}/W$.

The closed complement of $\tilde{C}_{\tau \cap \tau'}/W$ is the full subcategory

$$E := \{F \in \tilde{C}_{\tau \cap \tau'}: \text{ pr}_1 \colon W \times F \to W \text{ is an isomorphism}\}$$
$$= \{F \colon F(X) = * \text{ for all } X \text{ with } \epsilon_{\tau'}(X) = \emptyset_{\tau'}\}$$

of $\tilde{C}_{\tau \cap \tau'}$ [SGA4 IV.9.3]. The functor i_* takes values in E, since for $B \in \tilde{C}_{\tau'}$ one has

$$(i_*B)(X) = \text{Hom}_{\tilde{C}_{\tau'}}(\epsilon_{\tau'}(X), B) \quad (X \in C).$$

The claim is that $F \to i_*i^*F$ is an isomorphism for every $F \in E$, or in other words, that every $F \in E$ is a sheaf with respect to τ'. Once this is shown the proof of (ii) $\Rightarrow$ (i) will be complete. So let $F \in E$ and $X \in C$, and let R' be a τ'-covering sieve of X. One has to show that $F(X) \to \text{Hom}_{\hat{C}}(R', F)$ is bijective. By hypothesis (ii) there is a τ-covering sieve R of X consisting of objects which are covered by $\emptyset$ under τ'. Now $R \cup R'$ is a covering sieve for $\tau \cap \tau'$. Since F is a sheaf for $\tau \cap \tau'$, the upper horizontal map in

$$
\begin{array}{ccccc}
F(X) & \xrightarrow{\approx} & \text{Hom}_{\hat{C}}(R \cup R', F) & = & \varprojlim_{U \in \hat{C}/R \cup R'} F(U) \\
\| & & \downarrow & & \downarrow \\
F(X) & \longrightarrow & \text{Hom}_{\hat{C}}(R', F) & = & \varprojlim_{V \in \hat{C}/R'} F(V)
\end{array}
$$

is bijective. Now since $F \in E$ one has $F(U) = *$ for each $U \in C/R$. This implies that the right hand vertical arrow between the $\varprojlim$'s is bijective. Hence so is the lower arrow, as desired.

(i) $\Rightarrow$ (ii): From hypothesis (i) it follows that $j^*W = *$, i.e. $a_\tau(W)$ is the constant sheaf $*$. Since W is a sub(pre-)sheaf of $*$, it is a separated presheaf with respect to any topology. So $* = L_\tau(W)$. Let $X \in C$. Then

$$L_\tau W(X) = \varinjlim_{R \in J_\tau(X)^\circ} \mathrm{Hom}_{\widehat{C}}(R, W),$$

where $J_\tau(X)$ is the category of τ-covering sieves of X [SGA4 II.3]. So there is a τ-covering sieve R of X with $\mathrm{Hom}(R, W) \neq \emptyset$, i.e. with $W(U) \neq \emptyset$ for all $U \to X$ in R. This means $\epsilon_{\tau'}(U) = \emptyset$ for all these U. This is condition (ii). $\quad\square$

(2.3) Definition. Let X be a scheme. The topology on Et/X which is the intersection of the étale and the real étale topology will be denoted by b. Thus a family $\{U_\lambda \to U\}_\lambda$ in Et/X is a covering for the topology b if and only if it is *both* surjective and real surjective (1.1). For the site $(\mathrm{Et}/X, b)$ the notation X_b will be used. I will write

$$j = (j^*, j_*)\colon \widetilde{X}_{et} \longrightarrow \widetilde{X}_b \quad \text{and} \quad i = (i^*, i_*)\colon \widetilde{X}_{ret} \longrightarrow \widetilde{X}_b$$

for the canonical topos embeddings.

(2.4) Let us verify that Proposition (2.2) applies. For Y a scheme and p a prime number let $Y_{(p)}$ be the largest open subscheme of Y on which p is invertible. If ζ_{p^ν} is a primitive p^ν-th root of unity then

$$Y_{(p)} \times \mathrm{spec}\, \mathbb{Z}[\zeta_{p^\nu}] \longrightarrow Y_{(p)}$$

is an étale covering. So by taking the two maps

$$Y_{(2)}[\sqrt{-1}] \longrightarrow Y \quad \text{and} \quad Y_{(3)}[\zeta_3] \longrightarrow Y$$

one has covered Y in the étale sense by two schemes with empty real spectrum. From (2.2) one therefore gets

(2.4.1) Proposition. — $j\colon \widetilde{X}_{et} \to \widetilde{X}_b$ *is an open topos embedding, and* $i\colon \widetilde{X}_{ret} \to \widetilde{X}_b$ *is the embedding of the closed complement.* $\quad\square$

In order to display in more detail what this means, one has to introduce the following functor:

(2.5) Definition. The *glueing functor* is the functor $\rho := i^* j_* \colon \widetilde{X}_{et} \to \widetilde{X}_{ret}$. It is left exact, i.e. preserves finite inverse limits.

(2.6) The topos $\widetilde{X}_b$ can be expressed solely in terms of $\widetilde{X}_{et}$, $\widetilde{X}_{ret}$ and the glueing functor ρ, as follows (cf. [SGA4 IV.9.5.4]). Consider triples (B, A, ϕ), where $B \in \widetilde{X}_{ret}$, $A \in \widetilde{X}_{et}$ and $\phi \colon B \to \rho A$ is a morphism of sheaves on X_{ret}. These triples form in a natural way a category which is denoted $(\widetilde{X}_{ret}, \widetilde{X}_{et}, \rho)$. There is a natural functor from $\widetilde{X}_b$ to $(\widetilde{X}_{ret}, \widetilde{X}_{et}, \rho)$, namely

$$F \longmapsto \left(i^* F, \ j^* F, \ \phi_F \colon i^* F \to i^* j_* j^* F = \rho j^* F \right) \qquad (F \in \widetilde{X}_b). \qquad (1)$$

Here ϕ_F comes from the adjunction map $F \to j_* j^* F$. Another way of stating (2.4.1) is

(2.6.1) Proposition. — *The functor* (1) *is an equivalence between* $\widetilde{X}_b$ *and the category* $(\widetilde{X}_{ret}, \widetilde{X}_{et}, \rho)$. *A quasi-inverse is given by the functor which sends a triple* $(B, A, \phi \colon B \to \rho A)$ *to the fibre product (in* $\widetilde{X}_b$) *of*

$$i_* B \xrightarrow{\ i_*(\phi)\ } i_* \rho A = i_* i^* j_* A \xleftarrow{\ \text{adj}\ } j_* A. \qquad \square$$

(2.7) Corollary. — *The functor* $j^* \colon \widetilde{X}_b \to \widetilde{X}_{et}$ *has a left adjoint* $j_! \colon \widetilde{X}_{et} \to \widetilde{X}_b$, *the "empty extension".* $\qquad \square$

(2.8) Recall how these functors are described if one identifies $\widetilde{X}_b$ with $(\widetilde{X}_{ret}, \widetilde{X}_{et}, \rho)$ via the equivalence (2.6.1):

$$
\begin{aligned}
j_* \colon \widetilde{X}_{et} \to \widetilde{X}_b \quad &\text{is} \quad & A \quad &\longmapsto \quad (\rho A, A, \text{id} \colon \rho A \to \rho A) \\
j^* \colon \widetilde{X}_b \to \widetilde{X}_{et} \quad &\text{is} \quad & (B, A, \phi) \quad &\longmapsto \quad A \\
j_! \colon \widetilde{X}_{et} \to \widetilde{X}_b \quad &\text{is} \quad & A \quad &\longmapsto \quad (\emptyset, A, \emptyset \to \rho A) \\
i_* \colon \widetilde{X}_{ret} \to \widetilde{X}_b \quad &\text{is} \quad & B \quad &\longmapsto \quad (B, *, B \to *) \\
i^* \colon \widetilde{X}_b \to \widetilde{X}_{ret} \quad &\text{is} \quad & (B, A, \phi) \quad &\longmapsto \quad B.
\end{aligned}
$$

(2.9) Corollary (Abelian sheaves). —

a) *The functor* $j^* \colon \text{Ab}(X_b) \to \text{Ab}(X_{et})$ *has a left adjoint* $j_! \colon \text{Ab}(X_{et}) \to \text{Ab}(X_b)$, *the "extension by zero".* $j_!$ *is an exact additive functor.*

b) *The functor* $i_* \colon \text{Ab}(X_{ret}) \to \text{Ab}(X_b)$ *has a right adjoint* $i^! \colon \text{Ab}(X_b) \to \text{Ab}(X_{ret})$. *In particular, the additive functor* i_* *is exact.* $\qquad \square$

(2.9.1) In terms of $(\widetilde{X}_{ret}, \widetilde{X}_{et}, \rho)$ these functors are the following:

$$
\begin{aligned}
j_! \colon \text{Ab}(X_{et}) \to \text{Ab}(X_b) \quad &\text{is} \quad & A \quad &\longmapsto \quad (0, A, 0 \to \rho A) \\
i^! \colon \text{Ab}(X_b) \to \text{Ab}(X_{ret}) \quad &\text{is} \quad & (B, A, \phi \colon B \to \rho A) \quad &\longmapsto \quad \ker \phi.
\end{aligned}
$$

The usual warning is issued here that the two functors $j_!$ (for set-valued and abelian sheaves, respectively) do not coincide on abelian sheaves. No confusion should arise from keeping the same notation for both.

(2.10) Corollary. — *For every $F \in \mathrm{Ab}(X_b)$ there are natural exact sequences*

$$0 \longrightarrow j_! j^* F \longrightarrow F \longrightarrow i_* i^* F \longrightarrow 0 \qquad (2)$$

and

$$0 \longrightarrow i_* i^! F \longrightarrow F \longrightarrow j_* j^* F \qquad (3)$$

on X_b. In particular, for $A \in \mathrm{Ab}(X_{et})$ there is a natural exact sequence

$$0 \longrightarrow j_! A \longrightarrow j_* A \longrightarrow i_* \rho A \longrightarrow 0 \qquad (4)$$

on X_b. □

(2.7)–(2.10) are true in general for an open subtopos and its closed complement, cf. [SGA4, IV.9.5 and IV.14].

(2.11) Remarks. All three sites under consideration are living on the same category Et/X. It is a trivial but useful exercise to express the different functors associated with this glueing situation explicitly, by regarding all sheaves on the various sites as presheaves on Et/X:

(2.11.1) j_* and i_* do "nothing", i.e.

$$(j_* A)(U) = A(U) \quad \text{and} \quad (i_* B)(U) = B(U)$$

for $A \in \widetilde{X}_{et}$, $B \in \widetilde{X}_{ret}$ and $U \in \mathrm{Et}/X$.

(2.11.2) j^* is taking the associated X_{et}-sheaf, i^* is taking the associated X_{ret}-sheaf, both being defined on $\widetilde{X}_b$. Also ρ is taking the associated X_{ret}-sheaf, but ρ is defined on $\widetilde{X}_{et}$. Note that étale sheaves on X, when considered as presheaves on Et/X, are rarely separated with respect to ret, so one has to apply the functor L_{ret} twice to get ρ resp. i^* [SGA4 II.3]. Let $F \in \widetilde{X}_b$. One can make j^* more explicit: Since clearly $(j^* F)(U) = F(U)$ for every $U \in \mathrm{Et}/X$ with $U_r = \emptyset$, it follows that

$$(j^* F)(U) = \ker \left(\mathrm{id}, \sigma : F(U') \rightrightarrows F(U') \right) \qquad (5)$$

holds for all $F \in \widetilde{X}_b$ and $U \in \mathrm{Et}/X$ with $\frac{1}{2} \in \mathcal{O}(U)$. Here $U' := U \otimes_{\mathbb{Z}} \mathbb{Z}[\sqrt{-1}]$, and σ is the involution of U' over U. If F is an abelian sheaf on X_b and U is as before, then (5) implies that

$$(i^! F)(U) = \ker \left(F(U) \xrightarrow{\mathrm{res}} F(U') \right),$$

by the exact sequence (3). One can think of $i^! F$ as the subsheaf of F consisting of all sections with real support.

(2.11.3) Sometimes the fact that representable presheaves are not in general sheaves on X_{ret} requires a little caution. For example, the sheaf $\emptyset$ on X_{ret} (and hence also $i_*\emptyset$ on X_b) is given by

$$U \longmapsto \begin{cases} * & \text{if } U_r = \emptyset, \\ \emptyset & \text{if } U_r \neq \emptyset \end{cases} \qquad (U \in \text{Et}/X).$$

Any sheaf B on X_{ret} satisfies $B(U) = *$ for all $U \in \text{Et}/X$ with $U_r = \emptyset$.

(2.11.4) $j_!$ (the set-valued version) sends an étale sheaf A to the sheaf

$$j_!A = i_*(\emptyset) \times j_*A : \; U \longmapsto \begin{cases} A(U) & \text{if } U_r = \emptyset, \\ \emptyset & \text{if } U_r \neq \emptyset \end{cases} \qquad (U \in \text{Et}/X)$$

on X_b.

(2.11.5) The functor $j_!$ for abelian sheaves (the extension by zero) can be thought of as the functor which sends an abelian étale sheaf A to its sub-X_b-sheaf of all sections with "nonreal support". Indeed, the exact sequence (4) shows, for $U \in \text{Et}/X$, that $(j_!A)(U)$ is the subgroup of all $a \in A(U)$ which vanish on some real étale covering of U, or equivalently, on some open subscheme of U which contains all points with a formally real residue field. So

$$(j_!A)(U) = \Big\{ a \in A(U) : \exists \text{ closed subscheme } T \subset U \text{ with } T_r = \emptyset \text{ and } a|_{U-T} = 0 \Big\}$$

$$= \varinjlim_{\substack{T \subset U \text{ closed} \\ T_r = \emptyset}} H^0_T(U, A).$$

However $j_!$ does not send injective X_{et}-sheaves to acyclic X_b-sheaves in general, and so $H^n(X_b, j_!A)$ is different from $\varinjlim H^n_T(X, A)$ for $n \geq 1$.

(2.12) **Notation.** Let $t \in \{et, b, ret\}$. Then $a_t : (\text{Et}/X)^\wedge \to \tilde{X}_t$ denotes the functor "associated t-sheaf", and $\epsilon_t = a_t \circ h : \text{Et}/X \to \tilde{X}_t$ denotes the functor which sends $U \in \text{Et}/X$ to the t-sheaf associated with $h_U = \text{Hom}_X(-, U)$. Thus $\epsilon_{et}(U) = \epsilon_b(U) = h_U$, while $\epsilon_{ret}(U) = \rho h_U$.

(2.13) **Remark.** Let A be an étale sheaf on X. The sheaf $(\rho A)^\sharp$ on the topological space X_r which corresponds to ρA under the equivalence $\tilde{X}_{ret} \sim \tilde{X}_r$ (cf. Sect. 1) is the sheaf associated with the presheaf

$$W \longmapsto \varinjlim_{(U,s) \in I_W^o} A(U) \qquad (W \subset X_r \text{ open}), \qquad (6)$$

as follows from Theorem (1.14b). (Note: It is easy to see that the presheaf (6) on X_r is separated. In the particular case $A = \mathbb{G}_a$ this was observed in [R, Prop. 2.8].) In particular consider the case when A is representable by an étale X-scheme $f : Y \to X$, i.e. $A = h_Y$. Then $f_r : Y_r \to X_r$ is an espace étalé over X_r (1.8). Viewed as a sheaf on X_r, this is the sheaf $(\rho A)^\sharp$ which corresponds to $\rho A = \epsilon_{ret}(Y)$. Indeed,

(6) shows that there is an obvious morphism of sheaves on X_r from $(\rho A)^\sharp$ to Y_r; it is an isomorphism since the induced maps between the stalks are bijective. (This requires identification of the stalks of $(\rho A)^\sharp$ with the fibres of f_r. A description of the stalks of ρA which is valid for any sheaf A on X_{et} will be given in (3.7.2) below.)

(2.14) Example. The b-topology coincides with the étale topology if and only if X has no point with a formally real residue field. Thus the most basic proper example for the b-topology arises from $X = \operatorname{spec} R$ with R a real closed field. Let G be the Galois group of $R(\sqrt{-1})/R$, i.e. the group of order two. Then $\tilde{X}_b$ is equivalent to the category of all triples $(B, A, \phi\colon B \to A^G)$ with B a set, A a G-set and ϕ a map, where A^G denotes the elements of A fixed by G.

(2.15) Notation. If t is one of et, b, ret and M is a set, denote by $\underline{M}_t$ the constant sheaf on X_t with stalks M.

(2.15.1) Proposition. — *j_* and ρ preserve constant sheaves. That is: If M is a set, then $j_*\underline{M}_{et} = \underline{M}_b$ and $\rho\underline{M}_{et} = \underline{M}_{ret}$.*

For a generalization see the remark after (3.7.2).

Proof. $\underline{M}_t$ is the coproduct in $\tilde{X}_t$ of M copies of the constant sheaf $*$. There is a canonical sheaf map

$$\underline{M}_b = \coprod_M j_*(*) \longrightarrow j_*\left(\coprod_M *\right) = j_*(\underline{M}_{et})$$

(note $j_*(*) = *$), which becomes

$$\underline{M}_{ret} \longrightarrow \rho\underline{M}_{et} \tag{7}$$

after applying i^*. It suffices to check that (7) is an isomorphism. But this is clear by inspection of the stalks (or follows directly from Remark (2.13)). □

(2.15.2) If A is only a locally constant sheaf on X_{et} then in general the sheaf j_*A on X_b will *fail* to be locally constant. This is so already for $X = \operatorname{spec} R$, with R a real closed field: Here j_*A is locally constant iff A is constant.

A detailed discussion of locally constant and constructible sheaves on the three sites X_{et}, X_b, X_{ret} can be found in Sect. 17.

(2.16) Functorial behavior. For every morphism $f\colon Y \to X$ of schemes and each $t \in \{et,\, b,\, ret\}$, the base change functor $f^{-1}\colon \operatorname{Et}/X \to \operatorname{Et}/Y$ is a morphism $f_t\colon Y_t \to X_t$ of sites. The corresponding topos morphism from $\tilde{Y}_t$ to $\tilde{X}_t$ is written $f_t = (f_t^*, f_{t*})$. If $g\colon Z \to Y$ is a second morphism of schemes then $(f \circ g)_t$ and $f_t \circ g_t$ are naturally isomorphic. Note that if f is étale, f_t^* has a left adjoint which is denoted by $f_{t!}$, as usual. It sends a sheaf B on Y_t to the X_t-sheaf associated to the presheaf on Et/X

$$U \longmapsto \coprod_{b \in \mathrm{Hom}_X(U,Y)} B(U \xrightarrow{b} Y) \qquad (U \in \mathrm{Et}/X);$$

and a similar description holds for the abelian sheaves version of $f_{t!}$.

(2.17) The b-sites of a scheme and of its underlying reduced scheme are equivalent. More precisely, if $f: Y \to X$ is an entire surjective radical morphism then the induced morphism of sites $f_b: Y_b \to X_b$ is an equivalence of sites. This is either shown by the same argument as for the étale topology [SGA4 VIII.1], or is deduced directly from the étale case, since the analogous statement for the real étale sites is obvious.

(2.18) For later reference, here are some basic finiteness properties of the three topologies et, b, ret. Nothing exciting happens, it is the same story as with the étale topology.

(2.18.1) Definition. Let Et'/X be the full subcategory of Et/X consisting of the finitely presented étale X-schemes, i.e. those étale X-schemes which are quasi-compact and quasi-separated over X. Let Et'_{qc}/X denote the full subcategory of Et'/X consisting of all $U \to X$ in Et'/X for which U is quasi-compact. Let $\mathrm{Et}_{\mathrm{aff}}/X$ be the full subcategory of Et/X consisting of all étale morphisms $U \to X$ with U an affine scheme. Each of these categories carries the induced topologies et, b, ret from Et/X. The site $(\mathrm{Et}'/X, et)$ is called the *restricted étale site* in [SGA4 VII.3.2].

(2.18.2) Proposition. — *Let X be a scheme, let $t \in \{et, b, ret\}$.*
   a) *Et'/X, Et'_{qc}/X and $\mathrm{Et}_{\mathrm{aff}}/X$ are closed in Et/X under fibre products.*
   b) *If X is quasi-separated then the site morphisms $X_t = (\mathrm{Et}/X, t) \to (\mathrm{Et}'/X, t) \to (\mathrm{Et}'_{qc}/X, t)$ given by the inclusions of the categories induce equivalences of the associated toposes. For the site morphism $X_t \to (\mathrm{Et}_{\mathrm{aff}}/X, t)$ this is true without any restriction to X.*
   c) *The topos $\widetilde{X}_t$ is always algebraic, and in particular, it is locally coherent. If X is quasi-separated then $\widetilde{X}_t$ is quasi-separated. If X is in addition quasi-compact then $\widetilde{X}_t$ is coherent.*
   d) *For every $U \in \mathrm{Et}_{\mathrm{aff}}/X$ the object $\epsilon_t(U)$ of the topos $\widetilde{X}_t$ is coherent. If X is quasi-separated the same holds for every $U \in \mathrm{Et}'_{qc}/X$.*

Proof. See [SGA4 VI.2.3] for the notions of quasi-separated, algebraic and (locally) coherent toposes. For a) see [EGA I*, 6.1.5 and 6.1.9]; b) is an application of the Comparison Lemma, compare [SGA4 VII.3.2]. To prove c) note that every t-covering of a quasi-compact scheme has a finite subcovering. (For $t = b$ or $t = ret$

this uses that the real spectrum of a quasi-compact scheme is quasi-compact, and that every étale morphism induces an open map between the real spectra, (1.8).) Hence the site $\left(\mathrm{Et}_{\mathrm{aff}}/X, t\right)$ satisfies condition (ii) of [SGA4 VI.2.1], showing that $\widetilde{X}_t$ is locally coherent. That $\widetilde{X}_t$ is actually algebraic follows from the fact that for U, $V \in \mathrm{Et}_{\mathrm{aff}}/X$ the scheme $U \times_X V$ is quasi-separated [EGA I*, 6.1.9]. The remaining statements of c) are clear. If X is quasi-separated then also the site $\left(\mathrm{Et}_{qc}'/X, t\right)$ satisfies (ii) of [SGA4 VI.2.1], by a) and b). So d) is a consequence of [loc.cit.]. $\square$

The reason why the properties of a topos being algebraic or coherent are useful is that they imply very instrumental facts about filtering limits of sheaves. See the next section.

3. Limit theorems, stalks, and other basic facts

This rather technical section provides some basic tools for working with the b-topology. This includes the identification of the stalk functors. It is easily seen that the topos $\widetilde{X}_b$ has sufficiently many points; this allows to argue stalkwise in many instances, a feature which is very convenient. For example, stalkwise argumentation is applied at the end of this section to establish, in a relative situation, the basic commutation rules between the various functors between the three topologies.

It is familiar to everybody working with étale cohomology that there are two kinds of limit theorems which are indispensable for many arguments: Commutation of cohomology with filtering direct limits of sheaves, and commutation of cohomology with certain filtering inverse limits of schemes. This section starts with similar theorems for the real étale and the b-topology.

(3.1) Definition. Let X be a scheme, and let t be one of the topologies et, b, ret. The right derived functors of the global sections functor $\mathrm{Ab}(X_t) \to (\mathrm{Ab})$, $F \mapsto F(X)$ are denoted by $F \mapsto H^n(X_t, F)$, $n \geq 0$ (or sometimes also by $H^n_t(X, F)$ if this is typographically less awkward). If $X = \mathrm{spec}\, A$ is affine I may also write $H^n(A_t, F)$.

(3.2) Proposition. — *Let X be a quasi-separated scheme and fix $t \in \{et, b, ret\}$. Let I be a right filtering category, let $F: I \to \widetilde{X}_t$, $i \mapsto F_i$ be an inductive system of sheaves on X_t, and denote its direct limit by $\varinjlim_i F_i$. Let $U \to X$ be an étale X-scheme, and assume that U is quasi-compact and quasi-separated (thus $U \in \mathrm{Et}'_{qc}/X$).*
   a) *The canonical map $\varinjlim_i F_i(U) \longrightarrow (\varinjlim_i F_i)(U)$ is bijective.*
   b) *If F takes values in $\mathrm{Ab}(X_t)$, then the canonical map $\varinjlim_i H^n(U_t, F_i) \longrightarrow H^n(U_t, \varinjlim_i F_i)$ is bijective for every $n \geq 0$.*

Proof. The sheaf $\epsilon_t(U)$ is a coherent object of the topos $\widetilde{X}_t$, see (2.18.2d), so one may cite VI.1.23 and VI.5.3 of [SGA4]. $\qquad\square$

(3.3) Corollary. — *Let $f: Y \to X$ be a morphism between two quasi-separated schemes, and assume that f is quasi-compact. Let $t \in \{et, b, ret\}$. Then the direct image functor $f_{t*}: \widetilde{Y}_t \to \widetilde{X}_t$ and all higher direct image functors $\mathrm{R}^n f_{t*}: \mathrm{Ab}(Y_t) \to \mathrm{Ab}(X_t)$ $(n \geq 0)$ commute with filtering direct limits of sheaves.*

Proof. These are immediate consequences of Proposition (3.2). (Alternatively one could cite [SGA4 VI.5.1].) $\qquad\square$

(3.4) The other result one needs is the following. Let S be a scheme, I a left filtering category, and consider a functor $\mathfrak{X}: I \to (\text{Sch}/S)$, $i \mapsto X_i$. Suppose that all transition morphisms $X_j \to X_i$ are affine. Then the inverse limit $\varprojlim X_i =: X$ in the category of schemes (over S) exists, and all projection maps $p_i: X \to X_i$ are affine [EGA IV, 8.2.3]. Locally, i.e. if the X_i are affine, one has $X = \text{spec} \varinjlim \Gamma(X_i, \mathcal{O}_{X_i})$.

(3.4.1) **Proposition.** — *Let S, $\{X_i\}$ and X be as above. Suppose in addition that the schemes X_i are quasi-compact and quasi-separated. Fix $t \in \{et, b, ret\}$.*

a) *For any sheaf F on X_t the canonical sheaf map*

$$\varinjlim_{I^\circ} (p_i)_t^* (p_i)_{t*} F \longrightarrow F$$

is an isomorphism. If F is an abelian sheaf then

$$\varinjlim_{I^\circ} H^n\left(X_{it}, (p_i)_{t*} F\right) \longrightarrow H^n(X_t, F)$$

is an isomorphism for all $n \geq 0$.

b) *For any t-sheaf F_0 on S let F_i resp. F be the preimage of F_0 on X_i $(i \in I)$ resp. on X. Then $F(X) = \varinjlim_{I^\circ} F_i(X_i)$. If F_0 is an abelian sheaf then all maps*

$$\varinjlim_{I^\circ} H^n\left(X_{it}, F_i\right) \longrightarrow H^n(X_t, F)$$

are isomorphisms, $n \geq 0$.

Proof. In the étale case the proof can be found in VII.5 and VI.8 of [SGA4]. The proof in the other cases proceeds in exactly the same way. Since the proof is quite long and technical, I restrict to briefly indicating the main steps and pointing out what specific changes to make for $t = b$ or $t = ret$.

I use the language of fibered categories and toposes, as done in [loc.cit.]. (See also Sect. 10 for a review of some of these concepts, there however mostly specialized to the case where the base category I is a group.)

Let $\mathcal{G}$ be the category whose objects are the finitely presented étale morphisms of schemes (and whose arrows are obvious), and let $\mathcal{G} \to (\text{Sch})$ be the functor "target". This is a fibered category over (Sch) in which the fibre $\mathcal{G}_T$ over a scheme T is the restricted category of étale T-schemes, Et'/T (2.18.1). By putting the (induced) t-topology on each fibre $\mathcal{G}_T$, the functor $\mathcal{G} \to (\text{Sch})$ becomes a fibered site. Pulling it back by $\mathfrak{X}$ gives a fibered site $\mathcal{G}_{\mathfrak{X}} \to I$ over I. Let $\underline{\mathcal{G}}_{\mathfrak{X}}$ denote the direct limit of the fibered category $\mathcal{G}_{\mathfrak{X}} \to I$. So $\underline{\mathcal{G}}_{\mathfrak{X}}$ is obtained from $\mathcal{G}_{\mathfrak{X}}$ by inverting all cartesian arrows. One equips $\mathcal{G}_{\mathfrak{X}}$ with the total topology and $\underline{\mathcal{G}}_{\mathfrak{X}}$ with the coarsest topology which makes the canonical functor $\mathcal{G}_{\mathfrak{X}} \to \underline{\mathcal{G}}_{\mathfrak{X}}$ continuous. The site $\underline{\mathcal{G}}_{\mathfrak{X}}$ is called the inverse limit of the fibered site $\mathcal{G}_{\mathfrak{X}} \to I$, terminology justified by [SGA4 VI.8.2.3]. There is a unique functor

$$\underline{\mathcal{G}}_{\mathfrak{X}} \longrightarrow \mathcal{G}_X \tag{1}$$

through which the canonical functor $\mathcal{G}_{\mathfrak{X}} \to \mathcal{G}_X$ factors. One has to verify that (1) is an equivalence of sites. Then the assertions follow from the general results in [SGA4 VI.8] for inverse limits of fibered sites over I. (To be specific, a) follows from 8.5.2 and 8.7.4, b) from 8.5.7 and 8.7.5 of [loc.cit.].) The necessary hypotheses are satisfied thanks to (3.2) and (3.3).

In [SGA4 VII.5] it is explained why (1) is an equivalence of sites for the étale topologies. So it is an equivalence of categories anyway. To show that the topologies t are preserved one has to check the following: Let $i \in I$, and let $f_i: V_i \to U_i$ be a morphism in Et'/X_i. Suppose that the pullback $V_i \times_{X_i} X \to U_i \times_{X_i} X$ of f_i to X is t-surjective. Then show that there exists $j \xrightarrow{\alpha} i$ in I such that already the pullback of f_i by $\mathfrak{X}(\alpha): X_j \to X_i$ is t-surjective.

Of course the case $t = b$ follows from the conjunction of the cases $t = et$ and $t = ret$. Both the latter ones are easy consequences of constructibility properties of finitely presented morphisms of schemes (in case $t = ret$ for the real spectrum). To indicate why, let $u_i: U_i \to X_i$ by the structure map of U_i. Let $p: Y \to X_i$ be any scheme over X_i, and denote by f the pullback $V_i \times_{X_i} Y \to U_i \times_{X_i} Y$ of f_i by p. Then

$$f \text{ is surjective} \iff u_i\big(U_i - f_i(V_i)\big) \cap p(Y) = \emptyset, \tag{2}$$

and this remains true if "surjective" is replaced by "real surjective", and the maps and schemes on the right are replaced by their real spectrum counterparts. (The reason for (2) is the "amalgamation property", i.e. the fact that for any diagram $Z' \to Z \leftarrow Z''$ of schemes the map $|Z' \times_Z Z''| \to |Z'| \times_{|Z|} |Z''|$ of underlying sets is surjective; and in the real case the corresponding fact for the real spectra.) Now u_i and f_i are finitely presented, so $u_i(U_i - f_i(V_i))$ is a (locally closed) constructible subset of the spectral space X_i ("Chevalley's theorem"), and dito for real spectra. Moreover $p_i(X)$ is the filtering intersection of the $q_\alpha(X_j)$, where $q_\alpha := \mathfrak{X}(\alpha)$ and α ranges over all arrows $j \to i$ in I. Since the $q_\alpha(X_j)$ are pro-constructible in X_i one has $u_i(U_i - f_i(V_i)) \cap q_\alpha(X_j) = \emptyset$ for some α, and dito for real spectra. Thus f_j is surjective (resp. real surjective) for such α, by (2). $\qquad\square$

(3.5) The next topic will be a study of the fibre functors of the toposes $\widetilde{X}_t$. To begin with, here are some quite general remarks. Let $z = \operatorname{spec} k$ be the spectrum of a field and let $\alpha: z \to X$ be a morphism of schemes. An *étale neighborhood* of α is a factorization $z \to U \to X$ of α with U étale over X. The category $Nb(\alpha)$ of étale neighborhoods of α is left filtering and contains a cofinal subcategory which consists of affine étale neighborhoods. I will write

$$X^\alpha := X^z := \varprojlim_{U \in Nb(\alpha)} U, \tag{3}$$

which is a kind of étale localization of X with respect to α. The scheme X^α is

affine, and its ring of global sections

$$\mathcal{O}_{X,\alpha} := \Gamma(X^\alpha, \mathcal{O}_{X^\alpha}) = \varinjlim_{U \in Nb(\alpha)^\circ} \mathcal{O}_U(U)$$

is a local henselian ring. Note that α has a canonical factorization

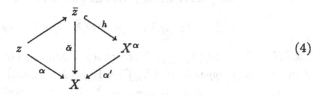

$$(4)$$

in which $h \colon \bar{z} \hookrightarrow X^\alpha$ is the inclusion of the closed point of X^α. Writing $x := \alpha(z) \in X$, the residue field $\kappa(\bar{z})$ of $\mathcal{O}_{X,\alpha}$ is the relative separable algebraic closure of $\kappa(x)$ in $\kappa(z)$; therefore $\mathcal{O}_{X,\alpha}$ lies between the henselization and the strict henselization of $\mathcal{O}_{X,x}$. For reference record

(3.5.1) Lemma. — *Let z be the spectrum of a field and $\alpha \colon z \to X$ a morphism of schemes. Factorize α as in (4).*

a) *For every étale sheaf A on X there are natural bijections*

$$(\alpha^\bullet A)(z) = \varinjlim_{U \in Nb(\alpha)^\circ} A(U) \xrightarrow{\sim} (\alpha'^* A)(X^\alpha) \xrightarrow{\sim} (\bar{\alpha}^* A)(\bar{z}) \xrightarrow{\sim} (\alpha^* A)(z).$$

In particular the natural map $(\alpha^\bullet A)(z) \to (\alpha^* A)(z)$ is bijective. ($\alpha^\bullet A$ is the presheaf inverse image of A, cf. (0.2.3).)

b) *For every abelian étale sheaf A on X the natural maps*

$$\varinjlim_{U \in Nb(\alpha)^\circ} H^n_{et}(U, A|_U) \longrightarrow H^n_{et}(X^\alpha, \alpha'^* A) \longrightarrow H^n_{et}(\bar{z}, \bar{\alpha}^* A) \quad (n \geq 0)$$

are isomorphisms. If $\kappa(z)$ is real closed or separably closed then also $H^n_{et}(\bar{z}, \bar{\alpha}^* A) \xrightarrow{\sim} H^n_{et}(z, \alpha^* A)$, $n \geq 0$.

Proof. This is of course completely well known, but I couldn't locate a convenient direct reference. The first equality in a) is clear, and the first arrows in a) resp. b) are bijective by particular cases of (3.4.1b). For any étale sheaf (resp. abelian étale sheaf) on X^α the inclusion $h \colon \bar{z} \hookrightarrow X^\alpha$ induces a bijection on global sections (resp. on all cohomology groups), [SGA4 VIII.8]. It remains therefore to look at $z \to \bar{z}$, which corresponds to a field extension $K \subset L$ which is relatively separably closed. Let L_s be a separable closure of L and K_s the separable algebraic closure of K in L_s. Then the restriction map $\operatorname{Gal}(L_s/L) \to \operatorname{Gal}(K_s/K)$ is surjective [Bo ch. V, § 10, no. 8], whence the last bijection in a). If K is real closed or separably closed this map is bijective, whence the last assertion in b). $\square$

(3.6) Now the question is considered which morphisms exist between the functors $\widetilde{X}_{et} \to$ (sets), $F \mapsto (\alpha^* F)(z)$, for different maps $\alpha \colon z \to X$ of the above kind. The following is an easy generalization of material in [SGA4 VIII.7].

Every X-scheme $f \colon Y \to X$ gives rise to a functor $\gamma_Y \colon \widetilde{X}_{et} \to$ (sets) which is characterized by $\gamma_Y(F) = \Gamma(Y, f^* F)$. These functors γ_Y are left exact and hence pro-representable [SGA4 I.8.3.3]. So one gets a functor

$$(\mathrm{Sch}/X) \longrightarrow \mathrm{pro}(\widetilde{X}_{et}), \quad Y \longmapsto \gamma_Y. \tag{5}$$

Recall (0.1.5) that $\mathrm{pro}(\widetilde{X}_{et})$, the category of pro-objects in $\widetilde{X}_{et}$, is a full subcategory of the category *opposite* to $(\widetilde{X}_{et})^{\vee} = \underline{\mathrm{Hom}}(\widetilde{X}_{et}, (\mathrm{sets}))$. Obviously the functor (5) extends the composite functor $\mathrm{Et}/X \overset{h}{\hookrightarrow} \widetilde{X}_{et} \overset{\iota}{\hookrightarrow} \mathrm{pro}(\widetilde{X}_{et})$, where ι is the canonical fully faithful functor and h the Yoneda embedding.

(3.6.1) **Proposition.** — *Let X be a scheme and Y, Z schemes over X. Assume that both Y and Z are filtering inverse limits of étale X-schemes which are quasi-compact and quasi-separated, with affine transition maps. Then the canonical map from* $\mathrm{Hom}_X(Y, Z)$ *to* $\mathrm{Hom}_{\mathrm{pro}(\widetilde{X}_{et})}(\gamma_Y, \gamma_Z) = \mathrm{Hom}_{(\widetilde{X}_{et})^{\vee}}(\gamma_Z, \gamma_Y)$ *induced by the functor (5) is bijective.*

Proof. Let $\{Y_i\}_I$ and $\{Z_j\}_J$ be filtering inverse systems in Et/X with the above properties such that $Y \cong \varprojlim Y_i$ and $Z \cong \varprojlim Z_j$. The first part of (3.4.1b) (for $t = et$) says that the canonical maps $\gamma_Y \to \{h_{Y_i}\}_I$ and $\gamma_Z \to \{h_{Z_j}\}_J$ in $\mathrm{pro}(\widetilde{X}_{et})$ are isomorphisms. (I am using obvious shorthand notations for pro-objects.) Thus the composite map

$$\mathrm{Hom}_{\mathrm{pro}(\widetilde{X}_{et})}\big(\{h_{Y_i}\}_I, \{h_{Z_j}\}_J\big) = \varprojlim_j \varprojlim_i \mathrm{Hom}_X(Y_i, Z_j)$$

$$\longrightarrow \varprojlim_j \mathrm{Hom}_X(Y, Z_j) = \mathrm{Hom}_X(Y, Z) \overset{(5)}{\longrightarrow} \mathrm{Hom}_{\mathrm{pro}(\widetilde{X}_{et})}(\gamma_Y, \gamma_Z)$$

is bijective. The first map is bijective since $\varinjlim_i \mathrm{Hom}_X(Y_i, Z_j) \to \mathrm{Hom}_X(Y, Z_j)$ is bijective for every j by [EGA IV, 8.8.2]. Hence the last map is bijective as well. $\square$

(3.6.2) **Corollary.** — *If $x \to X$ and $y \to X$ are morphisms of schemes with x, y spectra of fields, then* $\mathrm{Hom}_{\mathrm{pro}(\widetilde{X}_{et})}(\gamma_x, \gamma_y) = \mathrm{Hom}_{(\widetilde{X}_{et})^{\vee}}(\gamma_y, \gamma_x)$ *is in bijection with* $\mathrm{Hom}_X(x, X^y)$ *(cf. (3)).*

Proof. Since $\gamma_x \to \gamma_{X^x}$ and $\gamma_y \to \gamma_{X^y}$ are isomorphisms by Lemma (3.5.1a), the first set in the assertion is bijective to $\mathrm{Hom}_X(X^x, X^y)$, by (3.6.1). This latter set is bijective to $\mathrm{Hom}_X(x, X^y)$, by the henselian property of X^x. $\square$

Now the stalk (*alias* fibre) functors of each of the topologies et, b, ret on Et/X will be described. The following also serves the purpose of fixing some terminology. For the notion of points compare (0.2.4).

(3.7) Points of X_{ret}. By Sect. 1 the topos $\widetilde{X}_{ret}$ is equivalent to the category of sheaves on the real spectrum X_r. Hence $\underline{pt}(\widetilde{X}_{ret})$ is equivalent to $\underline{pt}(\widetilde{X}_r)$. The latter category is equivalent to the ordered set $(X_r, \succ)$ where $\succ$ is the specialization relation. (This is true for any sober topological space, i.e. space in which every non-empty irreducible closed subset has a unique generic point, [SGA4 IV.7.1].) To make these equivalences more explicit, fix an element $\xi \in X_r$ and represent it by a morphism $\alpha: x \to X$ where x is the spectrum of a real closed field. The global sections functor $F \mapsto F(x)$ on $\widetilde{x}_{ret}$ is an equivalence of categories $\widetilde{x}_{ret} \to$ (sets). Hence the functor

$$p_\alpha^*: \widetilde{X}_{ret} \longrightarrow \text{(sets)}, \quad p_\alpha^* B := B_\alpha := (\alpha_{ret}^* B)(x)$$

is obviously a fibre functor. It is clear that, under the equivalence of $\widetilde{X}_{ret}$ with $\widetilde{X}_r$, this p_α^* corresponds to the fibre functor on $\widetilde{X}_r$ given by ξ. (This follows also explicitly from (1.14).) Therefore I will also write p_ξ^* for p_α^*, and p_ξ for a point of $\widetilde{X}_{ret}$ associated with p_ξ^*.

(3.7.1) Since in x_{ret} there is only the trivial covering sieve for x, it is clear that for every presheaf Q on Et/x the map $Q(x) \to (a_{ret}Q)(x)$ is bijective. Since $\alpha_{ret}^* \circ a_{ret} = a_{ret} \circ \alpha^*$ holds (as functors from $(Et/X)\widehat{}$ to $\widetilde{x}_{ret}$) it follows for every presheaf P on Et/X that

$$p_\alpha^* \circ a_{ret}(P) = (\alpha_{ret}^* \circ a_{ret} P)(x) = (\alpha^* P)(x) = \varinjlim_{U \in Nb(\alpha)^\circ} P(U).$$

Using (3.5.1a) one gets in particular:

(3.7.2) Proposition. — *If $\xi \in X_r$ is represented by a morphism $\alpha: x \to X$, where x is the spectrum of a real closed field, then for every $A \in \widetilde{X}_{et}$ the stalk of ρA in ξ is*

$$(\rho A)_\xi = H^0(x_{et}, \alpha^* A). \qquad \square$$

As a corollary one sees from (3.7.2) that the functor ρ (and hence also j_*) preserves arbitrary coproducts ($\coprod$), resp. direct sums ($\bigoplus$) in the case of abelian sheaves. Indeed, this is obvious for $H^0(x_{et}, -)$.

(3.7.3) Recall that a *strictly real henselian* (or *local*) *ring* is a henselian local ring with a real closed residue field. (The terminology varies in the literature; also "real closed local ring" has been used.) A *strictly real local scheme* is the spectrum of such a ring.

Definition. Let X be a scheme.

a) A *real geometric point* (or simply *real point*) of X is a morphism of schemes $\alpha: x \to X$ where x is the spectrum of a real closed field.

b) Let a real geometric point $\alpha: x \to X$ of X be given, and let $\xi \in X_r$ be the element of the real spectrum represented by α. The X-scheme X^α of (3) is called the *strict real localization* of X at α (or x, or ξ), and is also denoted by X^x or X^ξ.

c) Given α as above, the global ring of sections of X^α is called the *strictly real local ring* of X at α (or x, or ξ), and is denoted by $\mathcal{O}_{X,\alpha}$ (or $\mathcal{O}_{X,x}$, or $\mathcal{O}_{X,\xi}$). If $X = \operatorname{spec} A$ is affine it is also customary to write A_ξ for $\mathcal{O}_{\operatorname{spec} A,\xi}$, and to call this ring the *strict real henselization* of A at ξ.

In the proof of the equivalence $\tilde{X}_{ret} \sim \tilde{X}_r$ as given in [CRC], the essential step is to establish rigidity of the strict real henselizations of a ring, cf. (1.4). The proof given in Sect. 1 did not use this rigidity property. Rather one can now turn things around and deduce rigidity from this proof:

(3.7.4) Corollary (to Section 1). — *Let A be a ring. Let $\xi, \eta \in \operatorname{sper} A$, and let A_ξ, A_η be the corresponding strict real henselizations of A. Then there is at most one A-homomorphism $A_\eta \to A_\xi$; and there is one if and only if $\xi \succ \eta$ in $\operatorname{sper} A$.*

Proof. Let $X := \operatorname{spec} A$. Let ξ and η be represented by real geometric points $x \to X$ and $y \to X$, respectively. Then

$$\operatorname{Hom}_{\underline{\mathrm{pt}}(\tilde{X}_{ret})}(p_\xi, p_\eta) = \operatorname{Hom}_{\mathrm{pro}(\tilde{X}_{et})}(\gamma_x, \gamma_y). \qquad (6)$$

By (3.6.1) the right hand set is $\operatorname{Hom}_X(X^x, X^y) = \operatorname{Hom}_A(A_\eta, A_\xi)$, while the left hand set is $*$ or $\emptyset$, depending on whether $\xi \succ \eta$ holds or not.

A rigorous proof for (6) goes as follows. Consider the diagram

$$
\begin{array}{ccc}
\underline{\mathrm{pt}}(\tilde{X}_{ret}) & \xrightarrow{\ b\ } & \mathrm{pro}(\mathrm{Et}/X) \\
{\scriptstyle a}\downarrow & & {\scriptstyle h^*}\uparrow\downarrow{\scriptstyle h_*} \\
\mathrm{pro}(\tilde{X}_{ret}) & \xrightarrow{\ \rho^*\ } & \mathrm{pro}(\tilde{X}_{et})
\end{array}
$$

in which the arrows are the following functors: a sends p to p^*, b sends p to $p^* \circ \epsilon_{ret} = p^* \circ \rho \circ \epsilon_{et}$, and ρ^* is the pro-adjoint of ρ [SGA4 I.8.11], i.e. is the restriction to the pro-categories of the functor $\rho^*: (\tilde{X}_{ret})^\vee \to (\tilde{X}_{et})^\vee$ induced by ρ. Moreover let $h = \epsilon_{et}: \mathrm{Et}/X \to \tilde{X}_{et}$ be the canonical fully faithful functor; then $h_* := \mathrm{pro}(h)$, and h^* is the pro-adjoint of h. Since h_* is fully faithful and h^* is *left* adjoint to h_* (not right adjoint, as stated in [loc.cit., I.8.11.5]!), it follows that $h^* \circ h_* \cong \mathrm{id}$.

Now b is fully faithful [SGA4 IV.4.9.4], and $b = h^* \circ \rho^* \circ a$ holds by definition. The image of $\rho^* \circ a$ lies (essentially) in the image of h_*: If p is a point of $\tilde{X}_{ret}$ represented by a real geometric point $x \to X$ of X, then

$$\rho^* \circ a(p) = p^* \circ \rho \cong \gamma_x \cong \gamma_{X^x}$$

by (3.7.2), and X^x is a $\varprojlim$ of étale X-schemes. Hence it follows that $\rho^* \circ a \cong h_* \circ b$, and that this functor is fully faithful, which proves (6). $\qquad\square$

(3.8) Points of X_{et}. To set up terminology I recall from [SGA4 VIII] the description of $\underline{pt}(\widetilde{X}_{et})$. A *geometric* (or *étale*) *point* of X is a morphism $\alpha \colon z \to X$ with z the spectrum of a separably closed field. Such α defines the fibre functor $p_\alpha^* := \gamma_z \colon \widetilde{X}_{et} \to$ (sets), $A \mapsto A_\alpha := (\alpha_{et}^* A)(z)$. For every presheaf P on Et/X one has (3.5.1a)

$$p_\alpha^* \circ a_{et}(P) = \left(\alpha_{et}^* \circ a_{et} P\right)(z) = (\alpha^* P)(z) = \varinjlim_{U \in Nb(\alpha)^\circ} P(U).$$

Every fibre functor of $\widetilde{X}_{et}$ is isomorphic to such a p_α^*; and p_α^* and p_β^* are isomorphic iff α and β have the same image point in X. So the isomorphism classes in $\underline{pt}(\widetilde{X}_{et})$ are in bijection with the underlying set of X. But the points of $\widetilde{X}_{et}$ admit nontrivial automorphisms in general; to be precise, the automorphism group of a point centered at $x \in X$ is isomorphic to the absolute Galois group of the residue field $\kappa(x)$. More generally, if α and β are geometric points of X then the morphisms $p_\alpha \to p_\beta$ between the associated points are in bijection with $\mathrm{Hom}_X(X^\alpha, X^\beta) \approx \mathrm{Hom}_X(\bar{x}_\alpha, X^\beta)$ where $\bar{x}_\alpha$ is the closed point of X^α.

(3.9) Points of X_b. Hence the points of $\widetilde{X}_b$ are the following, cf. [SGA4 IV.9.7]: The functors

$$\underline{pt}(\widetilde{X}_{et}) \longrightarrow \underline{pt}(\widetilde{X}_b) \longleftarrow \underline{pt}(\widetilde{X}_{ret})$$

induced by j resp. i are fully faithful, and the set of isomorphism classes of points of $\widetilde{X}_b$ is the disjoint union of the sets of isomorphism classes of points of $\widetilde{X}_{et}$ resp. $\widetilde{X}_{ret}$. To have a convenient way of speaking I will call a point of $\widetilde{X}_b$ either *étale* or *real*, depending on whether it is isomorphic to a point coming from $\widetilde{X}_{et}$ or from $\widetilde{X}_{ret}$. Since both $\widetilde{X}_{et}$ and $\widetilde{X}_{ret}$ have sufficiently many points the same is true for $\widetilde{X}_b$. This is convenient since it allows to argue stalkwise in many situations.

To avoid awkward expressions I will also abuse language and say for example, "Let $\alpha \colon z \to X$ be a point of X_b" instead of "Let $\alpha \colon z \to X$ be a morphism of schemes in which z is the spectrum of a field which is real or separably closed". Such a "point" α may be further qualified as real or étale depending on whether $\kappa(z)$ is real closed or separably closed. The point of $\widetilde{X}_b$ in the proper sense induced by α (i.e. the topos morphism (sets) $\to \widetilde{X}_b$) will be denoted by $p_\alpha = (p_\alpha^*, p_{\alpha*})$, and I will also write $F \mapsto F_\alpha$ or $F \mapsto F_z$ for the fibre functor $p_\alpha^* \colon \widetilde{X}_b \to$ (sets). Note that if α is étale (resp. real) then $p_\alpha = (p_\alpha^*, p_{\alpha*})$ may as well denote the corresponding point of $\widetilde{X}_{et}$ (resp. of $\widetilde{X}_{ret}$).

(3.9.1) To complete the description of the points of X_b we have to study the morphisms between real and étale points. Thus let $\xi \colon x \to X$ and $\eta \colon y \to X$ be points of X_b, with ξ real and η étale. Then $\mathrm{Hom}(p_\xi, p_\eta) = \emptyset$ in $\underline{pt}(\widetilde{X}_b)$ since for $A \in \widetilde{X}_{et}$ with $A_\eta \neq \emptyset$ there is no map from $(j_! A)_\eta = A_\eta$ to $(j_! A)_\xi = \emptyset$. Conversely, the morphisms $p_\eta \to p_\xi$ in $\underline{pt}(\widetilde{X}_b)$ are in natural bijection with the morphisms $p_\xi^* \circ \rho \to p_\eta^*$ of functors $\widetilde{X}_{et} \to$ (sets). This bijection is induced by the functors

$$\underline{pt}(\widetilde{X}_b)^\circ \longrightarrow (\widetilde{X}_b)^\vee \xrightarrow{\ -\circ j_*\ } (\widetilde{X}_{et})^\vee.$$

Indeed, if $\varphi: p_\xi^* \to p_\eta^*$ is a morphism of fibre functors of $\widetilde{X}_b$, then φ can be recovered from $\varphi * j_*$ (a morphism $p_\xi^* \circ j_* \to p_\eta^* \circ j_*$ of functors $\widetilde{X}_{et} \to$ (sets)) by the commutative diagram

$$
\begin{array}{ccc}
p_\xi^* & \xrightarrow{\quad \varphi \quad} & p_\eta^* \\
\text{adj}\Big\downarrow & & \Big\| \,\text{adj} \\
p_\xi^* * j_* j^* & \xrightarrow{\varphi * j_* j^*} & p_\eta^* * j_* j^*
\end{array}
$$

in $(\widetilde{X}_b)^\vee = \underline{\mathrm{Hom}}(\widetilde{X}_b, (\text{sets}))$, since the right vertical map is the identity. (These facts about the category $\underline{\mathrm{pt}}(\widetilde{X}_b)$ are true in any glued topos, cf. [SGA4 IV.9.7.2].) So from (3.6.2) it follows that

$$
\mathrm{Hom}_{\underline{\mathrm{pt}}(\widetilde{X}_b)}(p_\eta, p_\xi) \approx \mathrm{Hom}_X(X^\eta, X^\xi) \approx \mathrm{Hom}_X(y, X^\xi).
$$

In particular, this set is non-empty if and only if the specialization $\eta(y) \succ \xi(x)$ holds in X. Assume this is the case. Let $X' := X \otimes_{\mathbb{Z}} \mathbb{Z}[\sqrt{-1}]$, and let $\eta_1, \eta_2: y \rightrightarrows X'$ be the two (different!) liftings of η to X', moreover $\xi': x' := x[\sqrt{-1}] \to X'$ the base extension of ξ by $X' \to X$. Then the morphisms $p_\eta \to p_\xi$ of points of $\widetilde{X}_b$ are in bijection with the morphisms $p_{\eta_\nu} \to p_{\xi'}$ between the geometric points of X', for both $\nu = 1, 2$. This follows since $(X')^{\xi'} \cong X^\xi \times_X X'$ (as X'-schemes).

To summarize (part of) the above discussion:

(3.9.2) Proposition. — *Let X be a scheme. Fix $t \in \{et, b, ret\}$, and let $\alpha: z \to X$ be a t-point of X. Recall that a_t is the functor "associated t-sheaf".*

a) The corresponding fibre functor $p_\alpha^*: \widetilde{X}_t \to$ (sets) satisfies

$$
p_\alpha^* \circ a_t(P) = \varinjlim_{U \in \overline{Nb}(\alpha)^\circ} P(U) = (\alpha^\bullet P)(z) = (\alpha_t^* \circ a_t P)(z)
$$

for every presheaf P on Et/X. If A is a sheaf on X_{et} and α is real then also

$$
p_\alpha^*(\rho A) = p_\alpha^*(j_* A) = (\alpha_{et}^* A)(z).
$$

b) If $\beta: y \to X$ is a second t-point of X then $\mathrm{Hom}_{\underline{\mathrm{pt}}(\widetilde{X}_t)}(p_\beta, p_\alpha)$ ist identified with

$$
\mathrm{Hom}_X(X^\beta, X^\alpha) = \mathrm{Hom}_X(y, X^\alpha) = \mathrm{Hom}_{(\widetilde{X}_{et})^\vee}(p_\alpha^* \circ a_t, p_\beta^* \circ a_t).
$$

This set is empty if β is real and α is étale (and $t = b$). In particular, the functor $p \mapsto p^* \circ a_t$ from $\underline{\mathrm{pt}}(\widetilde{X}_t)^\circ$ to $\underline{\mathrm{Hom}}(\widetilde{X}_{et}, (\text{sets}))$ is fully faithful. $\square$

As an application I want now to record how the various functors between the three topologies are compatible with the change of the scheme:

(3.10) Proposition. — *Let $f: Y \to X$ be a morphism of schemes. The diagram*

$$\begin{array}{ccccc}
Y_{et} & \xrightarrow{\ j\ } & Y_b & \xleftarrow{\ i\ } & Y_{ret} \\
{\scriptstyle f_{et}}\downarrow & & {\scriptstyle f_b}\downarrow & & \downarrow{\scriptstyle f_{ret}} \\
X_{et} & \xrightarrow{\ j\ } & X_b & \xleftarrow{\ i\ } & X_{ret}
\end{array}$$

of sites and site morphisms commutes. Moreover the following commutation rules hold:

a) $f_b^* i_* = i_* f_{ret}^* \colon \widetilde{X}_{ret} \longrightarrow \widetilde{Y}_b$;

b) $f_b^* j_! = j_! f_{et}^* \colon \widetilde{X}_{et} \longrightarrow \widetilde{Y}_b$;

c) $f_b^* j_* = j_* f_{et}^* \colon \widetilde{X}_{et} \longrightarrow \widetilde{Y}_b$;

d) $f_{ret}^* \rho = \rho f_{et}^* \colon \widetilde{X}_{et} \longrightarrow \widetilde{Y}_{ret}$;

e) $j^* f_{b*} = f_{et*} j^* \colon \widetilde{Y}_b \longrightarrow \widetilde{X}_{et}$;

f) *if f is étale then also* $j^* f_{b!} = f_{et!} j^* \colon \widetilde{Y}_b \longrightarrow \widetilde{X}_{et}$.

The same rules are true when the respective functors and categories of set-valued sheaves are replaced by their abelian sheaves counterparts.

(3.10.1) Remark. The canonical morphisms

g) $i^* f_{b*} \longrightarrow f_{ret*} i^*$ (of functors $\widetilde{Y}_b \to \widetilde{X}_{ret}$),

h) $\rho f_{et*} \longrightarrow f_{ret*} \rho$ (of functors $\widetilde{Y}_{et} \to \widetilde{X}_{ret}$),

i) $j_! f_{et*} \longrightarrow f_{b*} j_!$ (of functors $\widetilde{Y}_{et} \to \widetilde{X}_b$)

are not in general isomorphisms.

Proof of (3.10). It is trivial that the diagram commutes. a) follows from $i^*(f_b^* i_*) = f_{ret}^* i^* i_* = i^*(i_* f_{ret}^*)$ and $j^*(f_b^* i_*) = f_{et}^* \emptyset = \emptyset = j^*(i_* f_{ret}^*)$. b) is shown similarly, and e) follows from b) by adjunction. c) and d) are easily seen to be equivalent, and c) implies f) by adjunction. So it remains to prove d), say. To check for $A \in \widetilde{X}_{et}$ that the canonical map

$$f_{ret}^* \rho A = i^* f_b^* j_* A \longrightarrow i^* j_* f_{et}^* A = \rho f_{et}^* A$$

is an isomorphism one may argue stalkwise. So let $\eta : y \to Y$ be a real point of Y. Then by (3.7.2) one has

$$\left(f_{ret}^* \rho A\right)_\eta = (\rho A)_{f \circ \eta} \cong \left((f \circ \eta)_{et}^* A\right)(y) = \left(\eta_{et}^* f_{et}^* A\right)(y) \cong \left(\rho f_{et}^* A\right)_\eta. \qquad \square$$

(3.11) Proposition. — *Let $f: Y \to X$ be a morphism of schemes which is quasi-compact and quasi-separated, and fix $t \in \{et, b, ret\}$. Let $\alpha: z \to X$ be a t-point of X, and let X^α be the corresponding local scheme (3.5). Let F be a sheaf on Y_t. For the stalks in α one has*

a) $(f_{t*} F)_\alpha \xrightarrow{\sim} \Gamma(X^\alpha \times_X Y, \tilde{F})$;

b) *if F is an abelian sheaf then* $(R^n f_{t*} F)_\alpha \xrightarrow{\sim} H_t^n(X^\alpha \times_X Y, \tilde{F})$ *for every* $n \geq 0$.

Here $\tilde{F} \in (X^\alpha \times_X Y)_t^{\sim}$ denotes the pullback of F under $\mathrm{pr}_2 \colon X^\alpha \times_X Y \to Y$, i.e. $\tilde{F} = (\mathrm{pr}_2)_t^* F$.

Proof. This follows in completely the same way as in the étale case [SGA4 VIII.5.2], using (3.4b) and the description of the points of $\widetilde{X}_t$ (3.9.2).   □

(3.11.1) Corollary. — *If $f: Y \to X$ is a finite morphism of schemes then the additive functor $f_{t*}: \mathrm{Ab}(Y_t) \to \mathrm{Ab}(X_t)$ is exact for $t = et, b, ret$. Moreover, for every sheaf F on Y_t and every t-point $\alpha: z \to X$ of X there is a canonical bijection*

$$(f_{t*}F)_\alpha \;\xrightarrow{\sim}\; \prod_{y \in z \times_X Y} F_y.$$

Proof (compare [SGA4 VIII.5.5]): Since $X^\alpha \times_X Y$ is finite over X^α it is a finite disjoint sum of strictly local or strictly real local schemes. So the assertions follow from (3.11) and

(3.11.2) Lemma. — *Let X be a strictly local or strictly real local scheme, and let $h: \bar{x} \hookrightarrow X$ be the inclusion of the closed point. Let $t \in \{et, b, ret\}$. For every t-sheaf F on X the map $F(X) \to (h_t^* F)(\bar{x})$ is bijective. For every abelian t-sheaf F on X one has $H^n(X_t, F) = 0$ for $n > 0$, unless X is strictly real local and $t = et$.*

Proof. The first assertion is a particular case of (3.9.2a) except when $t = et$ and $\kappa(\bar{x})$ is real closed, or $t = ret$ and $\kappa(\bar{x})$ is separably closed. The first of these two cases is clear [SGA4 VIII.8], and in the second one only has to remark that a strictly henselian ring has empty real spectrum. The assertion about cohomology follows from this since the global sections functor $\mathrm{Ab}(\bar{x}_t) \to (\mathrm{Ab})$ is exact in all cases except when $t = et$ and $\kappa(\bar{x})$ is real closed.   □

Next follow the basic exactness properties of the functors associated with the glueing situation. Below I always mean the abelian sheaves versions of the respective functors.

(3.12) Proposition. — *Let X be a scheme, and let $\pi: X' = X \otimes_{\mathbb{Z}} \mathbb{Z}[\sqrt{-1}] \to X$ be the projection to the first factor.*
a) *The functors $j_!$, j^*, i^*, i_* are all exact. The functors j_*, ρ, $i^!$ are left exact.*
b) *If $A \in \mathrm{Ab}(X_{et})$ then $R^n j_* A = i_* R^n \rho A$ for $n \geq 1$.*
c) *If $A \in \mathrm{Ab}(X_{et})$ and $\xi: x \to X$ is a real geometric point then $(R^n j_* A)_\xi = (R^n \rho A)_\xi = H_{et}^n(x, \xi_{et}^* A)$ for $n \geq 0$. In particular, $R^n j_* A$ and $R^n \rho A$ are $\mathbb{Z}/2$-sheaves (i.e. are killed by 2) for every $n \geq 1$.*
d) *The functors $j_* \pi_*: \mathrm{Ab}(X'_{et}) \to \mathrm{Ab}(X_b)$ and $\rho \pi_*: \mathrm{Ab}(X'_{et}) \to \mathrm{Ab}(X_{ret})$ are exact $(\pi_* := \pi_{et*})$.*

Proof. a) is clear. For b) note that $\mathrm{id} = R(j^* j_*) = j^* R j_*$. Thus $j^* R^n j_* = 0$, or equivalently $R^n j_* = i_* i^* R^n j_*$, for $n \geq 1$. But $i^* R j_* = R(i^* j_*) = R\rho$. c) $R^n j_* A$ is

the X_b-sheaf associated with the presheaf $U \mapsto H^n_{et}(U, A)$ on Et/X [SGA4 V.5.1].
Thus

$$(R^n j_* A)_\xi = \varinjlim_{U \in N b(\xi)^\circ} H^n_{et}(U, A) = H^n_{et}(x, \xi^*_{et} A),$$

the second equality by (3.5.1). As cohomology groups of G, the group of order two,
these groups are annihilated by 2 for $n \geq 1$. d) follows from c): For $S \in Ab(X'_{et})$
and ξ as before one has

$$\left(R^n (j_* \pi_*)(S) \right)_\xi = H^n_{et}(x, \xi^*_{et} \pi_* S) = H^n_{et}(x', \xi'^*_{et} S)$$

by finite base change, where $\xi' : x' \to X'$ is the base extension of ξ by π. Since x'
is the spectrum of an algebraically closed field these groups vanish for $n > 0$. $\qquad\square$

(3.12.1) Remark. The sequence of derived functors $R^n \rho$, $n = 1, 2, \dots$ is periodic.
To see this, note that there are natural cup product pairings $R^p \rho A \otimes R^q \rho B \to$
$R^{p+q} \rho(A \otimes B)$ for A, $B \in Ab(X_{et})$, and similarly for the $R^p j_*$. From c) of (3.12)
one sees that $R^n \rho \mathbb{Z} = \mathbb{Z}/2$ for even $n > 0$, and $R^n \rho \mathbb{Z} = 0$ for odd n. Again from
c) it follows that the cup pairings with $R^2 \rho \mathbb{Z}$ give isomorphisms

$$R^n \rho A \xrightarrow{\sim} R^{n+2} \rho A \quad (n \geq 1)$$

for every $A \in Ab(X_{et})$. If $2A = 0$ then cup product with $R^1 \rho(\mathbb{Z}/2) = \mathbb{Z}/2$ gives
isomorphisms $R^n \rho A \xrightarrow{\sim} R^{n+1} \rho A$ $(n \geq 1)$.

(3.12.2) Remark. There are commutation rules similar to those of (3.10) which
involve derived functors (for abelian sheaves). Let $f : Y \to X$ be a morphism of
schemes. If one derives e) one finds

$$j^* \circ R f_{b*} = R f_{et*} \circ j^*$$

since j^* has an exact left adjoint. (On the other hand, note that $i^* R f_{b*} \to R f_{ret*} i^*$
will rarely be an isomorphism!) It is also of interest to look at c) and d): From c)
of (3.12) it follows immediately that

$$f^*_{ret} \circ R\rho \xrightarrow{\sim} R\rho \circ f^*_{et},$$

since for every $A \in Ab(X_{et})$ and $n \geq 0$ the stalks of the map $f^*_{ret} R^n \rho A \to R^n \rho(f^*_{et} A)$
are isomorphisms. Note that this also implies

$$f^*_b \circ R j_* \xrightarrow{\sim} R j_* \circ f^*_{et}.$$

4. Some reminders on Weil restrictions

The purpose of this section is to collect some properties of the Weil restriction functor which will be applied in Sect. 5. In particular we need some existence criteria and have to know that Weil restriction preserves certain properties of morphisms. Part of the following material can be found in §7.6 of [BLR]. For the purpose of this book it seemed more appropriate, however, to include an independent presentation, which also contains some additional material not found in [BLR].

Recall that (Sch/X) is the category of schemes over X and $(\mathrm{Sch}/X)\hat{}$ the category of presheaves on (Sch/X).

(4.1) Definition.

a) Let $f: Y \to X$ be a morphism of schemes and let V be a Y-scheme. If the presheaf

$$U \longmapsto \mathrm{Hom}_Y(U \times_X Y, V)$$

on (Sch/X) is representable by an X-scheme R, then R is called the *Weil restriction* of V with respect to f, in symbols: $R = \mathrm{Res}_{Y/X}(V)$.

b) Let $A \to B$ be a ring homomorphism and let D be a B-algebra. If the functor

$$\{A\text{-algebras}\} \longrightarrow (\text{sets}), \quad C \longmapsto \mathrm{Hom}_{B\text{-alg}}(D, C \otimes_A B)$$

is representable by an A-algebra W, then W is called the *Weil restriction* of D with respect to B/A, in symbols: $W = \mathrm{res}_{B/A}(D)$.

(4.2) Remarks. Let $f: Y \to X$ be a morphism of schemes.

(4.2.1) In the literature the Weil restriction $\mathrm{Res}_{Y/X}(V)$ is often denoted by $\prod_{Y/X} V$ or $\prod_{Y/X}(V/Y)$.

(4.2.2) The defining property of the Weil restriction

$$\mathrm{Hom}_X(U, \mathrm{Res}_{Y/X}(V)) = \mathrm{Hom}_Y(U \times_X Y, V) \quad \text{for all } U \in (\mathrm{Sch}/X)$$

shows that $\mathrm{Res}_{Y/X}$, as far as it is defined, is right adjoint to the functor "base extension by $Y \to X$". Of course this picture comes from an adjunction in the true sense: The base extension functor $U \mapsto U_Y := U \times_X Y$ from (Sch/X) to (Sch/Y) is right adjoint to the functor which sends a Y-scheme $V \to Y$ to the X-scheme

$V \to Y \xrightarrow{f} X$. Denote the latter functor by $V \mapsto V|_X$. Correspondingly there is a quadruplet of adjoint functors

$$(\text{Sch}/Y)\hat{\ }$$

$$\Big\downarrow f_! \quad \Big\uparrow f^* \quad \Big\downarrow f_* \quad \Big\uparrow f^!$$

$$(\text{Sch}/X)\hat{\ }$$

where

$$(f^*P)(V) = P(V|_X) \qquad \text{for } P \in (\text{Sch}/X)\hat{\ }$$

and

$$(f_*Q)(U) = Q(U_Y) \qquad \text{for } Q \in (\text{Sch}/Y)\hat{\ },$$

see [SGA4 I.5]. On representable presheaves one has

$$f_!(h_V) = h_{V|_X} \quad \text{and} \quad f^*(h_U) = h_{U_Y},$$

for $V \in (\text{Sch}/Y)$ and $U \in (\text{Sch}/X)$. (Recall (0.1.3) that h denotes the Yoneda embedding.) Let V be a Y-scheme and consider the presheaf $f_*(h_V)$ on (Sch/X). By definition the Weil restriction $\text{Res}_{Y/X}(V)$ is the X-scheme representing this presheaf, if such an X-scheme exists. Note that

$$(f^!P)(V) = P(\text{Res}_{Y/X}V)$$

holds for $P \in (\text{Sch}/X)\hat{\ }$, provided that $\text{Res}_{Y/X}(V)$ exists. Also observe that one has $\text{Res}_{Y/X}(Y) = X$ in any case.

(4.2.3) Let V be a Y-scheme. If $\text{Res}_{Y/X}(V)$ exists then for every base change $X' \to X$ also $\text{Res}_{Y'/X'}(V')$ exists, where $Y' := Y \times_X X'$ and $V' := V \times_X X'$; and moreover $\text{Res}_{Y'/X'}(V') = \text{Res}_{Y/X}(V) \times_X X'$. Indeed, this is a purely formal consequence of the definition: Consider the diagram consisting of cartesian squares

$$
\begin{array}{ccccc}
V' & \longrightarrow & Y' & \xrightarrow{f'} & X' \\
\downarrow & & \downarrow g' & & \downarrow g \\
V & \longrightarrow & Y & \xrightarrow{f} & X.
\end{array}
$$

By definition $\text{Res}_{Y/X}(V) \times_X X'$ represents the presheaf $g^* f_*(h_V)$ on (Sch/X'). By base change $g^* f_* = f'_* g'^*$ (which is a formal triviality) this is the presheaf $f'_* g'^*(h_V) = f'_*(h_{V'})$.

(4.2.4) Let V be a scheme over Y. Since any representable presheaf on (Sch/X) is a sheaf for the Zariski topology (or even for the *fpqc* topology, which is finer), and since the base change functor

$$(\text{Sch}/X) \longrightarrow (\text{Sch}/Y), \quad U \longmapsto U \times_X Y$$

is continuous with respect to the Zariski (or *fpqc*) topologies (so $f_*(h_V)$ is a sheaf for these topologies), it suffices for an X-scheme R to satisfy (functorially)

$$\operatorname{Hom}_X(U, R) \approx \operatorname{Hom}_Y(U_Y, V)$$

for all X-schemes U which are *affine*, in order that $R = \operatorname{Res}_{Y/X}(V)$. In particular, definitions a) and b) of (4.1) are compatible in the sense that if $X = \operatorname{spec} A$, $Y = \operatorname{spec} B$ and $V = \operatorname{spec} D$ are affine, then $\operatorname{res}_{B/A}(D)$ exists if and only if $\operatorname{Res}_{Y/X}(V)$ exists and is affine, and $\operatorname{Res}_{Y/X}(V) = \operatorname{spec}(\operatorname{res}_{B/A}(D))$ in this case.

Moreover, $f_*(h_V)$ is representable iff it is representable Zariski locally on X. Therefore $\operatorname{Res}_{Y/X}(V)$ exists if and only if X has an open covering $\{X_\lambda\}$ such that $\operatorname{Res}_{Y_\lambda/X_\lambda}(V_\lambda)$ exists for every index λ, where Y_λ and V_λ are the base extensions of Y and V, respectively, by $X_\lambda \subset X$.

(4.2.5) Assume that U is an X-scheme for which $\operatorname{Res} U_Y := \operatorname{Res}_{Y/X}(U_Y)$ exists. If $Y \to X$ is faithfully flat and quasi-compact then the adjunction morphism $\alpha\colon U \to \operatorname{Res} U_Y$ is a closed embedding. Indeed, by descent it suffices to verify this for the base extension α_Y of α. The composite map

$$U_Y \xrightarrow{\alpha_Y} (\operatorname{Res} U_Y)_Y \xrightarrow{\lambda} U_Y$$

is the identity, where λ denotes the (other) adjunction map for U_Y. Hence α_Y is a closed embedding. (It is not necessary to require that U is separated, as done in [BLR, p. 198].)

(4.2.6) Assume $Y = Y_1 \amalg \cdots \amalg Y_n$ where the Y_i are X-schemes. For $V \in (\operatorname{Sch}/Y)$ write $V_i := V \times_Y Y_i$. If $\operatorname{Res}_{Y_i/X}(V_i)$ exists for $i = 1, \ldots, n$ then also $\operatorname{Res}_{Y/X}(V)$ exists, and

$$\operatorname{Res}_{Y/X}(V) = \operatorname{Res}_{Y_1/X}(V_1) \times_X \cdots \times_X \operatorname{Res}_{Y_n/X}(V_n).$$

In particular, if $Y \to X$ is a *trivial* covering of finite degree n (i.e. $Y \cong X \amalg \cdots \amalg X$ as X-schemes) then $\operatorname{Res}_{Y/X}(V)$ exists for every Y-scheme $V = V_1 \amalg \cdots \amalg V_n$, namely $\operatorname{Res}_{Y/X}(V) = V_1 \times_X \cdots \times_X V_n$. (Probably this is the reason for Grothendieck's notation $\prod_{Y/X}$.)

(4.3) In general the Weil restriction does not exist, not even in very elementary cases. In [FGA, p. 195-13] Grothendieck argues that $\operatorname{Res}_{Y/X}(V)$ can exist for "general" V only if $Y \to X$ is flat. Moreover he remarks that it is only reasonable to expect the existence of $\operatorname{Res}_{Y/X}(V)$ for sufficiently general V if $Y \to X$ is in addition proper. Under these restrictions he shows the existence of $\operatorname{Res}_{Y/X}(V)$ under fairly general conditions. Namely, if $Y \to X$ is proper, flat and finitely presented then $\operatorname{Res}_{Y/X}(V)$ exists for every quasi-projective Y-scheme V. He obtains this as a corollary to his construction of Hilbert schemes, cf. [FGA, exp. 221].

In this work only modest use will be made of Weil restrictions. In fact, only Weil restrictions with respect to projections $X \otimes_{\mathbb{Z}} \mathbb{Z}[\sqrt{-1}] \to X$ will be needed. In this case the Weil restrictions can be obtained in a much more elementary way. Therefore these constructions will be given directly here, without reference to Grothendieck's general theorem. More generally I will work under the hypothesis that $f : Y \to X$ is finite and locally free, the latter condition meaning that $f_* \mathcal{O}_Y$ is a locally free $\mathcal{O}_X$-Module. (If X is locally noetherian it is equivalent that f is finite and flat.)

A first supply of examples where the Weil restriction exists is provided by

(4.4) Proposition. — *Let $f : Y \to X$ be a finite morphism of schemes which is locally free. Then for every Y-scheme V which is affine over Y the Weil restriction $\mathrm{Res}_{Y/X}(V)$ exists and is affine over X. If in addition V is finitely presented (resp. étale, resp. unramified, resp. smooth) over Y then so is $\mathrm{Res}_{Y/X}(V)$ over X.*

(4.4.1) By a glueing argument (4.2.4) it suffices to prove the assertion Zariski locally with respect to the basis X. So one can assume that X is affine and $f_* \mathcal{O}_Y$ is free of finite rank over $\mathcal{O}_X$.

Hence let $A \to B$ be a ring homomorphism which makes B a free A-module of finite rank, n say. Fix a linear basis $b_1, \ldots, b_n$ of B over A. One is looking for a functor

$$\mathrm{res}_{B/A} : \{B\text{-algebras}\} \longrightarrow \{A\text{-algebras}\}$$

which is left adjoint to the base extension functor $C \longmapsto C \otimes_A B$.

Consider a fixed set $\{t_\alpha\}_{\alpha \in I}$ of indeterminants, and form the polynomial algebras $L = A[t_\alpha; \alpha \in I]$ over A and $M = L \otimes_A B = B[t_\alpha; \alpha \in I]$ over B. Write

$$L^{\otimes n} := \underbrace{L \otimes_A \cdots \otimes_A L}_{n \text{ factors}}.$$

Let $\epsilon : M \to L^{\otimes n} \otimes_A B$ be the B-algebra homomorphism defined by

$$\epsilon(t_\alpha) = \sum_{i=1}^{n} 1 \otimes \cdots \otimes \underset{\substack{\uparrow \\ i\text{-th position}}}{t_\alpha} \otimes \cdots \otimes 1 \otimes b_i, \qquad \alpha \in I.$$

Then for any A-algebra C the map

$$\begin{array}{ccc} \mathrm{Hom}_{A\text{-alg}}(L^{\otimes n}, C) & \longrightarrow & \mathrm{Hom}_{B\text{-alg}}(M, C \otimes_A B) \\ \varphi & \longmapsto & (\varphi \otimes 1) \circ \epsilon =: \psi \end{array} \tag{1}$$

is obviously bijective, where $\varphi \otimes 1$ denotes the map $L^{\otimes n} \otimes_A B \to C \otimes_A B$ induced by $\varphi : L^{\otimes n} \to C$. For $i = 1, \ldots, n$ let $\lambda_i : L^{\otimes n} \otimes_A B \to L^{\otimes n}$ be the $L^{\otimes n}$-linear map which satisfies $\lambda_i(1 \otimes b_j) = \delta_{ij}$, $j = 1, \ldots, n$. Then the identity

$$z = \lambda_1(z) \otimes b_1 + \cdots + \lambda_n(z) \otimes b_n \tag{2}$$

holds for every $z \in L^{\otimes n} \otimes_A B$.

Let φ and $\psi = (\varphi \otimes 1) \circ \epsilon$ be as in (1) and fix $f \in M$. From (2) for $z = \epsilon(f)$ one gets

$$\psi(f) = (\varphi \otimes 1)(\epsilon f) = \sum_{i=1}^{n} \varphi(\lambda_i(\epsilon f)) \otimes b_i.$$

So one sees that $\psi(f) = 0$ if and only if $\varphi(\lambda_i(\epsilon f)) = 0$ for $i = 1, \dots, n$. This discussion shows

(4.4.2) Lemma. — *Let B be a finite A-algebra which is free as an A-module with basis $b_1, \dots, b_n$. Let D be a B-algebra, and let*

$$D = B[t_\alpha;\ \alpha \in I] \big/ (f_\beta;\ \beta \in J)$$

be a presentation of D over B. Put $L = A[t_\alpha;\ \alpha \in I]$. Then the Weil restriction $W = \mathrm{res}_{B/A}(D)$ exists and can be presented over A as

$$W = \underbrace{\left(L \otimes_A \cdots \otimes_A L\right)}_{n\ \text{factors}} \big/ \left(\lambda_i(\epsilon f_\beta);\ \beta \in J,\ i = 1, \dots, n\right),$$

with $\epsilon \colon L \otimes_A B \longrightarrow L^{\otimes n} \otimes_A B$ and $\lambda_i \colon L^{\otimes n} \otimes_A B \longrightarrow L^{\otimes n}$ defined as above. In particular, if D is of finite type (resp. is finitely presented) over B then so is $\mathrm{res}_{B/A}(D)$ over A. □

(4.4.3) This proves the existence part of the proposition as well as the preservation of finite presentation. As to the other properties, recall the following definition [EGA IV, 17]: A presheaf P on (Sch/X) is said to be *formally étale* (resp. *formally unramified*, resp. *formally smooth*) if and only if, for every X-scheme W which is affine and every closed embedding $W_0 \hookrightarrow W$ defined by a nilpotent ideal, the restriction map $P(W) \to P(W_0)$ is bijective (resp. injective, resp. surjective). Depending on the point of view, it is either the definition or a theorem that an X-scheme U is étale (resp. unramified, resp. smooth) iff it is locally of finite presentation and the presheaf $\mathrm{Hom}_X(-, U)$ represented by U has the corresponding formal property. So by the first part of the proof it is sufficient to see that $f_* \colon (\mathrm{Sch}/Y)\widehat{\ } \to (\mathrm{Sch}/X)\widehat{\ }$ preserves the property of a presheaf to be formally étale (resp. ...). But this is a triviality (cf. (4.2.2)) since the base extension of $W_0 \hookrightarrow W$ by $Y \to X$ is again a closed nilpotent embedding of affine schemes. □

(4.5) Example. Let A be a ring and let $A' := A[i] = A \otimes_{\mathbb{Z}} \mathbb{Z}[\sqrt{-1}]$, i.e. A' is free as an A-module with basis $\{1, i\}$, and $i^2 = -1$. If f is a polynomial over A' in m variables $t = (t_1, \dots, t_m)$ one forms two polynomials

$$\mathrm{Re}\, f(u + iv) \quad \text{and} \quad \mathrm{Im}\, f(u + iv)$$

over A in $2m$ variables $u = (u_1, \ldots, u_m)$, $v = (v_1, \ldots, v_m)$, by substituting $t_j = u_j + iv_j$ and formally separating real and imaginary parts (including the coefficients). If an A'-algebra D is presented as

$$D = A'[t_\alpha; \; \alpha \in I] \big/ (f_\beta(t); \; \beta \in J),$$

the construction of (4.4.1) shows that $\operatorname{res}_{A'/A}(D)$ is generated as an A-algebra by elements u_α, v_α ($\alpha \in I$), subject only to the relations

$$\operatorname{Re} f_\beta(u + iv) = \operatorname{Im} f_\beta(u + iv) = 0, \quad \beta \in J.$$

More cases in which the Weil restriction exists are obtained using the following

(4.6) Lemma. — *Consider a cartesian square of schemes*

$$\begin{array}{ccc} Y' & \xrightarrow{\;g'\;} & Y \\ {\scriptstyle f'}\downarrow & & \downarrow{\scriptstyle f} \\ X' & \xrightarrow{\;g\;} & X \end{array} \tag{3}$$

in which g is an immersion (in the sense that every morphism of schemes $u \colon U \to X$ with $u(U) \subset g(X')$ factors uniquely through g) and f is surjective. Then $g_! f'_* = f_* g'_!$ holds, as functors from $(\mathrm{Sch}/Y')\hat{\ }$ to $(\mathrm{Sch}/X)\hat{\ }$ (cf. (4.2.2)).

This notion of immersion is slightly more general than that of [EGA I*, 4.2.1]. The lemma applies, for instance, to the morphism $g \colon \operatorname{spec} \mathcal{O}_{X,x} \to X$ for any $x \in X$, which in general is not an immersion in the sense of EGA.

By (4.2.2) the following corollary is a particular case of this lemma:

(4.6.1) Corollary. — *Let a diagram (3) as in (4.6) be given. If W' is a Y'-scheme for which $\operatorname{Res}_{Y'/X'}(W')$ exists then also $\operatorname{Res}_{Y/X}(W'|_Y)$ exists, and*

$$\operatorname{Res}_{Y/X}(W'|_Y) = \operatorname{Res}_{Y'/X'}(W')|_X$$

as X-schemes. $\qquad\qquad\qquad\qquad\qquad\qquad\qquad\qquad\qquad\qquad\qquad\qquad\qquad\square$

Proof of Lemma (4.6). Let $Q \in (\mathrm{Sch}/X')\hat{\ }$. First it is shown that $g_!(Q) = g_*(Q) \times h_{X'}$, as presheaves on (Sch/X). Note that $h_{X'}$ is a subpresheaf of the constant presheaf $*$ on (Sch/X). If $P \in (\mathrm{Sch}/X)\hat{\ }$, then to give a morphism $g_*(Q) \times h_{X'} \to P$ in $(\mathrm{Sch}/X)\hat{\ }$ is the same as to give maps $Q(U \times_X X') \to P(U)$ for all X-schemes U with $\operatorname{Hom}_X(U, X') \neq \emptyset$, such that these maps are compatible with restrictions. But whenever U is an X-scheme with $\operatorname{Hom}_X(U, X') \neq \emptyset$, then $\operatorname{pr}_1 \colon U \times_X X' \to U$ is an isomorphism. So the full subcategory of (Sch/X) of these U is equivalent to (Sch/X'), via $U \mapsto U \times_X X'$ resp. $V \mapsto V|_X$. Hence, to give $g_*(Q) \times h_{X'} \to P$ is the same thing as to give maps $Q(V) \to P(V|_X)$ for all X'-schemes V, compatibly with restrictions, or in other words, as to give a morphism $Q \to g^*P$ of presheaves on (Sch/X'). Thus $g_!(Q) = g_*(Q) \times h_{X'}$ is shown.

Since g' is an immersion (in the above sense) as well, one also has $g'_!(Q') = g'_*(Q') \times h_{Y'}$ in $(\mathrm{Sch}/Y')^\frown$, for every presheaf Q' on (Sch/Y'). To prove the lemma let Q' be a presheaf on (Sch/Y') and let U be an X-scheme. Then

$$(g_! f'_* Q')(U) = (g_* f'_* Q')(U) \times h_{X'}(U) = (f_* g'_* Q')(U) \times h_{X'}(U) = (f_* g'_! Q')(U).$$

Here the last equality results from $h_{X'} = f_*(h_{Y'})$. This in turn says for every X-scheme $t: T \to X$ that

$$\mathrm{Hom}_X(T, X') \neq \emptyset \quad \Longleftrightarrow \quad \mathrm{Hom}_Y(T_Y, Y') \neq \emptyset.$$

Of course "$\Rightarrow$" is trivial. Conversely suppose that $\mathrm{Hom}_Y(T_Y, Y') \neq \emptyset$, so that there is a solid arrow commutative diagram

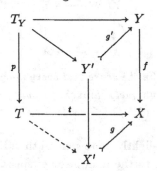

The question is whether the dotted arrow exists (leaving the diagram commutative), or equivalently, whether $t(T)$ is contained in $g(X')$. But since f was assumed to be surjective, the same is true for $p: T_Y \to T$, and the assertion follows. $\square$

(4.7) In order to construct $\mathrm{Res}_{Y/X}(V)$ in general, a natural approach is to cover V by open affine pieces V_λ for which the Weil restrictions over X exist, and then to try to glue the $\mathrm{Res}_{Y/X}(V_\lambda)$. In the following it is studied when this works. To begin with, $\mathrm{Res}_{Y/X}$ preserves open embeddings provided that $f: Y \to X$ is universally closed:

(4.7.1) Proposition. — *Let $f: Y \to X$ be a morphism of schemes which is universally closed, and let V be a Y-scheme. Assume that $\mathrm{Res}_{Y/X}(V)$ exists. Then $\mathrm{Res}_{Y/X}(V_0)$ exists for every open subscheme V_0 of V, and $\mathrm{Res}_{Y/X}(V_0) \to \mathrm{Res}_{Y/X}(V)$ is an open embedding.*

Proof. Put $R := \mathrm{Res}_{Y/X}(V)$. Let $\lambda: R_Y = R \times_X Y \to V$ be the adjunction map and $p: R_Y \to R$ the projection. Let $T := \lambda^{-1}(V - V_0)$, a closed subset of R_Y. Since p is closed, $R_0 := R - p(T)$ is an open subscheme of R. I claim that $R_0 = \mathrm{Res}_{Y/X}(V_0)$:

Let U be an X-scheme and $\alpha \in \mathrm{Hom}_X(U, R)$. The Y-morphism $U_Y \to V$ which by adjunction corresponds to α is $\lambda \circ \alpha_Y$:

$$
\begin{array}{ccccccc}
U_Y & \xrightarrow{\ \alpha_Y\ } & R_Y & \xrightarrow{\ \lambda\ } & V & \longrightarrow & Y \\
\downarrow & & \downarrow{\scriptstyle p} & & & & \downarrow{\scriptstyle f} \\
U & \xrightarrow{\ \alpha\ } & R & & & \longrightarrow & X
\end{array}
$$

Now $\lambda \circ \alpha_Y$ factorizes through $V_0 \subset V$ if and only if $T \cap \alpha_Y(U_Y) = \emptyset$. This in turn is equivalent to $p(T) \cap \alpha(U) = \emptyset$, i.e. to the condition that α factorizes through $R_0 \subset R$. Thus the functorial bijection $\mathrm{Hom}_X(U, R) \approx \mathrm{Hom}_Y(U_Y, V)$ restricts to a bijection $\mathrm{Hom}_X(U, R_0) \approx \mathrm{Hom}_Y(U_Y, V_0)$. $\qquad\square$

(4.7.2) Remark. However it is *not* true that the Weil restriction $\mathrm{Res}_{Y/X}$ preserves open coverings. To amplify this remark assume for simplicity that $f : Y \to X$ is universally closed. Let V be a Y-scheme and let $\mathfrak{V} = \{V_\lambda\}_{\lambda \in I}$ be a Zariski open covering of V. Assume that $\mathrm{Res}\, V_\lambda$ exists for every λ ($\mathrm{Res} := \mathrm{Res}_{Y/X}$). Since Res preserves fibre products it is clear from (4.7.1) that the family of X-schemes $\{\mathrm{Res}\, V_\lambda\}_\lambda$ satisfies the glueing condition, and hence can be glued to give an X-scheme R [EGA I*, 0.4.1.7]. But in general the presheaf represented by R is only a subpresheaf of $f_*(h_V)$. More precisely, an X-morphism $a : U \to R$ is the same as a Y-morphism $b : U_Y \to V$ with the following additional property: U has an open covering $\{U_\lambda\}_{\lambda \in I}$ such that $b((U_\lambda)_Y) \subset V_\lambda$ for every λ. I claim that $R = \mathrm{Res}\, V$ if and only if $\mathfrak{V}$ satisfies the following condition:

(†) *For every X-scheme x which is the spectrum of a field, every Y-morphism from $x_Y = x \times_X Y$ to V factors through one of the $V_\lambda \subset V$.*

It is clear that (†) holds if R represents $f_*(h_V)$. Conversely assume that (†) is satisfied. Let U be an X-scheme and $p : U_Y \to U$ the projection. Given $b \in \mathrm{Hom}_Y(U_Y, V)$ let $T_\lambda := U_Y - b^{-1}(V_\lambda)$ and $U_\lambda := U - p(T_\lambda)$ ($\lambda \in I$). Since p is closed, U_λ is an open subscheme of U. By construction $b((U_\lambda)_Y) \subset V_\lambda$. From (†) it follows that the union of the U_λ is all of U.

From these considerations one draws the following existence theorem for Weil restrictions [BLR, Thm. 4, p. 194], which is more general than Proposition (4.4):

(4.8) Proposition. — *Let Y be finite and locally free over X. Suppose V is a Y-scheme such that, for every $x \in X$, every finite subset of $V_x := \mathrm{spec}\, \kappa(x) \times_X V$ is contained in an open affine subscheme of V. Then $\mathrm{Res}_{Y/X}(V)$ exists.*

Proof. Let $\{V_\lambda\}$ be the family of all open affine subschemes of V whose image in X is contained in an open affine subscheme of X. All Weil restrictions $\mathrm{Res}_{Y/X}(V_\lambda)$ exist, by (4.4) and (4.6.1). The open covering $\{V_\lambda\}$ of V satisfies condition (†) above. Therefore $\mathrm{Res}_{Y/X}(V)$ exists, and can be obtained by glueing together the $\mathrm{Res}_{Y/X}(V_\lambda)$. $\qquad\square$

The proof shows that it suffices to know: For every $x \in X$, every subset of V_x of cardinality at most $[Y : X]_x$ ($:=$ local degree of Y over X at x) is contained in an open affine subscheme of V.

(4.8.1) Corollary. — *If Y/X is finite and locally free then $\mathrm{Res}_{Y/X}(V)$ exists for every Y-scheme V which is quasi-projective over Y.*   □

There exist examples where V is a complete Y-scheme of finite type but $\mathrm{Res}_{Y/X}(V)$ doesn't exist, even for Y/X a finite field extension. See the remark in [FGA, p. 221-27].

(4.9) Proposition. — *Let $Y \to X$ be finite and locally free. Let $h\colon W \to V$ be a morphism of schemes over Y, and suppose that both W and V satisfy the hypothesis of Proposition (4.8) (e.g. that W and V are quasi-projective over Y). For morphisms of schemes consider the properties of being*
>   (a) *quasi-compact,*
>   (b) *quasi-separated,*
>   (c) *separated,*
>   (d) *a closed embedding,*
>   (e) *an open embedding,*
>   (f) *affine,*
>   (g) *quasi-affine,*
>   (h) *(locally) of finite type,*
>   (i) *(locally) finitely presented,*
>   (j) *unramified,*
>   (k) *smooth,*
>   (l) *étale.*

If one of these properties holds for h then it also holds for $\mathrm{Res}_{Y/X}(h)\colon \mathrm{Res}_{Y/X}(W) \to \mathrm{Res}_{Y/X}(V)$.

Proof. (See [BLR, Prop. 5, p. 195] for weaker versions of some of these cases.) Write $\mathrm{Res} := \mathrm{Res}_{Y/X}$. Each of the properties (a)–(l) is local with respect to the base scheme. By the above construction of $\mathrm{Res}\, V$ one can assume that X (hence also Y) and V are affine. Thus $\mathrm{Res}\, V$ is affine as well.

To prove (a) suppose that W is quasi-compact. Let $n < \infty$ be an upper bound for the local degrees of Y over X. The hypothesis on W shows that the n-fold fibre product $W^n := W \times_X \cdots \times_X W$ is covered by the $W_\lambda \times_X \cdots \times_X W_\lambda$ where W_λ ranges over the open affine subschemes of W. Since $W \times_X \cdots \times_X W$ is quasi-compact there are finitely many open affine subschemes $W_1, \ldots, W_N$ of W such that

$$W^n = W \times_X \cdots \times_X W = \bigcup_{i=1}^{N} W_i \times_X \cdots \times_X W_i.$$

But then $\mathrm{Res}\,W$ is covered by the $\mathrm{Res}\,W_i$, $i = 1, \ldots, N$ (compare the remark after Proposition (4.8)), and hence is quasi-compact. Therefore $\mathrm{Res}(h)$ is quasi-compact [EGA I*, 6.1.5(v)].

(d) is a consequence of the affine construction (4.4.2), (e) has already been shown (4.7.1) and (f) follows directly from (4.4). (g) is a consequence of (e) and (f). (Use that the hypothesis of (4.8) carries over from V to every scheme affine over V.) (b) and (c) follow from (a) and (d), respectively, since Res preserves fibre products and also $W \times_V W$ satisfies the hypothesis of (4.8). The "local" versions of (h) and (i) are local on the source scheme, so (using (4.7.1)) one can assume here that also W is affine. Then the assertions follow from (4.4.2). The remaining cases of (h) and (i) follow from this and (a), (b). Finally (j), (k), (l) follow, using the "formal" definitions of these properties (see (4.4.3)), from (i) and the following formal

(4.9.1) Lemma. — *Let C, D be categories and $f^*: D \to C$, $f_*: C \to D$ functors such that f^* is left adjoint to f_*. Suppose that an arrow $\alpha: x' \to x$ in C and a commutative square*

$$
\begin{array}{ccc}
y' & \longrightarrow & f_* x' \\
\beta \downarrow & & \downarrow f_*(\alpha) \\
y & \longrightarrow & f_* x
\end{array}
\tag{4}
$$

in D are given. Let

$$
\begin{array}{ccc}
f^* y' & \longrightarrow & x' \\
f^*(\beta) \downarrow & & \downarrow \alpha \\
f^* y & \longrightarrow & x
\end{array}
\tag{5}
$$

be the commutative square in C deduced from (4) by adjunction. Then under the adjunction bijection $\mathrm{Hom}_D(y, f_* x') \approx \mathrm{Hom}_C(f^* y, x')$ the arrows which leave (4) resp. (5) commutative correspond to each other bijectively. □□

(4.10) In the next section it will be needed that the Weil restriction $\mathrm{Res}_{Y/X}$ preserves finiteness of morphisms. Unfortunately this can fail to be true if Y is ramified over X, cf. (5.14). So I make the stronger assumption now that $Y \to X$ is finite and étale, i.e. an étale covering.

Under this assumption it follows that $\mathrm{Res}_{Y/X}$ preserves all properties of morphisms which descend with respect to the étale topology. More precisely, let $\mathbb{P}$ be a property of scheme morphisms and consider the following conditions:

(P1) *(Isomorphisms)* $\mathbb{P}$ is true for every isomorphism.

(P2) *(Composition)* If g and f are two composable morphisms and $\mathbb{P}$ holds for each of them then $\mathbb{P}$ holds for $g \circ f$.

(P3) *(Base extension)* If $\mathbb{P}$ holds for $f: X \to S$ and $S' \to S$ is any morphism then $\mathbb{P}$ holds for the base extension $f_{S'}: X_{S'} \to S'$ of f.

(P4) (*Étale descent*) If $\{S_\lambda \to S\}$ is a surjective family of étale S-schemes,
   if $f : Y \to X$ is a morphism of schemes over S, and if $\mathbb{P}$ holds for each
   base extension $f_{S_\lambda} : Y_{S_\lambda} \to X_{S_\lambda}$ of f, then $\mathbb{P}$ holds for f.

(4.10.1) Proposition. — *Let $Y \to X$ be a finite étale morphism of schemes and $\mathbb{P}$ a property which satisfies (P1)–(P4) above. Let $h : W \to V$ be a morphism of Y-schemes for which $\mathbb{P}$ holds. Then $\mathbb{P}$ holds also for $\mathrm{Res}_{Y/X}(h) : \mathrm{Res}_{Y/X}(W) \to \mathrm{Res}_{Y/X}(V)$, provided that these two schemes exist.*

Proof. (Res $:= \mathrm{Res}_{Y/X}$) There are a surjective family $\{X_\lambda \to X\}_I$ of étale X-schemes and X_λ-isomorphisms $Y_\lambda := Y \times_X X_\lambda \xrightarrow{\sim} X_\lambda \amalg \cdots \amalg X_\lambda$ (n_λ-fold sum, $0 \le n_\lambda < \infty$, $\lambda \in I$). By *(P4)* it suffices to show that each base extension $(\mathrm{Res}\, h)_{X_\lambda}$ of $\mathrm{Res}(h)$ satisfies $\mathbb{P}$. But $(\mathrm{Res}\, h)_{X_\lambda}$ is the Weil restriction of $h_{Y_\lambda} : V_{Y_\lambda} \to W_{Y_\lambda}$ with respect to $Y_\lambda \to X_\lambda$. By *(P3)* one can therefore assume $Y = X \amalg \cdots \amalg X$ (n-fold sum). Then h is the sum of X-morphisms $h_i : V_i \to W_i$ ($i = 1, \ldots, n$) which satisfy $\mathbb{P}$, and $\mathrm{Res}(h)$ is their direct product over X (4.2.6). So $\mathrm{Res}(h)$ satisfies $\mathbb{P}$ as well. □

(4.10.2) All the properties (a)–(l) listed in (4.9) satisfy *(P1)–(P4)*. Other examples of such properties $\mathbb{P}$ are: being surjective, universally open, universally closed, proper, finite, quasi-finite, entire. These properties and those of (4.9) descend all with respect to the *fpqc* topology, see [EGA IV, 2.6, 2.7, 17.7].

(4.11) To conclude this section, here are some trivial remarks concerning the case when $Y \to X$ is actually a finite Galois covering.

(4.11.1) If a (discrete) group G acts on a scheme Y from the left then G acts on (Sch/Y) from the *left* by

$$\left(g, (V \xrightarrow{v} Y)\right) \longmapsto \left(g \circ v : V \xrightarrow{v} Y \xrightarrow{g} Y\right).$$

I write ${}^g V$, or also $V_{g^{-1}}$, for the Y-scheme $V \xrightarrow{v} Y \xrightarrow{g} Y$. Correspondingly G acts on $(\mathrm{Sch}/Y)^\wedge$ from the *right* by

$$(Q, g) \longmapsto \left(Q^g : V \mapsto Q({}^g V)\right).$$

For representable presheaves one has $(h_V)^g = h_{V_g}$.

(4.11.2) Recall that a morphism $f : Y \to X$ of schemes is a *finite Galois covering* with group G if the finite group G acts on Y over X (from the left) such that Y is a G-torsor over X for the *fpqc* topology. This means that f is faithfully flat and quasi-compact, and that the horizontal map in the commutative triangle

$$
\begin{array}{ccc}
G_Y = \coprod\limits_{g \in G} Y & \xrightarrow{\;\amalg_g(1,g)\;} & Y \times_X Y \\[4pt]
{\scriptstyle \amalg_g g}\!\searrow & & \swarrow{\scriptstyle \mathrm{pr}_2} \\[4pt]
 & Y &
\end{array}
$$

is an isomorphism. Note that this isomorphism condition says that the Y-scheme $\mathrm{pr}_2\colon Y\times_X Y \to Y$ is isomorphic to $\coprod_g Y_g$. Of course it then follows from descent that f is actually finite and étale.

(4.11.3) If $f\colon Y \to X$ is a finite Galois covering with group G then for every presheaf Q on (Sch/Y) which carries finite coproducts to direct products (for example, for every representable presheaf) there is a natural isomorphism

$$f^*f_*Q \xrightarrow{\sim} \prod_{g\in G} Q^g \tag{6}$$

of presheaves on (Sch/Y). Indeed, if V is a Y-scheme then $(f^*f_*Q)(V) = Q((V|_X)\times_X Y)$, and the Y-scheme $(V|_X)\times_X Y$ is isomorphic to $\coprod_g {}^gV$, as follows from the cartesian square

$$
\begin{array}{ccc}
\coprod_g Y & \xrightarrow{\ \amalg\, g\ } & Y \\
{\scriptstyle \amalg\,\mathrm{id}}\downarrow & & \downarrow{\scriptstyle f} \\
Y & \xrightarrow{\ f\ } & X.
\end{array}
$$

If in particular V is a Y-scheme for which $\mathrm{Res}_{Y/X}(V) =: \mathrm{Res}\,V$ exists, there is a canonical isomorphism

$$(\mathrm{Res}\,V)\times_X Y \xrightarrow{\sim} \prod_{g\in G} V_g \tag{7}$$

of Y-schemes (the right hand side is a multiple fibre product over Y). The adjunction map $(\mathrm{Res}\,V)\times_X Y \to V$ gets identified under (7) with the projection to the neutral element component of the product.

5. Real spectrum of X and étale site of $X[\sqrt{-1}]$

Let X be a scheme and put $X' := X \otimes_{\mathbb{Z}} \mathbb{Z}[\sqrt{-1}]$. The group G of order two acts on X' over X, and hence also on the étale site X'_{et}. As outlined in the introduction, it is basic for this work to see the étale site of X' with the G-action in close analogy to a topological G-space. In this picture the subspace of fixpoints of a G-space corresponds to the real spectrum of X. In this section a natural topos morphism ν from $\widetilde{X}_{ret}$ to $\widetilde{X}'_{et}$ will be studied. This ν is induced by the Weil restriction of X'/X, which (*cum grano salis*) is a site morphism from X_{ret} to X'_{et}.

In general ν is not a topos embedding. Nevertheless it plays in a very precise sense the role of the "inclusion of the fixtopos" of the Galois action on $\widetilde{X}'_{et}$ over X, if 2 is invertible on X. This will be shown in Sect. 11.1.

If a group acts on a Hausdorff space then the subspace of fixpoints is closed. Therefore, if one wants to gain some confidence in the analogy of $\widetilde{X}_{ret}$ with a kind of G-fixobject, one will ask whether the direct image functor $\nu_*: \mathrm{Ab}(X_{ret}) \to \mathrm{Ab}(X'_{et})$ is exact. This is indeed so if $\frac{1}{2} \in \mathcal{O}(X)$, and that is the main result of this section (Theorem (5.9)). An immediate consequence is that all sheaf cohomological dimensions of the real spectrum X_r are bounded above by the corresponding étale cohomological dimensions of $X[\sqrt{-1}]$.

(5.1) Notation. Throughout this book the notation $X' := X[\sqrt{-1}] := X \otimes_{\mathbb{Z}} \mathbb{Z}[\sqrt{-1}]$ will be used. Moreover I use the letter π to denote the projection $X' \to X$. For obvious reasons I mostly write $\pi = (\pi^*, \pi_*)$, instead of $\pi_{et} = (\pi^*_{et}, \pi_{et*})$, for the topos morphism from $\widetilde{X}'_{et}$ to $\widetilde{X}_{et}$ induced by π. Generally in this setup, $G = \{1, \sigma\}$ denotes the group of order 2 which acts on X' over X.

(5.2) Proposition. — *The composite functor $\rho \circ \pi_*: \widetilde{X}'_{et} \to \widetilde{X}_{ret}$ preserves all direct limits and all finite inverse limits.*

Proof. (Recall that ρ was defined in (2.5).) Since both π_* and ρ preserve finite inverse limits, the second claim is clear anyway. Since all stalk functors commute with both types of limits in question, the assertion may be checked stalkwise. For any étale sheaf S on X' and any real geometric point $\xi: x \to X$ of X (3.7.3) one has

$$(\rho \pi_* S)_\xi = (\xi^*_{et} \pi_{et*} S)(x) = (\xi'^*_{et} S)(x') = S_{\xi'}$$

where $\xi': x' = x[\sqrt{-1}] \to X'$ is the base extension of ξ by π. (The first equality holds by (3.7.2), the second follows from finite base change.) So the assertion is obvious. (The argument was used already in (3.12d).) $\qquad\square$

From the proposition it follows that $\rho\pi_*$ has a right adjoint, and hence that $\rho\pi_*$ is the *inverse* image functor of a topos morphism from $\widetilde{X}_{ret}$ to $\widetilde{X}'_{et}$:

(5.3) Definition. Define

$$\nu^* := \rho\pi_*\colon \widetilde{X}'_{et} \longrightarrow \widetilde{X}_{ret} \quad \text{and} \quad \nu_* := \text{a right adjoint of } \nu^*.$$

Then $\nu = (\nu^*, \nu_*)$ is a topos morphism from $\widetilde{X}_{ret}$ to $\widetilde{X}'_{et}$.

(5.4) It is useful to describe ν by a morphism of sites, ideally from X_{ret} to X'_{et}. Since π_* corresponds to the Weil restriction of X'/X, the latter will play a role. Since for a general étale X'-scheme V it is not clear whether $\text{Res}_{X'/X}(V)$ exists, one has to pass from Et/X' to a suitable subcategory. This can be done as follows. Let C' denote the site $\text{Et}_{\text{aff}}/X'$ (2.18) with the (induced) étale topology. By (2.18.2) the inclusion $\text{Et}_{\text{aff}}/X' \subset \text{Et}/X'$ induces an equivalence of the toposes $\widetilde{X}'_{et} \to \widetilde{C}'$. By (4.8) and (4.9) the Weil restriction is defined on all of $\text{Et}_{\text{aff}}/X'$ and is a functor

$$\text{Res}_{X'/X}\colon \text{Et}_{\text{aff}}/X' \longrightarrow \text{Et}/X.$$

This functor is a site morphism from X_{ret} to C'. Indeed, all fibre products exist in $\text{Et}_{\text{aff}}/X'$, and they are preserved by $\text{Res}_{X'/X}$. If V is any X'-scheme for which $\text{Res}_{X'/X}(V)$ exists, and if $x \to X$ is a real geometric point of X, then

$$\text{Hom}_X(x, \text{Res}_{X'/X}(V)) \approx \text{Hom}_{X'}(x', V),$$

and $x' = x \times_X X'$ is the spectrum of an *algebraically* closed field. Therefore it is obvious that $\text{Res}_{X'/X}$ carries surjective families of étale arrows to real surjective families. So $\text{Res}_{X'/X}$ is a site morphism from X_{ret} to C'.

Now the topos morphism induced by this site morphism is ν. This follows from

(5.4.1) Lemma. — *Let V be an étale X'-scheme. Assume that $\text{Res}_{X'/X}(V)$ exists and is étale over X (e.g. that $V \in \text{Et}_{\text{aff}}/X'$). Then*

$$(\nu_* B)(V) = B\big(\text{Res}_{X'/X}(V)\big)$$

holds for every sheaf B on X_{ret}.

Proof. Let $R := \text{Res}_{X'/X}(V)$. One has

$$(\nu_* B)(V) = \text{Hom}_{\widetilde{X}'_{et}}(h_V, \nu_* B) = \text{Hom}_{\widetilde{X}_{ret}}(\nu^* h_V, B)$$

$$= \text{Hom}_{\widetilde{X}_{ret}}(\rho\pi_* h_V, B) = \text{Hom}_{(\text{Et}/X)^\sim}(\pi_* h_V, B)$$

since ρ is the functor "associated *ret*-sheaf" (2.11.2). By the definition of the Weil restriction, $\pi_* h_V$ is the presheaf $\text{Hom}_X(-, R)$ represented by R. Thus by the Yoneda lemma, the last set is $B(R)$. $\qquad\square$

(5.5) Remarks and Examples. Let X be a scheme, and write Res $:=$ Res$_{X'/X}$ in the following.

(5.5.1) The effect of ν^* on representable sheaves is made "explicit" as follows. Let $V \to X'$ be an étale X'-scheme, and assume that $U := $ Res V exists and is étale over X (e.g. that V satisfies the condition of (4.8)). Then $U_r \to X_r$ is an espace étalé. Considered as a sheaf on X_r, this is the sheaf $(\nu^* h_V)^{\sharp}$ which corresponds to $\nu^* h_V$ (2.13).

This can be generalized: Assume that $V \to X'$ is an arbitrary étale X'-scheme. Let V_{cx} be the complex spectrum of V over X, cf. (5.6) below. Then $V_{cx} \to X_r$ is an espace étalé, and it corresponds to the sheaf $\nu^* h_V$ on X_{ret}. This follows from the first case by covering V with open affine subschemes.

The moral is therefore that for $\nu^* h_V$ it is irrelevant whether Res V exists or not; $\nu^* h_V$ is "the real spectrum of Res V", which is a well-defined espace étalé over X_r even if Res V doesn't exist (5.6.1).

(5.5.2) Consider the diagram of toposes and topos morphisms

$$
\begin{array}{ccc}
\widetilde{X}_{ret} & \xrightarrow{\;\nu\;} & \widetilde{X}'_{et} \\
{\scriptstyle i}\downarrow & & \downarrow{\scriptstyle \pi} \\
\widetilde{X}_b & \xleftarrow{\;j\;} & \widetilde{X}_{et}.
\end{array}
$$

Unless $X_r = \emptyset$ it does *not* commute, as can be seen by comparing $i^*(j_!A) = \emptyset$ and $\nu^*\pi^*j^*(j_!A) = \nu^*\pi^*A$ for $A \in \widetilde{X}_{et}$. But there is a natural morphism

$$
j\pi\nu \longrightarrow i
$$

between the two topos morphisms from $\widetilde{X}_{ret}$ to $\widetilde{X}_b$. On pullback functors it is given as

$$
i^* \xrightarrow{\;\text{adj}\;} i^*(j \circ \pi)_*(j \circ \pi)^* = \rho\pi_*\pi^*j^* = (j\pi\nu)^*. \tag{1}
$$

On direct image functors it is the composition

$$
j_*\pi_*\nu_* \xrightarrow{\;\text{adj}\;} i_*i^*j_*\pi_*\nu_* = i_*\nu^*\nu_* \xrightarrow{\;\text{adj}\;} i_*. \tag{2}
$$

I claim that for every $B \in \widetilde{X}_{ret}$ the adjunction map $\nu^*\nu_*B \to B$ is surjective. Note that this also implies that (2) is surjective on abelian sheaves (but not necessarily on set-valued sheaves).

The question is local on X, so assume that X is affine. If $U \in $ Et$/X$ is affine consider the composite map

$$
B(\text{Res}\, U_{X'}) = (\pi_*\nu_*B)(U) \xrightarrow{\;\rho\;} (\rho\pi_*\nu_*B)(U) = (\nu^*\nu_*B)(U) \xrightarrow{\;\text{adj}\;} B(U). \tag{3}
$$

It is induced by the adjunction morphism $\alpha\colon U \to $ Res $U_{X'}$ over X. Now α is a closed embedding (4.2.5), and since α is étale it is in fact a clopen embedding. Therefore (3) is surjective, and in particular, $\nu^*\nu_*B \to B$ is surjective.

Let B be an abelian sheaf on X_{ret}. Define $K \in \text{Ab}(X_b)$ by the exact sequence

$$0 \longrightarrow K \longrightarrow j_*\pi_*\nu_*B \overset{(2)}{\longrightarrow} i_*B \longrightarrow 0.$$

Later (5.9) it will be shown that ν_* is an exact functor on abelian sheaves, if 2 is invertible on X. Since also $j_*\pi_*$ is exact (3.12d) one sees from this that (2) induces isomorphisms of all cohomology groups $H^n(X_b, -)$, $n \geq 0$. In other words, $H^n(X_b, K) = 0$ for all n.

(5.5.3) The other adjunction map $S \to \nu_*\nu^*S$ for $S \in \widetilde{X}'_{et}$ is in general neither injective nor surjective. However, if every residue field of X is formally real then $S \to \nu_*\nu^*S$ is injective for every $S \in \widetilde{X}'_{et}$.

To verify this one may again assume that X is affine. Let V be an affine étale X'-scheme. The adjunction map $S(V) \to (\nu_*\nu^*S)(V)$ is the composition

$$S(V) \longrightarrow S((\text{Res } V)_{X'}) = (\pi_*S)(\text{Res } V) \overset{\rho}{\longrightarrow} (\rho\pi_*S)(\text{Res } V) \qquad (4)$$

in which the first arrow is induced by the adjunction morphism $\lambda \colon (\text{Res } V)_{X'} \to V$. Let a, $b \in S(V)$ be two sections which have the same image under (4). This means: For every real geometric point $\xi \colon x \to \text{Res } V$ of $\text{Res } V$ the pullbacks of a and b to $x' = x[\sqrt{-1}]$ coincide:

$$
\begin{array}{ccccccc}
x' & \overset{\xi'}{\longrightarrow} & (\text{Res } V)_{X'} & \overset{\lambda}{\longrightarrow} & V & \longrightarrow & X' \\
\downarrow & & \downarrow & & & & \downarrow{\scriptstyle\pi} \\
x & \overset{\xi}{\longrightarrow} & \text{Res } V & & & \longrightarrow & X.
\end{array}
$$

Since every residue field of X is real and $\text{Res } V \to X$ is real surjective it follows from

$$\text{Hom}_{X'}(x', V) \approx \text{Hom}_X(x, \text{Res } V)$$

that the map $(\text{Res } V)_r \longrightarrow V$, $[\xi] \longmapsto \lambda \circ \xi'(x')$ (the "complex support", see (5.7.4) below) is surjective. Hence a and b coincide in every geometric point of V, so $a = b$.

(5.5.4) Let $\mathcal{N} := \rho\mathbb{G}_{a,X}$ be the sheaf of (abstract) Nash functions on X_{ret} (cf. (1.15)). There is a natural isomorphism

$$\mathcal{N}_c := \mathcal{N} \otimes_{\mathbb{Z}} \mathbb{Z}[\sqrt{-1}] \overset{\sim}{\longrightarrow} \nu^*\mathbb{G}_{a,X'} \qquad (5)$$

of sheaves on X_{ret}, i.e. $\nu^*\mathbb{G}_{a,X'}$ is the sheaf $\mathcal{N}_c$ of "complex Nash functions". Indeed, composition of (1) with j_* gives a natural morphism of functors $\rho \to \nu^*\pi^*$, which here yields a sheaf morphism $\mathcal{N} \to \nu^*\mathbb{G}_{a,X'}$. Together with the global section $\sqrt{-1}$ of $\nu^*\mathbb{G}_{a,X'}$ this defines (5). Looking at stalks one sees that (5) is an isomorphism. Similarly $\nu^*\mathbb{G}_{m,X'}$ is isomorphic to $\mathcal{N}_c^*$, the sheaf of non-vanishing complex Nash functions.

(5.5.5) Let k be a real field and $X = \operatorname{spec} k$. The topos morphism ν can be described in this case as follows, see Sect. 9 for details. Let $\Gamma = \operatorname{Gal}(k_s/k)$ be the absolute Galois group of k, let Γ' be the subgroup which fixes $\sqrt{-1}$ and let T be the closed subspace of Γ consisting of all involutions. One can identify $\widetilde{X}'_{et}$ with the category of discrete Γ'-sets and $\widetilde{X}_{ret}$ with the category of ("continuous") Γ'-sheaves on the Γ'-space T (9.3). Under these identifications ν^* sends a Γ'-set N to the espace étalé $\operatorname{pr}_1: T \times N \to T$ with diagonal Γ'-action, and ν_* sends a Γ'-sheaf B on T to the discrete Γ'-set $H^0(T, B)$.

(5.5.6) Let $f: Y \to X$ be a morphism of schemes, and consider the pullback square

$$
\begin{array}{ccc}
Y' & \xrightarrow{\ f'\ } & X' \\
{\scriptstyle \pi'}\downarrow & & \downarrow{\scriptstyle \pi} \\
Y & \xrightarrow{\ f\ } & X.
\end{array}
$$

Then the diagram

$$
\begin{array}{ccc}
\widetilde{Y}'_{et} & \xrightarrow{\ f'_{et}\ } & \widetilde{X}'_{et} \\
{\scriptstyle \nu}\uparrow & & \uparrow{\scriptstyle \nu} \\
\widetilde{Y}_{ret} & \xrightarrow{\ f_{ret}\ } & \widetilde{X}_{ret}
\end{array}
$$

of toposes and topos morphism commutes, up to natural isomorphism, since (3.10)

$$
(\nu \circ f_{ret})^* = f_{ret}^* \rho \pi_{et*} = \rho f_{et}^* \pi_{et*} = \rho \pi'_{et*} f_{et}'^*.
$$

So in the cube (of toposes and topos morphisms)

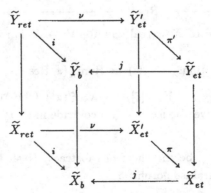

all side faces commute (but not in general the top and the bottom face).

(5.5.7) Assume that 2 is invertible on X. Then the direct image functor ν_* "descends" to $\widetilde{X}_{et}$, i.e. there is a natural functor $\mu: \widetilde{X}_{ret} \to \widetilde{X}_{et}$ such that

$$
\begin{array}{ccc}
\widetilde{X}_{ret} & \xrightarrow{\ \nu_*\ } & \widetilde{X}'_{et} \\
& {\scriptstyle \mu}\searrow \quad \nearrow {\scriptstyle \pi^*} & \\
& \widetilde{X}_{et} &
\end{array}
\qquad (6)
$$

commutes. It is therefore a curious fact that ν^* factors through π_* and ν_* factors through π^*.

There are several possible ways to see the factorization (6), one being the following. Let C be the site $\mathrm{Et}_{\mathrm{aff}}/X$ with the étale topology. Then $\widetilde{X}_{et} \to \widetilde{C}$ is an equivalence of categories, and the topos morphism $\pi \circ \nu$ from $\widetilde{X}_{ret}$ to $\widetilde{X}_{et}$ is induced by the functor $C \to \mathrm{Et}/X$, $U \mapsto \mathrm{Res}\, U_{X'}$, which is a site morphism from X_{ret} to C (5.4).

There is a natural action of the group $G = \{1, \sigma\}$ on the functor $\pi_*\pi^* \colon \widetilde{X}_{et} \to \widetilde{X}_{et}$. Indeed, for $A \in \widetilde{X}_{et}$ and $U \in \mathrm{Et}/X$ one has $(\pi_*\pi^* A)(U) = A(U \times_X X')$, and σ acts on this set through its action $1 \times \sigma$ on $U \times_X X'$. Hence G acts also on $\nu^*\pi^* = \rho\pi_*\pi^*$, and by adjunction it acts on $\pi_*\nu_*$. Define μ as the invariants of this last action, i.e. for $B \in \widetilde{X}_{ret}$ and $U \in \mathrm{Et}_{\mathrm{aff}}/X$ define

$$(\mu B)(U) := \left((\pi_*\nu_* B)(U) \right)^G = B(\mathrm{Res}\, U_{X'})^G.$$

Then μ is a functor $\widetilde{X}_{ret} \to \widetilde{X}_{et}$. I claim that $\nu_* B$ is naturally isomorphic to $\pi^*\mu B$.

For this let $V \xrightarrow{v} X'$ be an étale X'-scheme, with V affine, say. Then

$$(\pi^*\mu B)(V) = B\left(\mathrm{Res}(V|_X \times_X X') \right)^G. \tag{7}$$

Recall (4.11.3) that the X'-scheme $V|_X \times_X X'$ is canonically isomorphic to $V \coprod V_\sigma$, where V_σ is the X'-scheme $V \xrightarrow{v} X' \xrightarrow{\sigma} X'$. Hence

$$B\left(\mathrm{Res}(V|_X \times_X X') \right) \cong B(\mathrm{Res}\, V) \times B(\mathrm{Res}\, V_\sigma). \tag{8}$$

Via the canonical isomorphism $\mathrm{Res}\, V \cong \mathrm{Res}\, V_\sigma$ the group G acts on the right hand product in (8), by interchanging the two factors. The corresponding action on the left hand side coincides with the action with respect to which the invariants in (7) are formed. From this it is clear that $(\pi^*\mu B)(V) \cong B(\mathrm{Res}\, V) = (\nu_* B)(V)$.

An "explanation" of the factorization (6) in a general context will be given in (10.19.1).

It is convenient to introduce the following language:

(5.6) Definition. Let A be a ring, and write $A[i] := A \otimes_{\mathbb{Z}} \mathbb{Z}[\sqrt{-1}]$. If D is an $A[i]$-algebra, the topological space

$$\mathrm{spex}_A D := \mathrm{sper}\left(\mathrm{res}_{A[i]/A}(D) \right)$$

over $\mathrm{sper}\, A$ is called the *complex spectrum* of D (over A).

The complex spectrum has been studied under various aspects by several authors ([Cu], [Pu], [Hu2]). The above terminology is taken from [Hu2].

The definition globalizes easily:

(5.6.1) Lemma. — *Let X be a scheme, $X' = X[\sqrt{-1}]$, and let $V \to X'$ be an arbitrary X'-scheme. Let $\mathfrak{V} = \{V_\lambda\}$ be a covering of V by open subschemes V_λ such that $\mathrm{Res}_{X'/X}(V_\lambda) =: U_\lambda$ exists for every λ (e.g. take all V_λ affine). If U is the X-scheme obtained by glueing the U_λ (4.7.2) then the topological space U_r over X_r does not depend on the choice of $\mathfrak{V}$.*

(5.6.2) Definition. If V is an X'-scheme, the topological space U_r over X_r defined by Lemma (5.6.1) is called the *complex spectrum* of V (over X), and will be denoted by $V_{\mathrm{cx}/X}$ or V_{cx}.

Proof of the lemma. Let $\tilde{\mathfrak{V}} = \{\tilde{V}_\mu\}$ be a second covering of V by open subschemes. If $\tilde{\mathfrak{V}}$ is a refinement of $\mathfrak{V}$ then $\mathrm{Res}_{X'/X}(\tilde{V}_\mu) =: \tilde{U}_\mu$ exists for all μ, and the X-scheme $\tilde{U}$ obtained by glueing the $\tilde{U}_\mu$ is an open subscheme of U (4.7.1). It is immediate to see that $\tilde{U}_r = U_r$, from which the lemma follows.   □

(5.6.3) For fixed X the complex spectrum $V \mapsto V_{\mathrm{cx}/X}$ is a functor $(\mathrm{Sch}/X') \to (\mathrm{Top})/X_r$. This functor sends surjective families to surjective families. From (4.9) one sees that it sends étale morphisms to local homeomorphisms. If

$$
\begin{array}{ccc}
W_1 & \longrightarrow & W \\
\downarrow & & \downarrow \\
V_1 & \xrightarrow{\;f\;} & V
\end{array}
\tag{9}
$$

is a cartesian square of X'-schemes and if f is étale then the complex spectrum functor sends (9) to a cartesian square (of spaces over X_r), as follows from (4.9) and (1.7).

(5.7) Further remarks and examples. Fix a ring A and write $A[i] := A \otimes_{\mathbf{Z}} \mathbf{Z}[\sqrt{-1}]$, $i := 1 \otimes \sqrt{-1}$. The following are some remarks on the Weil restriction functor $\mathrm{res} := \mathrm{res}_{A[i]/A}$ and on the complex spectrum functor spex_A.

(5.7.1) (Canonical presentation of $\mathrm{res}\,D$) Let D be an $A[i]$-algebra. From the canonical presentation of D (which uses one generator for every single element of D) one derives the following presentation of $\mathrm{res}\,D$, compare (4.5): As an A-algebra, $\mathrm{res}\,D$ is generated by elements u_d, v_d ($d \in D$). The relations are generated by the following ones:

$$
\begin{aligned}
(1) \quad & u_a = a, \; v_a = 0, \\
& u_i = 0, \; v_i = 1, \\
(2) \quad & u_{d+d'} = u_d + u_{d'}, \quad v_{d+d'} = v_d + v_{d'}, \\
(3) \quad & u_{dd'} = u_d u_{d'} - v_d v_{d'}, \quad v_{dd'} = u_d v_{d'} + u_{d'} v_d,
\end{aligned}
$$

for $a \in A$ and $d, d' \in D$. This presentation, although highly redundant, is sometimes convenient because of its naturality. The adjunction mappings are

$$
D \longrightarrow (\mathrm{res}\,D) \otimes_A A[i], \quad d \longmapsto u_d \otimes 1 + v_d \otimes i,
\tag{10}
$$

for D an $A[i]$-algebra, and

$$\operatorname{res}(B \otimes_A A[i]) \longrightarrow B, \quad u_{b+ci} \longmapsto b \text{ and } v_{b+ci} \longmapsto c, \tag{11}$$

for B an A-algebra ($b, c \in B$). Let D be an $A[i]$-algebra and let $\tilde{D}$ be the $A[i]$-algebra obtained by twisting D with the automorphism $i \mapsto -i$ of $A[i]$ over A. The isomorphism (7) in (4.11.3) is made explicit as follows: The canonical homomorphism

$$D \otimes_{A[i]} \tilde{D} \longrightarrow (\operatorname{res} D) \otimes_A A[i]$$

of $A[i]$-algebras is

$$d \otimes 1 \longmapsto u_d \otimes 1 + v_d \otimes i, \quad 1 \otimes d \longmapsto u_d \otimes 1 - v_d \otimes i.$$

If 2 is a unit in A then this map has an inverse, given by

$$u_d \otimes 1 \longmapsto \tfrac{1}{2}(d \otimes 1 + 1 \otimes d), \quad v_d \otimes 1 \longmapsto \tfrac{1}{2i}(d \otimes 1 - 1 \otimes d).$$

(5.7.2) (Localization) Let D be an $A[i]$-algebra and let $\Sigma \subset D$ be a multiplicative subset. Then

$$\operatorname{res}(\Sigma^{-1} D) = S^{-1}\operatorname{res}(D)$$

where

$$S = \{u_d^2 + v_d^2 : d \in \Sigma\} \subset \operatorname{res} D.$$

Indeed, let B be an A-algebra and $\alpha : \operatorname{res} D \to B$ an A-algebra homomorphism. The corresponding map $\beta : D \to B \otimes_A A[i]$ sends d to $\alpha(u_d) + i\alpha(v_d)$. Since the "norm"

$$N : D \longrightarrow \operatorname{res} D, \quad N(d) = u_d^2 + v_d^2$$

is multiplicative, it follows for every $d \in D$ that $\beta(d)$ is a unit in $B \otimes_A A[i]$ if and only if $\alpha(N(d))$ is a unit in B. This proves the assertion.

(5.7.3) (Complex spectrum) Let an $A[i]$-algebra D be fixed. The complex spectrum $\operatorname{spex}_A D$ can be described in terms of D as follows (see also [Pu], [Hu2]). For every prime ideal $\mathfrak{q}$ of D fix an algebraic closure $\kappa(\mathfrak{q}) \hookrightarrow \bar{\kappa}(\mathfrak{q})$ of its residue field, and write $d \mapsto d_{\mathfrak{q}}$ for the canonical map $D \to \kappa(\mathfrak{q})$.

Consider pairs $(\mathfrak{q}, t)$ with $\mathfrak{q} \in \operatorname{spec} D$ and t a non-trivial involution of $\bar{\kappa}(\mathfrak{q})$ over A. Two pairs $(\mathfrak{q}, t)$ and $(\mathfrak{q}', t')$ are said to be equivalent iff $\mathfrak{q} = \mathfrak{q}'$ and there is an automorphism ψ of $\bar{\kappa}(\mathfrak{q})$ over D with $t\psi = \psi t'$. The set of equivalence classes $[\mathfrak{q}, t]$ of such pairs is in canonical bijection with the set $\operatorname{spex}_A D$, as follows: Given $(\mathfrak{q}, t)$, let $\bar{\kappa}(\mathfrak{q})^t$ be the subfield of $\bar{\kappa}(\mathfrak{q})$ consisting of the elements fixed by t, and define

$$\operatorname{Re}_t(\lambda) := \tfrac{1}{2}(\lambda + t(\lambda)), \quad \operatorname{Im}_t(\lambda) := \tfrac{1}{2i}(\lambda - t(\lambda))$$

for $\lambda \in \bar{\kappa}(\mathfrak{q})$. Then $\bar{\kappa}(\mathfrak{q})^t$ is a real closed subfield of $\bar{\kappa}(\mathfrak{q})$ which contains the image of A, and

$$\operatorname{res} D \longrightarrow \bar{\kappa}(\mathfrak{q})^t, \quad \begin{cases} u_d \longmapsto \operatorname{Re}_t(d_\mathfrak{q}) \\ v_d \longmapsto \operatorname{Im}_t(d_\mathfrak{q}) \end{cases}$$

is an A-algebra homomorphism. The point of $\operatorname{spex}_A D = \operatorname{sper}(\operatorname{res} D)$ which it represents corresponds to $(\mathfrak{q}, t)$.

An easy way to see this goes as follows: Let $\mathcal{A}$ be the category of all homomorphisms $\alpha \colon \operatorname{res} D \to K$ into real closed fields K, considered as a full subcategory of the category of $(\operatorname{res} D)$-algebras. Let $\mathcal{B}$ be the category of all pairs (β, t), where $\beta \colon D \to L$ is a homomorphism into an algebraically closed field L and t is a (proper) involution of L over A. The arrows in $\mathcal{B}$ are the involution-preserving D-homomorphisms. The categories $\mathcal{A}$ and $\mathcal{B}$ are naturally equivalent, by a construction similar to the above. As a *set*, the real spectrum $\operatorname{sper}(\operatorname{res} D)$ is the set of connected components of the category $\mathcal{A}$ (0.4.3), and hence also of $\mathcal{B}$. It is immediate that the connected components of $\mathcal{B}$ correspond to the equivalence classes $[\mathfrak{q}, t]$ defined above.

(5.7.4) (Complex support) Let D be an $A[i]$-algebra. From (5.7.3) one sees that there is a canonical map

$$\operatorname{supp}_c \colon \operatorname{spex}_A D \longrightarrow \operatorname{spec} D, \quad [\mathfrak{q}, t] \longmapsto \mathfrak{q}$$

which I call the *complex support*. This map supp_c is spectral. This is easily seen from the following description: Given $\xi \in \operatorname{spex}_A D$, let $\mathfrak{p} = \operatorname{supp} \xi \in \operatorname{spec}(\operatorname{res} D)$ be its usual support. Since the field $\kappa(\mathfrak{p})$ is formally real there is a unique prime ideal $\mathfrak{p}'$ lying over $\mathfrak{p}$ in $(\operatorname{res} D) \otimes_A A[i]$. The complex support $\operatorname{supp}_c(\xi)$ is the preimage of $\mathfrak{p}'$ under the adjunction map $D \to (\operatorname{res} D) \otimes_A A[i]$.

Of course this generalizes to schemes: If V is an X'-scheme then supp_c is a locally spectral map $V_{\operatorname{cx}/X} \to V$.

(5.7.5) (G-action on the complex spectrum) Let B be an A-algebra and write $B[i] := B \otimes_A A[i]$. The A-algebra $\operatorname{res}(B[i])$ carries a natural involution, which is explicitly given by

$$u_{b+ic} \longmapsto u_{b-ic}, \quad v_{b+ic} \longmapsto v_{-b+ic},$$

or equivalently, by $u_b \mapsto u_b$ and $v_b \mapsto -v_b$ ($b, c \in B$). (This is seen from the isomorphism of functors $\operatorname{Hom}_A(\operatorname{res} B[i], -) \approx \operatorname{Hom}_A(B, - \otimes_A A[i])$ and the obvious G-action on the latter functor.) The adjunction homomorphism $\operatorname{res}(B[i]) \to B$ is G-invariant.

Hence G acts on the complex spectrum $\operatorname{spex}_A B[i]$ (over $\operatorname{sper} A$), and

$$\operatorname{sper} B \hookrightarrow \operatorname{spex}_A B[i] \tag{12}$$

is a closed embedding whose image consists of points fixed by G. I claim that (12) actually identifies $\operatorname{sper} B$ with the space of fixpoints, i.e. that G acts freely outside sper B. Indeed, if $\alpha\colon \operatorname{res} B[i] \to R$ is a homomorphism into a real closed field R, then the point $[\alpha]$ in the real spectrum is fixed by G if and only if $\alpha(v_b) = 0$ for all $b \in B$. This means that $b \mapsto \alpha(u_b)$ is a homomorphism $\beta\colon B \to R$ and α is the pullback of β by the adjunction map (11). So $[\alpha] = [\beta]$ under (12).

This globalizes to schemes: If U is an X-scheme then G acts on the complex spectrum $(U')_{\mathrm{cx}}$ of $U' := U_{X'}$ over X. There is a canonical closed embedding $\alpha\colon U_r \hookrightarrow (U')_{\mathrm{cx}}$ of spaces over X_r. (It is derived from the local adjunction maps $\tilde{U} \to \operatorname{Res} \tilde{U}_{X'}$ which are closed embeddings, (4.2.5).) The map α identifies U_r with the space of G-fixpoints in $(U')_{\mathrm{cx}}$. (For a different interpretation, and generalization, of these facts see (6.4), in particular (6.4.3).)

(5.8) Let X be a scheme. For the proof of the main result below it is necessary to determine the stalks of direct image sheaves $\nu_* B$. So let B be a sheaf on X_{ret}, and let $\eta\colon y \to X' = X[\sqrt{-1}]$ be a geometric point of X'. Letting W range over the affine étale neighborhoods of η one finds

$$(\nu_* B)_\eta = \varinjlim_W (\nu_* B)(W) = \varinjlim_W B\big(\operatorname{Res}_{X'/X}(W)\big)$$

by (5.4.1). From the limit theorem (3.4.1b) it follows that

$$(\nu_* B)_\eta = \tilde{B}\Big(\varprojlim_W \operatorname{Res}_{X'/X}(W)\Big)$$

where $\tilde{B}$ is the pullback of B to the inverse limit. Since $\operatorname{Res}_{X'/X}$ preserves inverse limits this gives

$$(\nu_* B)_\eta = \tilde{B}\Big(\operatorname{Res}_{X'/X}(X'^\eta)\Big)$$

where $X'^\eta = \varprojlim W$ is the strict localization of X' at η (see (3.5)).

Let $x' := \eta(y)$ be the image point of η and put $x := \pi(x')$. Let $X_{(x)} := \operatorname{spec} \mathcal{O}_{X,x}$ be the Zariski local scheme of X at x, and write $X'_{(x)} := X_{(x)} \times_X X' = \operatorname{spec}\big(\mathcal{O}_{X,x} \otimes_{\mathbb{Z}} \mathbb{Z}[\sqrt{-1}]\big)$. By Corollary (4.6.1),

$$\operatorname{Res}_{X'/X}(X'^\eta) = \Big(\operatorname{Res}_{X'_{(x)}/X_{(x)}}(X'^\eta)\Big)\Big|_X$$

since $X_{(x)} \to X$ is an immersion in the sense of loc. cit. In summary one gets

(5.8.1) **Proposition.** — *Let X be a scheme. Let η be a geometric point of $X' = X[\sqrt{-1}]$ and let x be the image point of $\pi \circ \eta$ in X. Then*
a) $\operatorname{Res}_{X'/X}(X'^\eta)$ *is the X-scheme* $\operatorname{Res}_{X'_{(x)}/X_{(x)}}(X'^\eta) \longrightarrow X_{(x)} \longrightarrow X$, *where*
   $X_{(x)} = \operatorname{spec} \mathcal{O}_{X,x}$ *and* $X'_{(x)} = X_{(x)} \times_X X'$;
b) *for any sheaf B on X_{ret} one has canonically $(\nu_* B)_\eta \cong H^0(R_{\mathrm{ret}}, \tilde{B})$, where R is the X-scheme from a) and $\tilde{B}$ is the pullback of B to R.* $\square$

In other words, the stalk of $\nu_* B$ at a geometric point $\bar{x}'$ of X' is the set of sections of the pullback of B to $\operatorname{spex}_{\mathcal{O}_{X,x}}(\mathcal{O}_{X',\bar{x}'})$.

The main result of this section is

(5.9) Theorem. — *Assume that 2 is invertible on X. Then the direct image functor $\nu_*: \operatorname{Ab}(X_{ret}) \to \operatorname{Ab}(X'_{et})$ is exact.*

(5.10) By identification of the stalks (5.8.1) one is led to consider the following situation: Let A be a local ring, put $A' := A[i] := A \otimes_{\mathbb{Z}} \mathbb{Z}[i]$ and let $\mathfrak{m}$ be a maximal ideal of $A[i]$ (there are at most two). Let $\tilde{A}$ be a strict henselization of $A[i]$ in $\mathfrak{m}$. It suffices to show that the complex spectrum $\operatorname{spex}_A(\tilde{A})$ has vanishing sheaf cohomology in positive degrees.

Now $\tilde{A}$ is a filtering direct limit of $A[i]$-algebras which are local-étale over $A[i]_{\mathfrak{m}}$. (In [EGA IV, 18.6] the term *essentiellement étale* is used for what is here called local-étale.) Since $\operatorname{res}_{A[i]/A}$ commutes with such limits, the usual limit theorems for cohomology of the real spectrum (3.4.1) show that it suffices to replace $\tilde{A}$ by a local-étale $A[i]_{\mathfrak{m}}$-algebra. Unfortunately the argumentation for such algebras breaks down completely if the residue characteristic of A is 2, which is why I need the hypothesis on X in the theorem. I do not know whether the theorem is true without the restriction on the residue characteristics. It seems quite likely that this is the case but, alas, I was unable to prove it.

Every local-étale $A[i]_{\mathfrak{m}}$-algebra is a localization of a finite $A[i]$-algebra in some maximal ideal. Therefore the theorem will be an immediate consequence of Corollary (5.11.1) below:

(5.11) Proposition. — *Let A be a ring in which 2 is invertible. Let D be a finite $A[i]$-algebra and let E be a semilocalization of D. There are finitely many prime ideals $\mathfrak{p}_1, \ldots, \mathfrak{p}_r$ of $\operatorname{res} D := \operatorname{res}_{A[i]/A}(D)$ such that for every prime ideal $\mathfrak{p}$ of $\operatorname{res} D$ with $\sqrt{-1} \notin \kappa(\mathfrak{p})$:*

$$\mathfrak{p} \in \operatorname{spec}(\operatorname{res} E) \iff \mathfrak{p} \subset \mathfrak{p}_i \text{ for some } i \in \{1, \ldots, r\}.$$

Of course, $\operatorname{spec}(\operatorname{res} E)$ is identified here with a subset of $\operatorname{spec}(\operatorname{res} D)$, see Remark (5.7.2).

(5.11.1) Corollary. — *Let A be a ring containing $1/2$, D a finite $A[i]$-algebra and E a semilocalization of D. Then all sheaf cohomology of the complex spectrum $\operatorname{spex}_A E$ vanishes in degrees > 0.*

Proof. Choose $\mathfrak{p}_1, \ldots, \mathfrak{p}_r$ as in the proposition. The real spectrum of $\operatorname{res} E$ coincides with the real spectrum of the semilocalization of $\operatorname{res} D$ in $\mathfrak{p}_1, \ldots, \mathfrak{p}_r$. It is well known (see (19.2.1) for a proof) that the real spectrum of a semilocal ring has vanishing cohomology in positive degrees. $\square$

(5.12) Remarks and Examples.

(5.12.1) Consider the situation of the proposition. In general res E will definitely *not* be a semilocal ring. It is clear that res E is semilocal if and only if $(\mathrm{res}\,E)\otimes_A A[i]$ is semilocal. To state an example which is relevant for the following consider the case where

$$A \;=\; \text{localization of } \mathbb{R}[t_1,\ldots,t_n] \text{ in a } nonreal \text{ maximal ideal.}$$

So $A' = A[i] = A\otimes_{\mathbb{R}} \mathbb{C}$ has two maximal ideals $\mathfrak{m}_1$ and $\mathfrak{m}_2$. Let $E = (A')_{\mathfrak{m}_1}$. By Remark (5.7.1), $(\mathrm{res}\,E)\otimes_A A'$ is A'-isomorphic to $(A')_{\mathfrak{m}_1}\otimes_{A'}(A')_{\mathfrak{m}_2}$. The prime ideal spectrum of this ring consists of all irreducible Zariski closed subsets of $\mathbb{C}^n$ which pass through the two (different!) points which correspond to $\mathfrak{m}_1$ and $\mathfrak{m}_2$. So $(\mathrm{res}\,E)\otimes_A A'$ has certainly infinitely many maximal ideals if $n\geq 2$.

(5.12.2) If E is a semilocal $A[i]$-algebra, some finiteness condition on $A[i]\to E$ as in the corollary is necessary in order that the cohomology of $\mathrm{spex}_A E$ vanishes. Consider for example the case $A = \mathbb{R}$ and $E = \mathbb{C}[t]_{(t)}$, the local ring at the origin of the affine line over $A' = \mathbb{C}$. It is easy to see that the complex spectrum of E over $\mathbb{R}$ is canonically identified with the subspace

$$\left(\tilde{\mathbb{R}}^2 - \mathbb{R}^2\right)\cup\{0\}$$

of $\tilde{\mathbb{R}}^2 := \mathrm{sper}\,\mathbb{R}[t_1,t_2]$: All "rational" points in $\tilde{\mathbb{R}}^2$ are removed except the origin. This space has non-trivial cohomology; for example, $H^1(\mathrm{spex}_{\mathbb{R}} E, M) = \bigoplus_{\mathbb{R}^2-\{0\}} M$ for constant coefficients M.

(5.12.3) Let D be any $A[i]$-algebra, and consider the complex support map $\varphi_c := \mathrm{supp}_c\colon \mathrm{spex}_A D \to \mathrm{spec}\,D$ (5.7.4). For any abelian sheaf F on $\mathrm{spex}_A D$ one has

$$\left(\mathrm{R}^n\varphi_{c*}F\right)_{\mathfrak{q}} \;=\; H^n\!\left(\mathrm{spex}_A D_{\mathfrak{q}}, F\right)$$

for all $\mathfrak{q}\in\mathrm{spec}\,D$ ($n\geq 0$). Hence (5.11) implies that the direct image functor φ_{c*} is exact whenever D is finite over $A[i]$. On the other hand, (5.12.2) shows that in general φ_{c*} is not exact. One should compare this to (19.2), which says that φ_* (φ = usual support) is always exact.

(5.13) *Proof* of Proposition (5.11). Write $X := \mathrm{spec}\,A$, $X' := \mathrm{spec}\,A[i]$, $V := \mathrm{spec}\,D$ and $\mathrm{Res} = \mathrm{Res}_{X'/X}$. Since X'/X is finite étale and V/X' is finite, it follows from Proposition (4.10.1) (and (4.10.2)) that $\mathrm{Res}\,V \to X$, and hence also $(\mathrm{Res}\,V)_{X'} \to X'$, is a finite morphism. Since the latter map factors through the adjunction morphism $\lambda\colon(\mathrm{Res}\,V)_{X'} \to V$, also λ is finite. Let $\mathfrak{q}_1,\ldots,\mathfrak{q}_s \in V$ be the prime ideals of D in which E is the semilocalization, and let

$$\{\tilde{\mathfrak{p}}_1,\ldots,\tilde{\mathfrak{p}}_r\} \;:=\; \lambda^{-1}\{\mathfrak{q}_1,\ldots,\mathfrak{q}_s\} \subset (\mathrm{Res}\,V)_{X'}.$$

Let $p\colon (\operatorname{Res} V)_{X'} \to \operatorname{Res} V$ be the projection and put $\mathfrak{p}_i := p(\tilde{\mathfrak{p}}_i) \in \operatorname{Res} V = \operatorname{spec}(\operatorname{res} D)$:

$$(\operatorname{Res} V)_{X'} \xrightarrow{\ \lambda\ } V \longrightarrow X'$$

$$\operatorname{Res} V \longrightarrow X$$

with vertical map p on the left.

Writing $U := \operatorname{spec} E$ one has

$$\operatorname{Res} U = \operatorname{spec}(\operatorname{res} E) = \{\mathfrak{p} \in \operatorname{Res} V\colon \lambda(p^{-1}(\mathfrak{p})) \subset U\},$$

cf. the proof of (4.7.1). Let $\mathfrak{p} \in \operatorname{Res} V$ be a prime ideal of $\operatorname{res} D$ with $\sqrt{-1} \notin \kappa(\mathfrak{p})$ and let $\mathfrak{q} \in (\operatorname{Res} V)_{X'}$ be the unique point over $\mathfrak{p}$. One has to show that

$$\mathfrak{p} \subset \mathfrak{p}_i \text{ for some } i \iff \lambda(\mathfrak{q}) \subset \mathfrak{q}_j \text{ for some } j.$$

Assume $\lambda(\mathfrak{q}) \subset \mathfrak{q}_j$. Since λ is finite it has the going-up property. So $\mathfrak{q}$ is contained in one of the $\tilde{\mathfrak{p}}_i$, and accordingly $\mathfrak{p} = p(\mathfrak{q}) \subset p(\tilde{\mathfrak{p}}_i) = \mathfrak{p}_i$. Conversely assume $\mathfrak{p} \subset \mathfrak{p}_i = p(\tilde{\mathfrak{p}}_i)$. Since p is flat it has the going-down property. So there is a prime ideal $\tilde{\mathfrak{q}} \subset \tilde{\mathfrak{p}}_i$ with $\mathfrak{p} = p(\tilde{\mathfrak{q}})$. But $p^{-1}(\mathfrak{p}) = \{\mathfrak{q}\}$, so $\mathfrak{q} = \tilde{\mathfrak{q}} \subset \tilde{\mathfrak{p}}_i$, and hence $\lambda(\mathfrak{q}) \subset \lambda(\tilde{\mathfrak{p}}_i)$. This proves the proposition. $\square$

(5.14) Remark. It is true for any scheme X (i.e. without restrictions on the residue characteristics) and any finite X'-scheme V that $\operatorname{Res} V$ is finite over X if and only if $\lambda\colon (\operatorname{Res} V)_{X'} \to V$ is finite, since finiteness descends with respect to X'/X. Unfortunately, if X happens to have points of characteristic two, $\operatorname{Res} V \to X$ will in general be far from finite. It is this fact that causes the trouble with residue characteristic two in the proof of Theorem (5.9). For example, if A is a ring in which $2 = 0$ then $A' = A \otimes_{\mathbf{Z}} \mathbf{Z}[\sqrt{-1}]$ is A-isomorphic to $A[\varepsilon]/(\varepsilon^2)$. Consider $D = A'[t]/(t^n)$, a finite A'-algebra. Its Weil restriction is

$$\operatorname{res}_{A'/A}(D) = A[u,v]/(u^n, nu^{n-1}v)$$

which for $n > 1$ is not even quasi-finite over A. Rather it is geometrically a non-reduced version of the affine line over A.

One can show that, if X is local and v is a closed point of V such that $V \to X'$ is étale around v, the fibre $\lambda^{-1}(v)$ in $(\operatorname{Res} V)_{X'}$ is a finite set. But this is not sufficient to extend the proof of Proposition (5.11) to the case of residue characteristic 2.

(5.15) Remark. Finally in this section I want to mention a generalization of the topos morphism ν. It will be applied in Sect. 15.

Let X be a scheme and put $X' = X \otimes_{\mathbf{Z}} \mathbf{Z}[\sqrt{-1}]$ as usual. Fix an arbitrary X'-scheme Y, and let Y_{cx} be the complex spectrum of Y over X (5.6.2). If $V \to Y$

is an étale morphism then $V_{cx} \to Y_{cx}$ is an espace étalé, i.e. a sheaf on Y_{cx}. In this way a functor

$$\Phi^{-1} \colon \operatorname{Et}/Y \longrightarrow (Y_{cx})^{\sim}, \quad V \longmapsto (V_{cx} \to Y_{cx})$$

is defined. By (5.6.3) this is a morphism of sites from $(Y_{cx})^{\sim}$ to Y_{et}. Denote the corresponding topos morphism by

$$\Phi = (\Phi^*, \Phi_*) \colon (Y_{cx})^{\sim} \longrightarrow \widetilde{Y}_{et}.$$

Note that Y_{cx} is the real spectrum of $\operatorname{Res}_{X'/X}(Y)$, if this Weil restriction exists. The special case $Y = X'$ gives back the ν considered before. The topos morphism Φ is functorial in an obvious sense.

6. The fundamental long exact sequence

Let G be the group of order two. Assume that G acts on a paracompact topological space T, and let T^G be the subspace of fixpoints. For every abelian coefficient group M the cohomology sequence for the pair $\left(EG \times_G T, BG \times T^G\right)$ gives a long exact sequence

$$\cdots H^n_{\text{sing}}\left(T/G, T^G; M\right) \longrightarrow H^n_{G,\text{sing}}(T, M) \longrightarrow H^n_{\text{sing}}\left(BG \times T^G, M\right) \cdots \quad (1)$$

of singular cohomology groups, where the second group is G-equivariant singular cohomology. See (14.8) for more details.

The purpose of this section is to establish an analogous sequence for any scheme X with $\frac{1}{2} \in \mathcal{O}(X)$. As explained in Sect. 15, this generalizes the Cox sequence. To begin with, consider the étale topos of $X' = X[\sqrt{-1}]$ as a G-topos via the natural G-action on X' over X. Then $\widetilde{X}'_{et}(G)$, the topos of étale G-sheaves on X', is naturally equivalent to $\widetilde{X}_{et}$. As indicated in the introduction to the preceding section, the real topos $\widetilde{X}_r$ has to be regarded as the fixtopos of this G-action on $\widetilde{X}'_{et}$. Moreover the glued topos $\widetilde{X}_b$ (which was introduced in Sect. 2) is the analogue of the quotient space T/G, as will be explained in Sections 14 and 15. With these analogies in mind one can tentatively write down a long sequence of cohomology groups which should correspond to the groups in (1). The main result of this section (Theorem (6.6)) says that there is indeed such an exact sequence for every abelian sheaf A on X_{et}.

If A is a torsion sheaf whose torsion is odd, this sequence reduces to the usual long exact sequence associated with the glueing situation. Theorem (6.6) is therefore most useful when applied to 2-primary torsion sheaves.

The proper general setting for equivariant topos theory is the notion of G-toposes. See Sect. 10 for basic terminology and facts. Since the technicalities of this theory may appear somewhat deterring, I have inserted some reminders at the beginning of this section, specialized to the case of study here. In this way most of the following Sections 6–9 can be read independently from Sect. 10.

(6.1) Definitions and Notations. The following are some notations concerning G-objects in a topos. Let E be a topos and G an arbitrary group. Let BG denote here the category associated with G, so BG has a single object whose monoid of endomorphisms is G.

(6.1.1) One forms the topos $E(G) := \underline{\text{Hom}}(BG, E)$ of G-objects in E: The objects of $E(G)$ are objects x of E together with a left action of G on x (i.e. a group homomorphism $G \to \mathrm{Aut}_E(x)$), and the arrows are the action-preserving E-arrows. The category of abelian group objects in $E(G)$ is denoted by $\mathrm{Ab}_G(E)$; these are simply the left $\mathbb{Z}G$-module objects in E. In the language of G-toposes, $E(G) = \underrightarrow{\mathrm{Lim}}\mathrm{top}(\mathfrak{E}/G)$ where $\mathfrak{E}$ is the *trivial* G-topos E (Definition (10.6.5)).

(6.1.2) There are two natural topos morphisms $\pi_E = \pi\colon E \to E(G)$ and $r_E = r\colon E(G) \to E$, cf. (10.6.3). They have the following description. The inverse image functor π^* forgets the G-action, while π_* sends $x \in E$ to the direct product $\prod_G x$ with the natural G-action. The direct image functor $r_*\colon E(G) \to E$ is the formation of the G-invariants, while r^* puts the trivial G-action on objects. The composition $r \circ \pi$ is isomorphic to the identity. I will mostly write Γ_G or H_G^0 for the functor r_*. On abelian group objects this is a left exact additive functor $\Gamma_G\colon \mathrm{Ab}_G(E) \to \mathrm{Ab}(E)$. Its right derived functors will be written $A \mapsto \mathrm{H}_G^n(A) := \mathrm{R}^n\Gamma_G(A)$, $n \geq 0$.

(6.1.3) For $x \in E$ consider the left exact additive functor

$$\mathrm{Ab}_G(E) \longrightarrow (\mathrm{Ab}), \quad A \longmapsto \Gamma_G(x, A) := (\Gamma_G A)(x) = A(x)^G = \mathrm{Hom}_E(x, A)^G.$$

Its right derived functors

$$A \longmapsto H_G^n(x, A) := \mathrm{R}^n\Gamma_G(x, -)(A)$$

are the *G-equivariant cohomology groups* of x with coefficients A. There are two natural ways of writing the functor $\Gamma_G(x, -)$ as a composition; they yield spectral sequences which are functorial in $A \in \mathrm{Ab}_G(E)$, have E_2 terms

$$E_2^{pq} = H^p(x, \mathrm{H}_G^q(A)) \quad \text{resp.} \quad E_2^{pq} = H^p(G, H^q(x, A))$$

$(p, q \geq 0)$ and converge against $H_G^{p+q}(x, A)$. If $x = *$ one writes $\Gamma_G(E, A)$ and $H_G^n(E, A)$ for $\Gamma_G(*, A)$ and $H_G^n(*, A)$, respectively. Observe that $H_G^n(E, A) = \mathrm{Ext}_{\mathbb{Z}G}^n(E; \mathbb{Z}, A)$.

(6.1.4) If C is a site and $E = \widetilde{C}$ is the topos of sheaves on C then the above notations are extended in the canonical fashion: One writes $\mathrm{Ab}_G(C)$ for $\mathrm{Ab}_G(\widetilde{C})$, and $H_G^n(x, A)$ for $H_G^n(\epsilon x, A)$ ($x \in C$), resp. $H_G^n(C, A)$ for $H_G^n(\widetilde{C}, A)$. For $F \in \widetilde{C}(G)$ and $U \in C$ one has $(\mathrm{H}_G^0 F)(U) = F(U)^G = H_G^0(U, F)$.

A very useful fact is

(6.2) **Proposition.** — *Let E be a topos, and let G be a finitely generated group. Let p^* be a fibre functor of E. Then*

$$p^* \mathrm{H}_G^n(A) = H^n(G, p^* A)$$

holds for every $A \in \mathrm{Ab}_G(E)$ and every $n \geq 0$.

Proof. Like any fibre functor, p^* is pro-representable [SGA4 IV.6.8.5]. So there are a left filtering category I and a functor $i \mapsto x_i$ from I to E such that $p^*x \cong \varinjlim_i \mathrm{Hom}_E(x_i, x)$ holds functorially in $x \in E$. Since for every $A \in E(G)$ and $x \in E$ one has $\mathrm{Hom}_E(x, \Gamma_G A) = \mathrm{Hom}_E(x, A)^G$, and since for a finitely generated group G the functor $M \mapsto M^G$ from (G-sets) to (sets) preserves filtering direct limits, the proposition is true for $n = 0$ (even with set-valued sheaves). From this it follows that, to prove it for all n, it suffices to show that

$$p^*\colon \mathrm{Ab}_G(E) \longrightarrow (G\text{-mod})$$

carries injective objects of $\mathrm{Ab}_G(E)$ to acyclic G-modules. So let $J \in \mathrm{Ab}_G(E)$ be injective. For every $x \in E$ the G-module $J(x) = \mathrm{Hom}_E(x, J)$ is injective. (Quite generally, for any not necessarily commutative ring Λ the functor $A \mapsto A(x)$ from the category of (left) Λ-module objects in E to the category of (left) Λ-modules has an exact left adjoint and hence preserves injective objects; compare [SGA4 IV.11].) So $p^*J = \varinjlim I_i$ is a filtering direct limit of injective G-modules I_i, and hence $H^n(G, p^*J) = \varinjlim H^n(G, I_i) = 0$ for $n \geq 1$. — Observe that if G is finite (or more generally, if the group ring $\mathbb{Z}G$ is left noetherian) the G-module p^*J itself is injective, as is any filtering direct limit $\varinjlim I_i$ of injective G-modules: It suffices to show that any G-map from some left ideal L of the group ring $\mathbb{Z}G$ to the limit extends to all of $\mathbb{Z}G$ [Bo ch. X §1 Prop. 10]. Since L is finitely generated one sees that $L \to \varinjlim I_i$ factors through I_{i_0} for some index i_0, whence the assertion. $\square$

(6.3) Let A be an abelian sheaf on C. The following remarks relate the G-equivariant cohomology of r^*A (i.e. of the sheaf A with the trivial G-action) to the usual cohomology of A. With an eye to later applications, a more general situation is actually covered, namely also twisted versions of r^*A.

Let C be a site, G a group and k a commutative ring. Suppose that a character $\chi\colon G \to k^*$ is fixed. If M is any k-module, write M_χ for the kG-module M on which G operates through the character χ, i.e. by $(g, m) \mapsto \chi(g).m$. If A is any sheaf of k-modules on C, define the sheaf of kG-modules A_χ similarly. In particular, $r^*A = A_{\chi_0}$ where χ_0 is the trivial character.

Fix a sheaf A of k-modules on C. Let $L_\bullet \to k$ be a resolution of the trivial kG-module k by projective kG-modules L_p, and let $A \to I^\bullet$ be a resolution of A by injective sheaves I^q of k-modules. Then $H_G^*(C, A_\chi)$ is the hyper-G-cohomology of the complex $H^0(C, I_\chi^\bullet) = H^0(C, I^\bullet)_\chi$ of kG-modules. So $H_G^*(C, A_\chi)$ is the total cohomology of the double complex

$$\mathrm{Hom}_{kG}\bigl(L_\bullet,\, H^0(C, I^\bullet)_\chi\bigr) = \mathrm{Hom}_k\bigl(L_\bullet \otimes_{kG} k_\chi,\, H^0(C, I^\bullet)\bigr).$$

If k is a principal ideal domain, the Künneth formula (e.g. [HiSt, V.3.1]) applies to give

(6.3.1) Proposition. — *Let k be a principal ideal domain and $\chi\colon G \to k^*$ a character. For any sheaf A of k-modules on C there are natural exact sequences*

$$0 \longrightarrow \bigoplus_{p+q=n-1} \operatorname{Ext}_k^1\!\big(H_p(G, k_\chi), H^q(C, A)\big) \longrightarrow H^n_G(C, A_\chi)$$

$$\longrightarrow \bigoplus_{p+q=n} \operatorname{Hom}_k\!\big(H_p(G, k_\chi), H^q(C, A)\big) \longrightarrow 0,$$

$n \geq 0$, and these sequences split (non-canonically). ☐

Essentially, this proposition is a particular case of a general result of Grothendieck [Gr1, §4.3, Cor. 2]. However Grothendieck does not mention the important fact that these sequences split.

Specializing now to the case where k is a field and $\chi = \chi_0$ is the trivial character, one gets

(6.3.2) Corollary. — *If k is a field and A is a sheaf of k-vector spaces on C, the canonical map $H^*(G, k) \otimes_k H^*(C, A) \to H^*_G(C, r^*A)$ is an isomorphism of graded abelian groups.*

Indeed, its composition with the isomorphism (6.3.1)

$$H^*_G(C, r^*A) \xrightarrow{\ \sim\ } \operatorname{Hom}_k\!\big(H_*(G, k), H^*(C, A)\big)$$

is the canonical map, which over fields is well known to be an isomorphism.

For example, if $|G| = 2$, the cohomology ring $H^*_G(C, \mathbb{Z}/2)$ is canonically isomorphic to the graded polynomial ring $H^*(C, \mathbb{Z}/2)[\gamma]$, where $\deg(\gamma) = 1$.

(6.4) Let now X be a scheme on which 2 is invertible, and let G be the group of order two. The G-action on $X' = X[\sqrt{-1}]$ over X induces a (non-trivial!) G-action on $\widetilde{X}'_{et}$ which makes $\widetilde{X}'_{et}$ (the fibre of) a G-topos (Example (10.11)). Hence the topos $\widetilde{X}'_{et}(G)$ of G-sheaves on X'_{et} is defined (10.5), (10.6.1). Since $\frac{1}{2} \in \mathcal{O}(X)$ there is a canonical equivalence of $\widetilde{X}'_{et}(G)$ with $\widetilde{X}_{et}$ which identifies the canonical topos morphism $\pi\colon \widetilde{X}'_{et} \to \widetilde{X}'_{et}(G)$ with the topos morphism induced by the scheme morphism $\pi\colon X' \to X$. See Example (10.12).

(6.4.1) On X the group G acts trivially, so (6.1) and (6.2) apply to G-sheaves on the sites X_{ret}, X_b and X_{et}. Consider the sequence of topos morphisms

$$\widetilde{X}_{ret} \xrightarrow{\ \nu\ } \widetilde{X}'_{et} \xrightarrow{\ \pi\ } \widetilde{X}_{et}, \tag{2}$$

where ν was defined in (5.3). Both ν and π are "G-invariant". Thus it is plausible that (2) gives rise to the following diagram of topos morphisms, cf. (10.19):

$$
\begin{array}{ccccc}
\widetilde{X}_{ret} & \xrightarrow{\ \nu\ } & \widetilde{X}'_{et} & \xrightarrow{\ \pi\ } & \widetilde{X}_{et} \\
\uparrow{\scriptstyle r_{ret}}\,\big\uparrow{\scriptstyle g_{ret}} & & \big\downarrow{\scriptstyle \pi} & & \big\downarrow{\scriptstyle g_{et}}\,\big\uparrow{\scriptstyle r_{et}} \\
\widetilde{X}_{ret}(G) & \xrightarrow{\ \nu(G)\ } & \widetilde{X}_{et} \sim \widetilde{X}'_{et}(G) & \xrightarrow{\ \pi(G)\ } & \widetilde{X}_{et}(G).
\end{array} \tag{3}
$$

The notations of this diagram will be used in what follows, so here is an explanation: I have written

$$q_{ret} := \pi_{\widetilde{X}_{ret}}, \quad q_{et} := \pi_{\widetilde{X}_{et}} \quad \text{and} \quad r_{ret} := r_{\widetilde{X}_{ret}}, \quad r_{et} := r_{\widetilde{X}_{et}}$$

in the notation of (6.1.2). Recall that the functors r_* are the respective functors Γ_G (formation of the sheaf of G-invariants), while the functors r^* put the trivial G-action on sheaves. The functors q^* forget the G-action, while q_* sends F to $F \times F$ with the involution which interchanges the two factors. The topos morphisms $\nu(G)$ resp. $\pi(G)$ are induced by ν resp. π (10.6.2); see also (6.4.3) below.

The solid arrows part of (3) commutes. In addition there are a handful of relations between the various functors associated with (3). Some of them are listed in (10.19.1). The following will be used in the sequel:

(6.4.2) Lemma. —
a) $r_{ret} \circ q_{ret} = \text{id}$ *and* $r_{et} \circ q_{et} = r_{et} \circ \pi(G) = \text{id}$;
b) $\rho \cong (r_{ret})_* \circ \nu(G)^* = \Gamma_G \circ \nu(G)^*$;
c) $\nu(G)^* \cong \rho(G) \circ \pi(G)_*$, *where* $\rho(G): \widetilde{X}_{et}(G) \to \widetilde{X}_{ret}(G)$ *denotes the functor induced by* ρ.

Proof. a) and b) are in (10.19.1), and c) follows from $\nu^* = \rho \circ \pi_*$. But see also the following remark (6.4.3). $\qquad\qquad\qquad\qquad\qquad\qquad\qquad\qquad\qquad\qquad\qquad\qquad\square$

(6.4.3) Explanation. Here is a word of explanation on $\pi(G)$ and $\nu(G)$. The involution on X' over X induces an involution on the functor $\pi_*\pi^*: \widetilde{X}_{et} \to \widetilde{X}_{et}$. The direct image $\pi(G)_*$ sends a sheaf A on X_{et} to the sheaf $\pi_*\pi^*A$ with this involution (Example (10.12)). The adjunction morphism $A \to \pi_*\pi^*A$ is injective, and its image is $\Gamma_G(\pi_*\pi^*A) = \Gamma_G \circ \pi(G)_*A$. This is easily verified, and proves the last equality in a).

From this one also obtains a more explicit description of $\nu(G)$, by c). The inverse image functor $\nu(G)^*$ sends $A \in \widetilde{X}_{et}$ to the G-sheaf $\rho\pi_*\pi^*A$ on X_{ret} (the involution on this sheaf is induced from the involution on $\pi_*\pi^*A$). Since ρ is left exact, the identity $\Gamma_G \circ \nu(G)^* = \rho$ of b) follows from $\Gamma_G \circ \pi(G)_* = \text{id}$.

If $A = h_U$ is a representable sheaf (with $U \in \text{Et}/X$) then $\nu(G)^*h_U$ is the complex spectrum $(U_{X'})_{cx}$ of $U_{X'}$ over X — an espace étalé over X_r with a natural G-action which was described in (5.7.5). There it was also mentioned that the G-fixpoints of this space coincide with U_r; this is precisely what (6.4.2b) says in the case $A = h_U$.

One sees that it is possible to define $\pi(G)$ and $\nu(G)$ by the descriptions just given, and to verify rules like (6.4.2) by direct arguments. In so far the general view point of G-toposes of Sect. 10 is not needed for the following. But if one wants to reach a better "philosophical" understanding, the notion of G-toposes forms the right general setup, and it has the advantage to apply to other situations as well.

(6.5) Proposition. — *Assume $\frac{1}{2} \in \mathcal{O}(X)$. For every sheaf A on X_{et} there is a natural isomorphism $\mathbf{H}_G^0(j_!\pi(G)_*A) \cong j_!A$ of sheaves on X_b. If A is an abelian sheaf then $\mathbf{H}_G^n(j_!\pi(G)_*A) = 0$ for $n \geq 1$, and hence there are natural isomorphisms*

$$H_G^n\Big(X_b, j_!\pi(G)_*A\Big) \cong H^n(X_b, j_!A) \tag{4}$$

for $n \geq 0$.

Note that the $j_!$ in $j_!\pi(G)_*A$ denotes the functor $j_!$ on G-sheaves induced by the usual $j_!$. In general I will note distinguish these functors notationally.

Proof. Since $\Gamma_G \circ \pi(G)_*$ is the identity (6.4.2a) and the functor $j_!$ preserves fibre products, it is clear that $\Gamma_G\big(j_!\pi(G)_*A\big) = j_!A$. It suffices to show $\mathbf{H}_G^n(j_!\pi(G)_*A) = 0$ for $n \geq 1$, since (4) will follow from this by applying the first spectral sequence in (6.1.3) to $j_!\pi(G)_*A$. This is done by inspecting the stalks, so let $\alpha : z \to X$ be a b-point of X (3.9). By Proposition (6.2),

$$\Big(\mathbf{H}_G^n(j_!\pi(G)_*A)\Big)_\alpha = H^n\Big(G, \big(j_!\pi(G)_*A\big)_\alpha\Big).$$

If α is real the right hand group is trivially zero. So assume that α is étale. Then $\big(j_!\pi(G)_*A\big)_\alpha = \big(\pi(G)_*A\big)_\alpha$ is the group $(\pi_*\pi^*A)_\alpha \cong A_\alpha \times A_\alpha$ on which G acts by interchanging the two components: $(a,b) \mapsto (b,a)$. So $\big(j_!\pi(G)_*A\big)_\alpha = A_\alpha \otimes \mathbb{Z}G$ is an induced G-module, and thus has trivial cohomology. $\qquad\square$

(6.6) Theorem. — *Let X be a scheme on which 2 is invertible, and let A be an abelian sheaf on X_{et}. The short exact sequence*

$$0 \longrightarrow j_!\pi(G)_*A \longrightarrow j_*\pi(G)_*A \longrightarrow i_*\nu(G)^*A \longrightarrow 0 \tag{5}$$

in $\mathrm{Ab}_G(X_b)$ gives rise to the long exact sequence of cohomology groups

$$\cdots \longrightarrow H^n(X_b, j_!A) \longrightarrow H^n(X_{et}, A) \longrightarrow H_G^n\Big(X_{ret}, \nu(G)^*A\Big) \longrightarrow \cdots \tag{6}$$

This sequence is functorial in A. Moreover every morphism $f : Y \to X$ of schemes induces a morphism of long exact sequences from (6) to

$$\cdots \longrightarrow H^n(Y_b, j_!f_{et}^*A) \longrightarrow H^n(Y_{et}, f_{et}^*A) \longrightarrow H_G^n\Big(Y_{ret}, \nu(G)^*f_{et}^*A\Big) \longrightarrow \cdots$$

Proof. This is a corollary to results obtained earlier: The sequence (5) is the sequence (4) from (2.10), applied to the G-sheaf $\pi(G)_*A$ on X_{et}. The last term is identified by (6.4.2c). Application of the functor $\mathbf{H}_G^0(X_b, -)$ to (5) yields the long exact sequence

$$\cdots H_G^n\Big(X_b, j_!\pi(G)_*A\Big) \longrightarrow H_G^n\Big(X_b, j_*\pi(G)_*A\Big) \longrightarrow H_G^n\Big(X_b, i_*\nu(G)^*A\Big) \cdots$$

This is the sequence (6): The identification of the first term is given in (6.5). Since $\Gamma_G \circ j_*\pi(G)_* = j_* \circ \Gamma_G\pi(G)_* = j_*$ and $j_*\pi(G)_*$ is exact by (3.12d), the middle term is $H^n(X_{et}, A)$. The identification of the last term follows from exactness of i_*. To see functoriality of (6) with respect to a scheme morphism f apply f_b^* to the sequence (5). From the commutation rules in (3.10) it follows that the sequence so obtained is the sequence (5) for the étale sheaf f_{et}^*A on Y. $\qquad\square$

(6.6.1) Corollary. — *If M is an abelian group then $\nu(G)^* \underline{M}_{et}$ is the constant sheaf $\underline{M}_{ret}$ on X_{ret} with the trivial G-action. Hence (6) yields a long exact sequence*

$$\cdots \longrightarrow H^n(X_b, j_! M) \longrightarrow H^n(X_{et}, M) \longrightarrow H_G^n(X_{ret}, M) \longrightarrow \cdots \qquad (7)$$

Proof. It is clear that $\nu(G)^* \underline{M}_{et}$ is the constant abelian group object M in the topos $\widetilde{X}_{ret}(G)$, i.e. the inverse image of the abelian group M with respect to the final topos morphism f from $\widetilde{X}_{ret}(G)$ to (sets). But f can be factored in the topos morphisms

$$\widetilde{X}_{ret}(G) \xrightarrow{r_{ret}} \widetilde{X}_{ret} \longrightarrow \text{(sets)},$$

so the assertion is obvious since r_{ret}^* puts the trivial G-operation on sheaves. $\square$

(6.6.2) Remark. If $A \in \mathrm{Ab}(X_{et})$ is an odd torsion sheaf then the fundamental long exact sequence (6) simplifies to

$$\cdots H^n(X_b, j_! A) \longrightarrow H^n(X_{et}, A) \longrightarrow H^n(X_{ret}, \rho A) \cdots \qquad (8)$$

Indeed, if B is an abelian G-sheaf on X_{ret} which is odd torsion then $\mathbf{H}_G^n(B) = 0$ for $n > 0$ by (6.2), and so the first spectral sequence in (6.1.3) shows $H_G^*(X_r, B) = H^*(X_r, \Gamma_G B)$. Since $\Gamma_G \circ \nu(G)^*(A) = \rho A$ (6.4.2b) this gives the identification of the third term in (8). Of course, (8) is simply the long exact sequence derived from the short exact sequence $0 \to j_! A \to j_* A \to i_* \rho A \to 0$ in $\mathrm{Ab}(X_b)$, since $\mathrm{R}^n j_* A = 0$ for $n \geq 1$ by (3.12).

(6.6.3) The homomorphisms

$$h \colon H^n(X_{et}, A) \longrightarrow H_G^n(X_{ret}, \nu(G)^* A) \qquad (9)$$

of (6) should be regarded as sort of comparison maps. The basic idea is that these maps tend to be isomorphisms in high degrees. The exact sequence (6) shows that the obstructions against (9) being isomorphisms are cohomology groups of the glued site X_b. Therefore it is interesting to study the cohomology of X_b more closely, in particular the cohomological dimensions of X_b. This will be started in Sect. 7.

(6.6.4) A generalization of the material presented here to G-toposes of quite arbitrary nature will be sketched in Sect. 14. There one finds a detailed discussion of the long exact sequence, and how it generalizes facts from topological equivariant theory. In particular it will be made clear (15.5.1) that if X is an algebraic variety over $\mathbb{R}$ and A is constant with finite stalks, the sequence (6) is canonically isomorphic to the Cox sequence (cf. the Introduction).

(6.7) Remark. ($\frac{1}{2} \in \mathcal{O}(X)$) Let $A \in \mathrm{Ab}(X_{et})$. The homomorphisms (9) in the exact sequence (6) are nothing but pullback of cohomology classes by $\nu(G)^*$. Since $\pi^*\nu(G)_* = \nu_* q_{ret}^*$ (10.6.2) and ν_* is exact (Theorem 5.9) it follows that $\nu(G)_*$ is exact. So (9) can also be read as the map induced in *étale* cohomology by the adjunction morphism

$$\alpha: A \longrightarrow \nu(G)_* \nu(G)^* A. \tag{10}$$

The latter map, however, has non-trivial kernel *and* cokernel in general. This already suggests that it might be easier to deal with (9) when it is included in the long exact sequence (6).

Assume however that every residue field of X is formally real; for example, X may be the spectrum of a real field. Then the map (10) is injective for every $A \in \mathrm{Ab}(X_{et})$. Indeed, it is equivalent that its restriction to X'_{et} is injective. But this is

$$\pi^* A \longrightarrow \pi^* \nu(G)_* \nu(G)^* A = \nu_* q_{ret}^* \nu(G)^* A = \nu_* \nu^* \pi^* A$$

which was shown to be injective in (5.5.3). Let $P(A)$ denote the cokernel of α, so that one has a short exact sequence of étale sheaves on X, functorial in A:

$$0 \longrightarrow A \xrightarrow{\alpha} \nu(G)_* \nu(G)^* A \longrightarrow P(A) \longrightarrow 0. \tag{11}$$

Then there are canonical isomorphisms

$$H^n(X_b, j_! A) \cong H^{n-1}(X_{et}, P(A)), \quad n \in \mathbb{Z}, \tag{12}$$

which are again functorial in A.

This is not surprising since both sides figure in long exact sequences which otherwise coincide (on any two of three typical terms). To give a rigorous proof recall that $j_* \pi(G)_*: \mathrm{Ab}(X_{et}) \to \mathrm{Ab}_G(X_b)$ is an exact functor and that $H^0_G(X_b, -) \circ j_* \pi(G)_* = H^0(X_{et}, -)$. By general adjunction reasons there is a factorization β in the following diagram in $\mathrm{Ab}_G(X_b)$, and β is natural in A:

$$
\begin{array}{ccccccccc}
0 & \longrightarrow & j_*\pi(G)_* A & \longrightarrow & j_*\pi(G)_*\nu(G)_*\nu(G)^* A & \longrightarrow & j_*\pi(G)_* P(A) & \to & 0 \\
& & \| & & \Big\downarrow{\scriptstyle\beta} & & & & \\
0 & \to & j_!\pi(G)_* A & \longrightarrow & j_*\pi(G)_* A & \longrightarrow & i_*\nu(G)^* A & \longrightarrow & 0.
\end{array}
$$

Here the top row is the transformation of (11) by $j_*\pi(G)_*$, and the bottom row is (5). (To see the factorization β, apply i^* to the square in the diagram and use $\rho\pi(G)_* = \nu(G)^*$.) Both rows are exact, and moreover β induces an isomorphism in G-equivariant cohomology on X_b. The long exact sequence obtained by applying $\Gamma_G(X_b, -)$ to the top row is the long exact sequence for (11) on X_{et}, while the bottom row gives (6). From this one deduces the assertion.

Next I give the generalization of the long exact sequence (6) to a relative situation:

(6.8) Corollary. — *Let $f\colon Y \to X$ be a morphism of schemes, and suppose that 2 is invertible on Y. Then for every $A \in \mathrm{Ab}(Y_{et})$ there is a natural long exact sequence*

$$\cdots \longrightarrow R^n f_{b*}(j_! A) \longrightarrow R^n(j_* f_{et*}) A \longrightarrow i_* R^n f_{ret*}^G \circ \nu(G)^* A \longrightarrow \cdots \qquad (13)$$

of abelian sheaves on X_b.

Here $f_{ret*}^G \colon \mathrm{Ab}_G(Y_{ret}) \to \mathrm{Ab}(X_{ret})$ denotes the composite functor $f_{ret*} \circ \Gamma_G = \Gamma_G \circ f_{ret}(G)_*$. Compare with Corollary (6.9.2) below.

Proof. For every étale X-scheme U the exact sequence (6) on $U_Y = U \times_X Y$ for $\tilde{A} := A\big|_{U_Y}$ reads

$$\cdots \longrightarrow H_b^n(U_Y, j_! \tilde{A}) \longrightarrow H_{et}^n(U_Y, \tilde{A}) \longrightarrow H_{G,ret}^n\big(U_Y, \nu(G)^* \tilde{A}\big) \longrightarrow \cdots$$

If one varies U in Et/X one obtains a long exact sequence of presheaves on Et/X. Take the associated sheaves with respect to the topology b on Et/X. The resulting long sequence in $\mathrm{Ab}(X_b)$ is exact, and is easily identified with the sequence (13). $\square$

The following easy observation has some nice applications, cf. below and Sections 19, 20.2:

(6.9) Proposition. — *Let 2 be invertible on X. The additive functor $\nu(G)^*$ from $\mathrm{Ab}(X_{et})$ to $\mathrm{Ab}_G(X_{ret})$ carries injective sheaves to Γ_G-acyclic sheaves.*

Proof. Let $A \in \mathrm{Ab}(X_{et})$, and let $\xi \colon x \to X$ be a real geometric point of X. By Proposition (6.2)

$$\Big(\mathbf{H}_G^n(\nu(G)^* A)\Big)_\xi = H^n\Big(G, \big(\nu(G)^* A\big)_\xi\Big) \underset{(*)}{=} H_{et}^n(x, \xi_{et}^* A).$$

Identify $\tilde{x}_{et}$ with (G-sets). The reason for $(*)$ is that the G-modules $\big(\nu(G)^* A\big)_\xi$ and $\xi_{et}^* A$ are isomorphic. Hence it follows that

$$\Big(\mathbf{H}_G^n(\nu(G)^* A)\Big)_\xi = \varinjlim_{U \in \overline{Nb}(\xi)^\circ} H_{et}^n(U, A\big|_U),$$

by (3.5.1b). If A is injective then so is $A\big|_U$ for every U étale over X, and so the above stalks vanish for $n > 0$. $\square$

This fact has some interesting consequences:

(6.9.1) Corollary. — *Let 2 be invertible on X. Then $R\Gamma_G \circ \nu(G)^* = R\rho$, i.e. for every $A \in \mathrm{Ab}(X_{et})$ and $n \geq 0$ one has $R^n \rho A = \mathbf{H}_G^n(\nu(G)^* A)$.*

Proof. Immediate from Proposition (6.9) since $\rho = \Gamma_G \circ \nu(G)^*$ (6.4.2b). $\square$

(6.9.2) Corollary. — *If $f: Y \to X$ is a morphism of schemes over $\mathbb{Z}[\frac{1}{2}]$ then there is an isomorphism*

$$Rf^G_{ret*} \circ \nu(G)^* \cong Rf_{ret*} \circ R\rho$$

of functors $D^+(Y_{et}) \to D^+(X_{ret})$ (cf. (6.8) for the definition of f^G_{ret*}). *In particular, for any scheme X with $\frac{1}{2} \in \mathcal{O}(X)$ and any abelian sheaf A on X_{et} there is a cohomological spectral sequence*

$$E_2^{pq} = H^p(X_{ret}, R^q \rho A) \quad \Longrightarrow \quad H_G^{p+q}\Big(X_{ret}, \nu(G)^* A\Big), \tag{14}$$

functorial in A. The edge homomorphisms

$$H^n(X_{ret}, \rho A) \longrightarrow H_G^n\Big(X_{ret}, \nu(G)^* A\Big) \qquad (n \geq 0)$$

are isomorphisms modulo 2-torsion, i.e. are isomorphisms after tensorizing with $\mathbb{Z}[\frac{1}{2}]$.

Proof. The first assertion from (6.9.1) since $Rf^G_{ret*} \circ \nu(G)^* = Rf_{ret*} \circ R\Gamma_G \circ \nu(G)^*$. Since $R^q \rho A$ is killed by 2 for $q \geq 1$ (3.12c) one has $2E_2^{pq} = 0$ for $q \geq 1$, which implies the last statement about the edge homomorphisms. $\qquad \square$

The spectral sequence (14) calculates the equivariant cohomology groups on X_{ret} which enter into the basic long exact sequence (6), in terms of ordinary sheaf cohomology on the real spectrum. For some applications see Sections 19 and 20.2.

(6.10) There is a second interpretation of the long exact sequence (6) and the spectral sequence (14). This is an application of the above corollaries. Recall that if T is a topological space, $U \subset T$ is an open subspace and $Z = T - U$ is its closed complement, one has the long exact sequence of relative cohomology [Ht]

$$\cdots \longrightarrow H_Z^n(T, F) \longrightarrow H^n(T, F) \longrightarrow H^n(U, F) \longrightarrow \cdots$$

for $F \in Ab(T)$, where $H_Z^*(T, F)$ is cohomology of F with support in Z (or "relative" cohomology of T mod U). This sequence and related ones exist more generally in any glued topos. We observe now that (6) can be identified with such a relative cohomology sequence in X_b.

(6.10.1) Let X be any scheme (with characteristic 2 points allowed). Recall (Sect. 2) that the functor i_* on abelian sheaves admits a right adjoint $i^!: Ab(X_b) \to Ab(X_{ret})$ which is described by the exact sequence

$$0 \longrightarrow i_* i^! F \longrightarrow F \longrightarrow j_* j^* F$$

for $F \in Ab(X_b)$. It is customary in such a situation to write $H^0_{X_{ret}}(X_b, F) := H^0(X_b, i_* i^! F) = (i^! F)(X)$ and $H^n_{X_{ret}}(X_b, F) := R^n H^0_{X_{ret}}(X_b, -)(F)$, $n \geq 0$ [Ht], [SGA4 V.6]. Moreover the sheafified versions of these groups are written $\mathbf{H}^n_{X_{ret}}(F) := i_* R^n i^! F$, $n \geq 0$.

The following facts hold in complete generality [SGA4 V.6]:

(6.10.2) Proposition. — *Let X be a scheme and $F \in \mathrm{Ab}(X_b)$.*
a) *There are a natural exact sequence in $\mathrm{Ab}(X_b)$*

$$0 \longrightarrow H^0_{X_{ret}}(F) \longrightarrow F \longrightarrow j_* j^* F \longrightarrow H^1_{X_{ret}}(F) \longrightarrow 0$$

and natural isomorphisms

$$i_* R^n \rho(j^* F) = R^n j_*(j^* F) \xrightarrow{\;\sim\;} H^{n+1}_{X_{ret}}(F)$$

for $n \geq 1$.
b) *There is a natural long exact sequence*

$$0 \longrightarrow H^0_{X_{ret}}(X_b, F) \longrightarrow F(X) \longrightarrow (j^* F)(X) \longrightarrow \cdots$$
$$\cdots \longrightarrow H^n_{X_{ret}}(X_b, F) \longrightarrow H^n(X_b, F) \longrightarrow H^n(X_{et}, j^* F) \longrightarrow \cdots$$

c) *There is a natural spectral sequence ($p,\ q \geq 0$)*

$$E_2^{pq} = H^p(X_b, H^q_{X_{ret}}(F)) \implies H^{p+q}_{X_{ret}}(X_b, F). \qquad \square$$

If one puts $F = j_! A$ in b), with $A \in \mathrm{Ab}(X_{et})$, the resulting sequence coincides with (6) in two of three typical positions. This leads one to conjecture that

$$H^{n+1}_{X_{ret}}(X_b, j_! A) \cong H^n_G\left(X_{ret}, \nu(G)^* A\right)$$

should hold if $\frac{1}{2} \in \mathcal{O}(X)$. This is indeed the case:

(6.10.3) Corollary. — *Let X be any scheme.*
a) *$i^! \circ j_! = 0$ and $R^n \rho \xrightarrow{\sim} R^{n+1} i^! \circ j_!$ for all $n \geq 0$. In short,*

$$R\rho \cong [1] \circ Ri^! \circ j_!$$

as functors $D^+(X_{et}) \to D^+(X_{ret})$.
b) *If 2 is invertible on X then for $A \in \mathrm{Ab}(X_{et})$ there are natural isomorphisms ($n \in \mathbb{Z}$)*

$$H^n_G\left(X_{ret}, \nu(G)^* A\right) \xrightarrow{\;\sim\;} H^{n+1}_{X_{ret}}(X_b, j_! A).$$

Proof. a) Let $A \in \mathrm{Ab}(X_{et})$. From (6.10.2a), applied to $F = j_! A$, it follos that $i^! j_! A = 0$ and $R^n \rho A \xrightarrow{\sim} R^{n+1} i^!(j_! A)$ for $n \geq 0$; whence a). (This is still a general fact.) If $\frac{1}{2} \in \mathcal{O}(X)$ then (6.9.2) gives

$$R\Gamma_G(X_{ret}, -) \circ \nu(G)^* \cong R\Gamma(X_{ret}, -) \circ R\rho \underset{a)}{\cong} R\Gamma(X_{ret}, -) \circ [1] \circ Ri^! \circ j_!,$$

which is b). $\square$

(6.10.4) The corollary also identifies the spectral sequence (14) of (6.9.2) with the spectral sequence (6.10.2c) for cohomology with support, the latter for $F = j_! A$ and with a dimension shift by 1 in q.

(6.11) **Remark.** In this section X was mostly supposed to be a scheme without points of characteristic 2. Here are some complementary remarks on the case when 2 is not invertible on X.

If X has a point of characteristic two then the canonical topos morphism $\phi\colon \widetilde{X}'_{et}(G) \to \widetilde{X}_{et}$ (Example (10.11)) is no longer an equivalence. Instead of (3) one has therefore only a diagram

$$
\begin{array}{ccccc}
\widetilde{X}_{ret} & \xrightarrow{\ \nu\ } & \widetilde{X}'_{et} & \xrightarrow{\ \pi\ } & \widetilde{X}_{et} \\
{\scriptstyle r_{ret}}\Big\uparrow\Big\downarrow{\scriptstyle q_{ret}} & & p\Big\downarrow & & {\scriptstyle q_{et}}\Big\downarrow\Big\uparrow{\scriptstyle r_{et}} \\
\widetilde{X}_{ret}(G) & \xrightarrow{\ \nu(G)\ } & \widetilde{X}'_{et}(G) & \xrightarrow{\ \pi(G)\ } & \widetilde{X}_{et}(G) \\
& \searrow{\scriptstyle \bar{\nu}} & \phi\Big\downarrow & & \\
& & \widetilde{X}_{et}. & &
\end{array}
\tag{15}
$$

in which $\phi \circ p = \pi$. Define $\bar{\nu} := \phi \circ \nu(G)$, a topos morphism from $\widetilde{X}_{ret}(G)$ to $\widetilde{X}_{et}$. Roughly, if one replaces $\nu(G)^*$ by $\bar{\nu}^*$, those assertions of this section remain true which do not involve the topos morphism $\pi(G)$. More precisely, this holds for (6.4.2b), and then for (6.9), (6.9.1), (6.9.2) and (6.10.3).

7. Cohomological dimension of X_b, I: Reduction to the field case

Let X be a scheme on which 2 is invertible, and let $X' = X[\sqrt{-1}]$. In the preceding section it was shown that étale cohomology of X is related to G-equivariant cohomology of the real spectrum X_r through a long exact sequence, in which also sheaf cohomology of X_b figures. The comparison homomorphisms $H^n(X_{et}, A) \to H^n_G(X_{ret}, \nu(G)^*A)$ being the principal object of interest, one is thus led to a closer study of b-cohomology. This is taken up now.

The goal is to show that the cohomological ℓ-dimension of X_b (ℓ a prime) exceeds that of X'_{et} by at most one. Modulo exactness of ν_* (which was established in Sect. 5) this is quite trivial for odd ℓ, so the interesting case is $\ell = 2$. Here one can make a stronger statement: If A is an étale $\mathbb{Z}/2\mathbb{Z}$-sheaf on X then one can pass each of the three groups

$$H^n(X_{et}, A), \quad H^n_G(X_{ret}, \nu(G)^*A) \quad \text{and} \quad H^n(X_b, j_!A)$$

to the direct limit $n \to \infty$ in a natural way. The corresponding limit groups are related by an exact triangle (7.12.1). The main result (Theorem (7.17)) states that for the b-topology (third of the above groups) this limit group vanishes. Hence $H^n(X_{et}, A)$ and $H^n_G(X_{ret}, \nu(G)^*A)$ are isomorphic "in the limit $n \to \infty$". If $\mathrm{cd}_2(X'_{et})$ is finite, equal to d, one gets these isomorphisms already for $n > d$. The hypotheses needed for this theorem are very weak, it is only required that X be quasi-compact and quasi-separated (and $\frac{1}{2} \in \mathcal{O}(X)$, of course).

This result is non-trivial already for fields. However in this section the field case will be taken for granted, and it will be shown how to deduce the general case from it. The field case will be settled in Sect. 9, after some preparations in Sect. 8. The plan of the proof in this section is the following: First the theorem is proved by inductive arguments for schemes of finite type over spec $\mathbb{Z}$, using the field case. By limit arguments one extends this to finitely presented schemes over any base ring. Then a glueing argument gives the general case. The section will close with a series of immediate corollaries.

After some trivialities on torsion sheaves, this section starts with a collection of more or less easy comparison results between the cohomological dimensions of the various sites, before the proof of the main theorem is taken up.

(7.1) Definition. Let C be a site and ℓ a prime.

a) An abelian sheaf A on C is said to be ℓ-*primary torsion* if $\lim\limits_{\nu \to \infty} (_{\ell^\nu} A) \to A$ is an isomorphism of sheaves [SGA4 IX.1]. A is called *torsion* if $\lim\limits_{n \to \infty} (_n A) \to A$ is an isomorphism. (Recall that $_n A$ is the kernel of multiplication by n on A.)

b) The *cohomological ℓ-dimension* $\mathrm{cd}_\ell(C)$ of C is defined to be the largest integer n for which there is an ℓ-primary torsion sheaf A on C with $H^n(C, A) \neq 0$. If no such n exists one writes $\mathrm{cd}_\ell(C) = \infty$. I write $\mathrm{cd}(C)$ (with no prime specified) for the supremum of all integers $n \geq 0$ for which there is some abelian sheaf A (not necessarily torsion) such that $H^n(C, A) \neq 0$.

Of course these definitions apply in particular to any topological space T, such a space being identified with a site in the usual way.

(7.1.1) Remarks. Let E be a topos.

1. If E has sufficiently many points then $A \in \mathrm{Ab}(E)$ is ℓ-primary torsion if and only if every stalk of A is an ℓ-primary torsion group.

2. Assume that the topos E is locally coherent [SGA4 VI]. For example this holds for the topos $\widetilde{X}_t$ ($t \in \{et, b, ret\}$) of any scheme X, (2.18.2). For $A \in \mathrm{Ab}(E)$ to be ℓ-primary torsion, it is necessary and sufficient that $A(Y) = \mathrm{Hom}(Y, A)$ be an ℓ-primary torsion group for every quasi-compact $Y \in E$. For the sufficiency part it is enough to know this for Y ranging over some generating family of quasi-compact objects. Moreover, if E is coherent (e.g. $E = \widetilde{X}_t$ for X quasi-compact and quasi-separated) then the natural maps

$$\lim_{\nu \to \infty} H^n(E, _{\ell^\nu} A) \longrightarrow H^n(E, A), \quad n \geq 0,$$

are isomorphisms for each $A \in \mathrm{Ab}(E)$ which is ℓ-primary torsion [SGA4 IX.1.2]. In particular the $H^n(E, A)$ are ℓ-primary torsion groups for such A. Using the short exact sequences $0 \longrightarrow _{\ell^{\nu-1}} A \longrightarrow _{\ell^\nu} A \longrightarrow (_{\ell^\nu} A)/(_{\ell^{\nu-1}} A) \longrightarrow 0$ it follows immediately by induction that in a coherent topos E one has

$$\mathrm{cd}_\ell(E) = \sup\Big\{ n \geq 0 : \text{ there is a } \mathbb{Z}/\ell\text{-sheaf } A \in \mathrm{Ab}(E) \text{ with } H^n(E, A) \neq 0 \Big\}.$$

(7.1.2) Lemma. — *Let X, Y be schemes, and let $f : Y \to X$ be a morphism which is quasi-compact and quasi-separated. Each of the following additive functors maps ℓ-primary torsion sheaves to ℓ-primary torsion sheaves:*

    (i) *$j_!$, j^*, j_*, i^*, i_*, $i^!$, $R^n\rho$ for $n \geq 0$ (all are functors on X);*

    (ii) *ν^*, $\nu_* : \mathrm{Ab}(X_{ret}) \leftrightarrows \mathrm{Ab}(X'_{et})$;*

    (iii) *f_t^*, $R^n f_{t*} : \mathrm{Ab}(Y_t) \leftrightarrows \mathrm{Ab}(X_t)$, for $t \in \{et, b, ret\}$ and $n \geq 0$.*

Proof. The assertion holds trivially for any (left) exact additive functor which has a right adjoint. Equally clear are $i^!$ (a subfunctor of i^*) and j_* (look at the stalks). $R^n\rho$ for $n \geq 1$ is zero on odd torsion sheaves and is always annihilated by 2 (3.12c). For ν_* and $R^n f_{t*}$ the claim follows again from inspecting the stalks, using (5.8.1) and (3.11). $\square$

I want to compare the cohomological dimensions of the sites X_{et}, X'_{et}, X_b and X_r. Recall that $X' = X \otimes_{\mathbb{Z}} \mathbb{Z}[\sqrt{-1}]$ and $\pi: X' \to X$ is the projection. To start with, here are the most immediate observations:

(7.2) Proposition. — *Let X be any scheme and ℓ any prime.*
a) *The following inequalities hold:*

$$\mathrm{cd}_\ell(X_r) \leq \mathrm{cd}_\ell(X_b), \quad \mathrm{cd}_\ell(X'_{et}) \leq \mathrm{cd}_\ell(X_{et}), \quad \mathrm{cd}_\ell(X'_{et}) \leq \mathrm{cd}_\ell(X_b).$$

b) *If ℓ is odd then $\mathrm{cd}_\ell(X_{et}) = \mathrm{cd}_\ell(X'_{et})$.*
c) *If $\mathrm{cd}_2(X_{et}) < \infty$ and 2 is invertible on X then also $\mathrm{cd}_2(X_{et}) = \mathrm{cd}_2(X'_{et})$.*

Proof. a) follows from (7.1.2) since i_*, π_* and $j_* \pi_*$ (3.12d) are exact functors. Assume first that 2 is invertible on X. I drop all subscripts "et" in the remainder of this proof. One uses the trace of the étale covering π in a well known way: For $A \in \mathrm{Ab}(X)$ the composition $A \longrightarrow \pi_* \pi^* A \xrightarrow{\mathrm{tr}} A$ is multiplication by 2. The assertion of b) is immediate since $H^*(X, A)$ is uniquely 2-divisible, for A an odd torsion sheaf; and c) follows by using exact sequences $0 \to B \longrightarrow \pi_* \pi^* A \xrightarrow{\mathrm{tr}} A \to 0$ and downward induction. To prove b) in general let X be an arbitrary scheme and consider the cartesian squares

$$
\begin{array}{ccccc}
U' & \xrightarrow{\ g'\ } & X' & \xleftarrow{\ f'\ } & Z' \\
{\scriptstyle p}\downarrow & & {\scriptstyle \pi}\downarrow & & \downarrow \\
U & \xrightarrow{\ g\ } & X & \xleftarrow{\ f\ } & Z
\end{array}
$$

where Z is the closed reduced subscheme of char 2 points, $U = X - Z$, and f, g are the inclusions. Let $A \in \mathrm{Ab}(U)$ be an ℓ-primary torsion sheaf, ℓ odd. Applying $g_!$ to $A \longrightarrow p_* p^* A \xrightarrow{\mathrm{tr}} A$ (whose composition is 2) and using $g_! p_* = \pi_* g'_!$ one finds $H^q(X, g_! A) = 0$ for $q > \mathrm{cd}_\ell(X')$, by the same argument as before in the case 2 invertible. Moreover $\mathrm{cd}_\ell(Z) = \mathrm{cd}_\ell(Z')$ since $Z = (Z')_{\mathrm{red}}$, and obviously $\mathrm{cd}_\ell(Z') \leq \mathrm{cd}_\ell(X')$. So the assertion follows from the exact sequence $0 \to g_! g^* F \to F \to f_* f^* F \to 0$ for $F \in \mathrm{Ab}(X)$. $\qquad\square$

The argument in the last part of the proof was inspired by [SGA4 IX.5]. — The following is an immediate corollary to the main result of Sect. 5 (Theorem (5.9)), but nevertheless I state it as a theorem:

(7.3) Theorem. — *If X is any scheme on which 2 is invertible then $\mathrm{cd}_\ell(X_r) \leq \mathrm{cd}_\ell(X'_{et})$ for all primes ℓ.* $\qquad\square$

Of course this is true for any scheme as soon as one knows that ν_* is exact.

(7.4) Proposition. — *Let ℓ be an odd prime. If X is any scheme then*

$$\mathrm{cd}_\ell(X_b) = m_\ell \text{ or } 1 + m_\ell$$

with $m_\ell := \sup\{\mathrm{cd}_\ell(X_{et}), \mathrm{cd}_\ell(X_r)\}$. Moreover $\mathrm{cd}_\ell(X_b) = 1 + m_\ell < \infty$ is possible only if $\infty > \mathrm{cd}_\ell(X_r) \geq \mathrm{cd}_\ell(X_{et}) = \mathrm{cd}_\ell(X'_{et})$.

(7.4.1) Corollary. — *Assume that* 2 *is invertible on* X, *and let* ℓ *be an odd prime. Then* $\mathrm{cd}_\ell(X_b)$ *is either* $\mathrm{cd}_\ell(X_{et})$ *or* $1 + \mathrm{cd}_\ell(X_{et})$. *The latter case is possible only if* $\mathrm{cd}_\ell(X_r) = \mathrm{cd}_\ell(X_{et})$ *(or* $\mathrm{cd}_\ell(X_{et}) = \infty$ *).*

Proof. The corollary is immediate from the proposition by (7.2b) and Theorem (7.3). To prove the proposition let $A \in \mathrm{Ab}(X_{et})$ be an odd torsion sheaf. Then $R^n j_* A = 0$ for $n \geq 1$ (3.12) which shows $H^*(X_b, j_* A) = H^*(X_{et}, A)$. Thus there is a long exact sequence

$$\cdots \; H^n(X_b, j_! A) \longrightarrow H^n(X_{et}, A) \longrightarrow H^n(X_{ret}, \rho A) \; \cdots \tag{1}$$

From (1) together with the exact sequence

$$\cdots \; H^n(X_b, j_! j^* F) \longrightarrow H^n(X_b, F) \longrightarrow H^n(X_{ret}, i^* F) \; \cdots$$

(which holds for every $F \in \mathrm{Ab}(X_b)$) one reads off the assertion. □

It can actually happen that $\mathrm{cd}_\ell(X_b) = 1 + \mathrm{cd}_\ell(X_{et}) < \infty$, see (9.8.2). Of course the corollary should hold without the hypothesis on residue characteristics, since conjecturally ν_* is always exact.

(7.5) Remark. If X is any scheme with non-empty real spectrum then $\mathrm{cd}_2(X_{et}) = \infty$. Indeed, assume that 2 is invertible on X, and let $s = (-1) \in H^1(X_{et}, \mathbb{Z}/2)$ be the image of $-1 \in \mathcal{O}(X)^*$ under the boundary $\mathcal{O}(X)^* \to H^1(X_{et}, \mu_2)$ of the Kummer sequence $1 \to \mu_2 \to \mathbb{G}_m \xrightarrow{2} \mathbb{G}_m \to 1$. Let $\xi \colon \mathrm{spec}\, R \to X$ be a morphism with R a real closed field. Then $\xi^*(s^n) \neq 0$, and hence $s^n \neq 0$, for all $n \geq 0$. If 2 is not necessarily invertible one may argue similarly, using that the restriction map $H^n_{et}(\mathbb{Z}, \mathbb{Z}/2) \to H^n_{et}(\mathbb{R}, \mathbb{Z}/2)$ is surjective for $n \gg 0$ (for $n \geq 3$, in fact, and bijective for $n > 3$). Thus there are classes $s_n \in H^n_{et}(X, \mathbb{Z}/2)$, $n \gg 0$, which survive to all real points. The asserted surjectivity is deduced, for example, from the Leray spectral sequence for $\mathrm{spec}\,\mathbb{Q} \to \mathrm{spec}\,\mathbb{Z}$.

However one may ask whether this is the only possible obstruction to the equality $\mathrm{cd}_2(X_{et}) = \mathrm{cd}_2(X'_{et})$. In other words, does it follow from $\mathrm{cd}_2(X'_{et}) < \infty$ and $X_r = \emptyset$ that $\mathrm{cd}_2(X_{et}) < \infty$? This is closely related to a similar question in the cohomology theory of profinite groups which was settled by Serre in 1965 [Se2]. The main result of this section will answer this question in the affirmative, see Corollary (7.21) below.

Also the following result should be mentioned when discussing cohomological dimensions:

(7.6) Theorem. — *Let* X *be any scheme, let* X_{zar} *be the site associated with its underlying topological space. Then* $\mathrm{cd}(X_r) \leq \mathrm{cd}(X_{zar})$ *and* $\mathrm{cd}_\ell(X_r) \leq \mathrm{cd}_\ell(X_{zar})$ *for all* ℓ. *Moreover, if* X *is quasi-separated and is a union of countably many quasi-compact subspaces then* $\mathrm{cd}(X_{zar}) \leq \dim X$.

Proof. The direct image functor φ_* of the support mapping $\varphi: X_r \to X$ is exact. This will be proved in (19.2). It clearly implies that all cohomological dimensions of X_r are bounded above by those of X_{zar}. The inequality $\mathrm{cd}(X_{zar}) \leq \dim X$ for X as above is proved in [Sch1, Cor. 4.6]. $\qquad\Box$

The determination of the cohomological ℓ-dimension of X_b is the goal of this section. For odd ℓ this has already been achieved, so it is only the case $\ell = 2$ which remains. The aim is to show $\mathrm{cd}_2(X_b) \leq 1 + \mathrm{cd}_2(X'_{et})$. As indicated before, I will now assume that this is true for X the spectrum of a field (more precisely, I will use Corollary (9.8)). Then the next step is to treat finitely generated schemes:

(7.7) Theorem. — *Let X be a scheme of finite type over* $\mathrm{spec}\,\mathbb{Z}$*, and let d be its (Krull) dimension.*

a) $\mathrm{cd}_\ell(X_b) \leq 2d + 1$ *for every prime ℓ.*

b) *More precisely: Let F be an abelian sheaf on X_b such that j^*F is torsion. Let $s \geq 0$ be an integer such that $\dim \overline{\{x\}} \leq s$ holds for each $x \in X$ in which F has a nonzero (real or étale) stalk. Then $H^n(X_b, F) = 0$ for $n > 2s + 1$.*

This proposition has an almost exact counterpart in the étale theory: The analogous statements are true when b is replaced by et, *under the condition that the real spectrum of X is empty*: [SGA4 X.6.2]. (Of course this is actually included in (7.7) as a particular case.) Also the proof given here is very close to the proof of [loc.cit.], but it uses the latter result.

Proof. If $d = 0$ then X_{red} is a finite sum of spectra of finite fields; so $X_b = X_{et}$, and the assertion is well known. By noetherian induction one may therefore assume that b) has been proved for all proper closed subschemes of X. Let F be a sheaf as in b). There exists a directed system $\{Z_\alpha\}$ of closed subschemes Z_α of X of dimensions $\leq s$ such that $\bigcup_\alpha Z_\alpha$ contains all points of X in which F has a nonzero stalk. Write $f_\alpha: Z_\alpha \hookrightarrow X$ for the inclusion of Z_α. The direct image $(f_\alpha)_{b*}$ (on abelian sheaves) has a right adjoint $(f_\alpha)_b^!$. The natural sheaf map

$$\varinjlim_\alpha (f_\alpha)_{b*}(f_\alpha)_b^! F \longrightarrow F$$

is an isomorphism. Indeed, it is trivially injective. And it is surjective since X noetherian implies that every subset of $\bigcup_\alpha Z_\alpha$ which is closed in X is contained in one of the Z_α. So

$$\varinjlim_\alpha H_b^n\Big(Z_\alpha, (f_\alpha)_b^! F\Big) \xrightarrow{\sim} H_b^n(X, F)$$

for all $n \geq 0$ (3.2). If $s < d$ then the assertion to prove follows from the inductive hypothesis. So it is sufficient to prove $H^n(X_b, F) = 0$ for $n > 2d + 1$ and $F \in \mathrm{Ab}(X_b)$, j^*F torsion.

The exact sequence $0 \to j_! j^* F \to F \to i_* i^* F \to 0$ shows that it suffices to treat the cases $F = j_! A$ (with $A \in \mathrm{Ab}(X_{et})$ torsion) and $F = i_* B$ (with

$B \in \mathrm{Ab}(X_{ret})$. Since i_* is exact and $\mathrm{cd}(X_r) \leq \dim(X \times_{\mathrm{spec}\,Z} \mathrm{spec}\,\mathbb{Q}) < d$ (7.6) one is left with $F = j_! A$. Since A is the direct sum of its primary components it is sufficient to assume that A is ℓ-primary torsion for some prime ℓ. If ℓ is odd then $\mathrm{cd}_\ell(X_{et}) \leq 2d + 1$ by the étale case of the proposition [SGA4 X.6.2], so $\mathrm{cd}_\ell(X_b) \leq 2d + 1$ by (7.4). Hence one can assume $\ell = 2$.

Let $f_\nu : x_\nu \hookrightarrow X$ ($\nu = 1, \ldots, N$) be the inclusions of the generic points of the irreducible components of X. There is an exact sequence

$$0 \longrightarrow A' \longrightarrow A \longrightarrow \bigoplus_\nu f_{\nu*} f_\nu^* A \longrightarrow A'' \longrightarrow 0 \qquad (2)$$

in $\mathrm{Ab}(X_{et})$ such that A' and A'' have zero stalks in the x_ν. (Here and below some subscripts "et" are omitted.) By the inductive hypothesis, the b-cohomology of $j_! A'$ and $j_! A''$ vanishes in degrees $\geq 2d$. It suffices therefore to prove that $H_b^n(X, j_! f_{\nu*} f_\nu^* A)$ vanishes for $n > 2d+1$, $\nu = 1, \ldots, N$. Now for any $B \in \mathrm{Ab}(X_{et})$ the exact sequence

$$0 \longrightarrow j_! B \longrightarrow j_* B \longrightarrow i_* \rho B \longrightarrow 0,$$

combined with $\mathrm{cd}(X_{ret}) < d$, shows that $H^n(X_b, j_! B) = H^n(X_b, j_* B)$ for $n > d$. So one can replace $j_!$ by j_*. Let X_ν be the reduced closure of $\{x_\nu\}$ in X and write $x_\nu \xrightarrow{g_\nu} X_\nu \xrightarrow{h_\nu} X$ for the inclusions. Since $j_* f_{\nu*} = (h_\nu)_{b*} j_* g_{\nu*}$ and $(h_\nu)_{b*}$ is exact, one has

$$H_b^*\Big(X, j_* f_{\nu*} f_\nu^* A\Big) = H_b^*\Big(X_\nu, j_* g_{\nu*} g_\nu^*(A|_{X_\nu})\Big),$$

which shows that one can replace X by $X_\nu = \overline{\{x_\nu\}}$ and A by $A|_{X_\nu}$.

So one can assume: X is integral, of dimension d, $f : x \hookrightarrow X$ is the inclusion of the generic point, and $S \in \mathrm{Ab}(x_{et})$ is a 2-primary torsion sheaf. One has to show that the group

$$H^n(X_b, j_* f_* S) = H^n(X_b, f_{b*} j_* S)$$

vanishes for $n > 2d + 1$. Only the case $\mathrm{char}\,x = 0$ has to be considered since otherwise $X_b = X_{et}$. Write $F := j_* S$, a sheaf on x_b, and consider the Leray spectral sequence for f_b:

$$E_2^{pq} = H^p(X_b, R^q f_{b*} F) \implies H^{p+q}(x_b, F).$$

The function field $\kappa(x)$ of X has transcendence degree $d-1$ over $\mathbb{Q}$. So $\mathrm{cd}_2\big(x[\sqrt{-1}]_{et}\big) = d+1$ [Se4, II, Prop. 11, 13], and thus $\mathrm{cd}_2(x_b) = d+1$ by (9.8). One has to show $E_2^{n0} = 0$ for $n > 2d+1$, so it suffices that $E_2^{pq} = 0$ if $p + q \geq 2d+1$ and $q \geq 1$.

Let $\alpha : z \to X$ be a b-point of X. Then by (3.11)

$$(R^q f_{b*} F)_\alpha = H_b^q(X^\alpha \times_X x, \tilde{F})$$

with $\tilde{F}$ = pullback of F under $\mathrm{pr}_2 : X^\alpha \times_X x \to x$, and $X^\alpha = \mathrm{spec}\, \mathcal{O}_{X,\alpha}$ the "localization" of X in α. So $X^\alpha \times_X x = \mathrm{spec}\, \Omega$ where Ω is the total ring of fractions of $\mathcal{O}_{X,\alpha}$. Put

$$\tilde{\mathcal{O}} := \begin{cases} \mathcal{O}_{X,\alpha} & \text{if } \alpha \text{ is étale,} \\ \mathcal{O}_{X,\alpha}[i] & \text{if } \alpha \text{ is real.} \end{cases}$$

Then $\tilde{\mathcal{O}}$ is a strictly henselian ring with total ring of fractions

$$\tilde{\Omega} := \begin{cases} \Omega & \text{if } \alpha \text{ is étale,} \\ \Omega[i] & \text{if } \alpha \text{ is real.} \end{cases}$$

From [SGA4 X.3.2] it follows that

$$\mathrm{cd}_2\Big((\mathrm{spec}\,\tilde{\Omega})_{et}\Big) \le \dim \tilde{\mathcal{O}} = \dim \mathcal{O}_{X,\alpha},$$

and $\dim \mathcal{O}_{X,\alpha} =: c$ is the codimension in X of the closure $\overline{\{\alpha(z)\}}$. Again from (9.8) one has therefore $\mathrm{cd}_2\big((\mathrm{spec}\,\Omega)_b\big) \le c$. (This also holds for $c = 0$ since then Ω is a field which is real closed or separably closed.) So one concludes that

$$\big(R^q f_{b*} F\big)_\alpha = 0 \quad \text{for} \quad q > c = d - \dim \overline{\{\alpha(z)\}}.$$

Altogether this shows that $R^q f_{b*} F = 0$ for $q > d$, and that $R^q f_{b*} F$ has support in dimension $\le d - q$ if $q \le d$. So the inductive hypothesis gives for $q \ge 1$ that $E_2^{pq} = 0$ if $q > d$ or $p > 2(d-q)+1$, in particular if $p + q > 2d+1$. This completes the proof. $\square$

(7.8) The following exposition (until Lemma (7.14.2) inclusively) does not make use of Theorem (7.7). Let from now on X always be a scheme on which 2 is invertible, $X' = X \otimes_{\mathbb{Z}} \mathbb{Z}[\sqrt{-1}]$ as usual, and let $G = \{1, \sigma\}$ be the group of order two. It is well known that the right adjoint π_* of $\pi^* : \mathrm{Ab}(X_{et}) \to \mathrm{Ab}(X'_{et})$ is also left adjoint to π^*. (This is more generally true for every finite étale morphism π [SGA4 IX.5.].) The adjunction map on X for the adjunction (π_*, π^*) is called the *trace map*:

$$\mathrm{tr} : \pi_* \pi^* \longrightarrow \mathrm{id}.$$

Recall that locally (in the étale sense, namely over X') the sheaf $\pi_* \pi^* A$ is isomorphic to $A \times A$, and the trace is (locally) the sum of the two components. I will occasionally write "adj" for the other adjunction map $\mathrm{id} \longrightarrow \pi_* \pi^*$. The composition $\mathrm{id} \xrightarrow{\mathrm{adj}} \pi_* \pi^* \xrightarrow{\mathrm{tr}} \mathrm{id}$ is multiplication by $2 = |G| = [X' : X]$.

(7.8.1) In the following write $\mathrm{Mod}_{\mathbb{Z}/2}(X_{et})$ for the abelian category of $\mathbb{Z}/2$-sheaves on X_{et} (i.e. sheaves which are annihilated by 2). Note that $H^*(X_{et}, A)$ is a graded module over $H^*(X_{et}, \mathbb{Z}/2)$ for every $\mathbb{Z}/2$-sheaf A. If A is a $\mathbb{Z}/2$-sheaf on X_{et} then the short sequence

$$\mathbf{E}(A): \quad 0 \longrightarrow A \xrightarrow{\mathrm{adj}} \pi_* \pi^* A \xrightarrow{\mathrm{tr}} A \longrightarrow 0$$

is *exact*. Actually $\mathbf{E}(A)$ is isomorphic to $\mathbf{E}(\mathbb{Z}/2) \otimes A$. I will abbreviate $\mathbf{E}(\mathbb{Z}/2)$ by $\mathbf{E}$. These sequences will be used very much in the following.

(7.8.2) Next three functors from $\mathrm{Mod}_{Z/2}(X_{et})$ to (Ab) are introduced. They reflect the limit behavior of cohomology with 2-torsion coefficients on the sites X_{et}, X_b and X_{ret}, where "limit" refers to passing to arbitrarily high degrees in a suitable way:

Since the functors $j_!$ and $\nu(G)^*$ are exact, they transform $\mathbf{E}(A)$ into short exact sequences in $\mathrm{Ab}(X_b)$ resp. $\mathrm{Ab}_G(X_{ret})$. Applying the respective global sections functors to the exact sequences $j_!\mathbf{E}(A)$, $\mathbf{E}(A)$ and $\nu(G)^*\mathbf{E}(A)$ yields in particular boundary maps

$$H^n(X_b, j_!A) \longrightarrow H^{n+1}(X_b, j_!A),$$
$$H^n(X_{et}, A) \longrightarrow H^{n+1}(X_{et}, A), \tag{3}$$
$$H_G^n(X_{ret}, \nu(G)^*A) \longrightarrow H_G^{n+1}(X_{ret}, \nu(G)^*A),$$

which are functorial in $A \in \mathrm{Mod}_{Z/2}(X_{et})$. In each of the three cases we form the limit for $n \to \infty$:

(7.8.3) Definition. For $A \in \mathrm{Mod}_{Z/2}(X_{et})$ put

$$L_b(A) := \varinjlim H^n(X_b, j_!A),$$

$$L_{et}(A) := \varinjlim H^n(X_{et}, A),$$

$$L_{ret}^G(A) := \varinjlim H_G^n(X_{ret}, \nu(G)^*A),$$

each limit being formed for $n \to \infty$ with (3) as the transition maps. This defines three functors L_b, L_{et}, L_{ret}^G from $\mathrm{Mod}_{Z/2}(X_{et})$ to (Ab). If the underlying scheme X is to be emphasized I may also write $L_b(X, A)$ etc.

(7.9) Since $\mathbf{E}(A)$ is isomorphic to $\mathbf{E} \otimes A$, the boundary maps $\partial\colon H^n(X_{et}, A) \to H^{n+1}(X_{et}, A)$ of (3) satisfy

$$\partial(\lambda \cdot \alpha) = \partial(\lambda) \cdot \alpha \tag{4}$$

for $\lambda \in H^p(X_{et}, \mathbb{Z}/2)$ and $\alpha \in H^q(X_{et}, A)$, where $\lambda \cdot \alpha$ is the cup pairing

$$H^p(X_{et}, \mathbb{Z}/2) \times H^q(X_{et}, A) \longrightarrow H^{p+q}(X_{et}, A)$$

[Mi, V.1.16, p. 171]. (One need not care about signs since all groups are $\mathbb{Z}/2$-modules.) Similar formulas hold for the other two boundary maps in (3). To make this more precise, observe that both $j_!$ and $\nu(G)^*$ preserve tensor products; i.e.

$$j_!(M \otimes N) = j_!(M) \otimes j_!(N) \quad \text{and} \quad \nu(G)^*(M \otimes N) = \nu(G)^*(M) \otimes \nu(G)^*(N)$$

hold for M, $N \in \mathrm{Ab}(X_{et})$. (For $\nu(G)^*$ this is clear [SGA4 IV.13.4], and for $j_!$ it follows from applying j^* and i^* to $j_!M \otimes j_!N$.) So

$$\mathbf{E}(A) = \mathbf{E} \otimes A, \qquad j_!\mathbf{E}(A) = j_!\mathbf{E} \otimes j_!A, \qquad \nu(G)^*\mathbf{E}(A) = \nu(G)^*\mathbf{E} \otimes \nu(G)^*A$$

hold for $A \in \mathrm{Mod}_{\mathbb{Z}/2}(X_{et})$. (All tensor products are taken over $\mathbb{Z}$.)

Hence formula (4) holds also for the first resp. third of the boundary maps (3), with $\lambda \in H^p(X_b, j_!\mathbb{Z}/2)$ and $\alpha \in H^q(X_b, j_!A)$ in the first case resp. $\lambda \in H^p_G(X_{ret}, \mathbb{Z}/2)$ and $\alpha \in H^q_G(X_{ret}, \nu(G)^*A)$ in the third.

At least for $t = et$ and $t = ret$ one can make the boundary maps (3) more explicit. This will be done next, in (7.10) and (7.11).

(7.10) Consider the Kummer sequence $1 \to \mu_2 \to \mathbb{G}_m \xrightarrow{2} \mathbb{G}_m \to 1$ on X_{et}. It is exact since 2 is invertible. Moreover μ_2 is equal to the constant sheaf $\mathbb{Z}/2$. If $u \in \mathcal{O}(X)^*$ is a global unit let $(u) \in H^1(X_{et}, \mu_2) = H^1(X_{et}, \mathbb{Z}/2)$ be the image of u under the boundary $H^0(X_{et}, \mathbb{G}_m) \to H^1(X_{et}, \mu_2)$ of the Kummer sequence.

The next proposition is certainly folklore, but for lack of a reference I include a proof.

(7.10.1) Proposition. — *Let $A \in \mathrm{Mod}_{\mathbb{Z}/2}(X_{et})$. The transition maps*

$$\partial\colon H^n(X_{et}, A) \longrightarrow H^{n+1}(X_{et}, A)$$

of (3) are multiplication (= cup product) with the class $s := (-1) \in H^1(X_{et}, \mathbb{Z}/2)$.

Proof. I drop all subscripts "*et*" in the course of this proof. One can replace the above Kummer sequence by

$$\mathbf{K}\colon \quad 1 \longrightarrow \mu_2 \longrightarrow \mu_4 \longrightarrow \mu_2 \longrightarrow 1.$$

Denote the boundaries $H^n(X, \mu_2) \to H^{n+1}(X, \mu_2)$ of $\mathbf{K}$ by ∂' and the boundaries $H^n(X, \mathbb{Z}/2) \to H^{n+1}(X, \mathbb{Z}/2)$ of $\mathbf{E}$ by ∂. Formula (4) shows that the transition maps in question are cup product with $\partial(1)$. So one must check that $\partial(1) = \partial'(-1)$ holds in $H^1(X, \mathbb{Z}/2) = H^1(X, \mu_2)$. (Of course, under the identification of $\mathbb{Z}/2$ with μ_2 the global section 1 becomes -1 !) This can be done using Čech cohomology. Consider the étale covering $\mathfrak{U} = \{X' \xrightarrow{\pi} X\}$ of X. Both sequences $\mathbf{E}$ and $\mathbf{K}$ are *presheaf* exact when restricted to Et/X'. For $A \in \mathrm{Ab}(X)$ the Čech complex with respect to $\mathfrak{U}$ is

$$C^\bullet(\mathfrak{U}, A)\colon \quad 0 \longrightarrow A(X') \longrightarrow A(X' \times_X X') \longrightarrow \cdots$$

Now $X' \times_X X'$ is X-isomorphic to $X' \amalg X'$, and under this isomorphism the projections $\mathrm{pr}_1, \mathrm{pr}_2\colon X' \times_X X' \rightrightarrows X'$ become $\mathrm{id} \amalg \mathrm{id}$ and $\mathrm{id} \amalg \sigma\colon X' \amalg X' \rightrightarrows X'$. Using these identifications the Čech complex becomes

$$C^\bullet(\mathfrak{U}, A)\colon \quad 0 \longrightarrow A(X') \longrightarrow A(X') \times A(X') \longrightarrow \cdots,$$

and the first differential sends $a \in A(X')$ to $(0, a - \sigma a) \in A(X') \times A(X')$ (writing A additively). One checks immediately for the Kummer sequence $\mathbf{K}$ that the boundary $\partial'(-1) \in \check{H}^1(X, \mu_2) = H^1(X, \mu_2)$ is represented by the Čech cocycle $(1, -1) \in \mu_2(X') \times \mu_2(X')$. For the sequence $\mathbf{E}$ a slightly longer verification gives the same result, namely that $\partial(1) \in H^1(X, \mathbb{Z}/2)$ is represented by the Čech cocycle $(0, 1) \in \mathbb{Z}/2(X') \times \mathbb{Z}/2(X')$. $\qquad\square$

(7.10.2) Let $A = \bigoplus_{d \geq 0} A_d$ be a $\mathbb{Z}_+$-graded ring and let $s \in A_1$ be a homogeneous element of degree 1. Consider the homogeneous localization $A_{(s)}$, i.e. the degree 0 component in the localization of A by $\{1, s, s^2, \dots\}$ (the latter being a graded ring in a canonical way). Then it is obvious that $A_{(s)}$ is the direct limit of the sequence $A_0 \xrightarrow{s} A_1 \xrightarrow{s} A_2 \xrightarrow{s} \cdots$, or also that $A_{(s)} = A/(s-1)A$. If M is a graded A-module one forms in the same way the homogeneous localization $M_{(s)}$ of M. Then $M_{(s)} = M \otimes_A A_{(s)}$ is the direct limit of $M_0 \xrightarrow{s} M_1 \xrightarrow{s} M_2 \xrightarrow{s} \cdots$.

(7.10.3) **Corollary.** — *The ring $L_{et}(\mathbb{Z}/2) = \lim\limits_{n \to \infty} H^n(X_{et}, \mathbb{Z}/2)$ is the homogeneous localization $H^*(X_{et}, \mathbb{Z}/2)_{(s)}$ of the cohomology ring $H^*(X_{et}, \mathbb{Z}/2)$ with respect to $s = (-1) \in H^1(X_{et}, \mathbb{Z}/2)$. More generally, for every $A \in \mathrm{Mod}_{\mathbb{Z}/2}(X_{et})$ the group $L_{et}(A)$ is the homogeneous localization $H^*(X_{et}, A)_{(s)}$. In particular, $L_{et}(A)$ is an $L_{et}(\mathbb{Z}/2)$-module in a natural way.* $\square$

The group $H^1(X_{et}, \mathbb{Z}/2)$ classifies the étale coverings of X of degree 2. Under this correspondence the element $s = (-1)$ is associated with the covering $\pi \colon X' \to X$.

(7.11) Recall (6.6.1) that $\nu(G)^*\mathbb{Z}/2$ is the constant sheaf $\mathbb{Z}/2$ on X_{ret} with the trivial G-action. By (6.3.2), its cohomology ring is the polynomial ring

$$H^*_G(X_{ret}, \mathbb{Z}/2) = H^*(X_{ret}, \mathbb{Z}/2)[\gamma]$$

in which $\deg \gamma = 1$ and γ is the image of the global section 1 under the boundary $\partial \colon H^0(X_{ret}, \mathbb{Z}/2) = H^0_G(X_{ret}, \mathbb{Z}/2) \to H^1_G(X_{ret}, \mathbb{Z}/2)$ (3). (So $H^n_G(X_{ret}, \mathbb{Z}/2) = \bigoplus_{i=0}^n H^i(X_{ret}, \mathbb{Z}/2) \cdot \gamma^{n-i}$.) In analogy to (7.10.1) one gets therefore

(7.11.1) **Proposition.** — *For $A \in \mathrm{Mod}_{\mathbb{Z}/2}(X_{et})$ the transition maps*

$$\partial \colon H^n_G(X_{ret}, \nu(G)^*A) \longrightarrow H^{n+1}_G(X_{ret}, \nu(G)^*A)$$

of (3) are multiplication with $\gamma \in H^1_G(X_{ret}, \mathbb{Z}/2)$. $\square$

(7.11.2) **Corollary.** — *$L^G_{ret}(\mathbb{Z}/2)$ is canonically identified with the full cohomology ring*

$$H^*(X_{ret}, \mathbb{Z}/2) = \bigoplus_{n \geq 0} H^n(X_{ret}, \mathbb{Z}/2)$$

of the real spectrum; and $L^G_{ret}(A)$ is a module over this ring, for $A \in \mathrm{Mod}_{\mathbb{Z}/2}(X_{et})$. $\square$

(7.11.3) Remark. The "base change" $\nu(G)^*\pi_* \to q_{ret*}\nu^*$ is an isomorphism, as is seen by applying q^*_{ret} to both sides (or from (10.19.1b)). (See (6.4.1) for notations) Thus the G-sheaf $\nu(G)^*\pi_*\pi^*A$ on X_{ret} (which figures in the exact sequence $\nu(G)^*\mathbf{E}(A)$) is the sheaf $\nu^*\pi^*A \times \nu^*\pi^*A$, on which G acts by interchanging the two factors. Since q_{ret*} is exact one has canonically

$$H^n_G\big(X_{ret},\ \nu(G)^*\pi_*\pi^*A\big) \cong H^n\big(X_{ret}, \nu^*\pi^*A\big) = H^n\big(X_{ret},\ \rho\pi_*\pi^*A\big)$$

for $n \geq 0$. In particular, the left hand group is zero for $n > \operatorname{cd}(X_r)$. Note that $\rho\pi_*\pi^*A$ is the underlying sheaf of the G-sheaf $\nu(G)^*A$.

(7.11.4) For the b-topology there does not seem to be such a convenient description of the differentials (3). The reason is that $j_!\mathbb{Z}/2$ fails to be a sheaf of rings *with units* on X_b, and so the cohomology "ring" $H^*(X_b, j_!\mathbb{Z}/2)$ need not have a unit. For example, if X is connected with non-empty real spectrum then $H^0(X_b, j_!\mathbb{Z}/2) = 0$.

(7.12) Consider now the basic exact sequence (see (6.6))

$$0 \longrightarrow j_!\pi(G)_*A \longrightarrow j_*\pi(G)_*A \longrightarrow i_*\nu(G)^*A \longrightarrow 0 \qquad (5)$$

in $\operatorname{Ab}_G(X_b)$. It is functorial in $A \in \operatorname{Ab}(X_{et})$. Together with $\mathbf{E}(A)$ it yields a commutative 3-by-3 square in $\operatorname{Ab}_G(X_b)$ which again is functorial in A:

$$
\begin{array}{ccccccccc}
& & 0 & & 0 & & 0 & & \\
& & \downarrow & & \downarrow & & \downarrow & & \\
0 & \longrightarrow & j_!\pi(G)_*A & \longrightarrow & j_!\pi(G)_*\pi_*\pi^*A & \longrightarrow & j_!\pi(G)_*A & \longrightarrow & 0 \\
& & \downarrow & & \downarrow & & \downarrow & & \\
0 & \longrightarrow & j_*\pi(G)_*A & \longrightarrow & j_*\pi(G)_*\pi_*\pi^*A & \longrightarrow & j_*\pi(G)_*A & \longrightarrow & 0 \qquad (6) \\
& & \downarrow & & \downarrow & & \downarrow & & \\
0 & \longrightarrow & i_*\nu(G)^*A & \longrightarrow & i_*\nu(G)^*\pi_*\pi^*A & \longrightarrow & i_*\nu(G)^*A & \longrightarrow & 0 \\
& & \downarrow & & \downarrow & & \downarrow & & \\
& & 0 & & 0 & & 0 & &
\end{array}
$$

For A a $\mathbb{Z}/2$-sheaf all rows and lines are exact since $j_!\pi(G)_*$, $j_*\pi(G)_*$ and $i_*\nu(G)^*$ are exact functors (3.12d).

(7.12.1) Proposition. — *For* $A \in \operatorname{Mod}_{\mathbb{Z}/2}(X_{et})$ *the 3-by-3 square* (6) *induces an exact triangle*

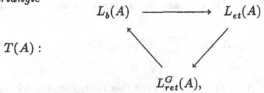

$$T(A):$$

and this triangle depends functorially on A.

Proof. Application of $\Gamma_G(X_b, -)$ yields a long exact sequence for each row and for each column of (6). The terms of these sequences have been identified in (6.6). The groups $L_b(A)$, $L_{et}(A)$, $L_{ret}^G(A)$ are the direct limits with respect to the respective horizontal connecting homomorphisms. The arrows $L_b(A) \to L_{et}(A) \to L_{ret}^G(A)$ are obvious from (6), they are induced by the morphisms from the first to the second and from the second to the third row. To get a map $L_{ret}^G(A) \to L_b(A)$ one must check that the horizontal $(=: \partial')$ and vertical $(=: \partial'')$ boundary operators commute, i.e. that

$$
\begin{array}{ccc}
H_G^n(X_{ret}, \nu(G)^*A) & \xrightarrow{\ \partial'\ } & H_G^{n+1}(X_{ret}, \nu(G)^*A) \\
\downarrow{\scriptstyle\partial''} & & \downarrow{\scriptstyle\partial''} \\
H^{n+1}(X_b, j_!A) & \xrightarrow{\ \partial'\ } & H^{n+2}(X_b, j_!A)
\end{array}
$$

commutes. Actually this diagram *anti*-commutes, i.e. $\partial'\partial'' + \partial''\partial' = 0$ [CE, III §4]; but this is exactly what is needed since all groups are $\mathbb{Z}/2$-modules. Finally any two consecutive arrows in the triangle $T(A)$ can be written as a direct limit of exact sequences $\bullet \to \bullet \to \bullet$. So the triangle is exact. $\qquad\Box$

(7.12.2) Proposition. — *Let L_t be one of the functors L_b, L_{et}, L_{ret}^G.*

a) *For every short exact sequence $0 \to A' \to A \to A'' \to 0$ in $\mathrm{Mod}_{\mathbb{Z}/2}(X_{et})$ there is a natural map $L_t(A'') \to L_t(A')$; and the triangle*

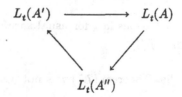

$$
\begin{array}{ccc}
L_t(A') & \xrightarrow{\hspace{2cm}} & L_t(A) \\
& & \\
& L_t(A'') &
\end{array}
$$

is exact.

b) *If X is quasi-compact and quasi-separated then L_t commutes with filtering direct limits in $\mathrm{Mod}_{\mathbb{Z}/2}(X_{et})$.*

c) *If $f: Y \to X$ is a morphism of schemes and $A \in \mathrm{Mod}_{\mathbb{Z}/2}(X_{et})$ then the pullback of cohomology classes by f_t^* induces a natural map $L_t(X, A) \to L_t(Y, f^*A)$.*

Proof. a) follows from an obvious exact 3×3-diagram in a similar way as (7.12.1). For b) use (3.2). For c) it suffices to remark that $j_!E(f^*A) = f_b^*j_!E(A)$ and $\nu(G)^*E(f^*A) = f_{ret}^*\nu(G)^*E(A)$, compare (3.10). $\qquad\Box$

(7.13) Proposition. — *Assume that $d := \mathrm{cd}_2(X_{et}')$ is finite. Then for every $A \in \mathrm{Mod}_{\mathbb{Z}/2}(X_{et})$ the canonical map*

$$
H^n(X_b, j_!A) \longrightarrow L_b(A)
$$

is surjective for $n = d+1$ and is an isomorphism for $n > d+1$. If $\mathrm{cd}_2(X_r) < \mathrm{cd}_2(X_{et}')$ then this map is surjective for $n = d$ and is an isomorphism for $n > d$.

Proof. Since $H^n(X_b, j_* \pi_* \pi^* A) = H^n(X'_{et}, \pi^* A)$ vanishes for $n > d$ it follows from the exact sequence $0 \to j_! \to j_* \to i_* \rho \to 0$ that $H^n(X_b, j_! \pi_* \pi^* A) = 0$ for $n > d + 1$, using $\mathrm{cd}_2(X_{ret}) \leq \mathrm{cd}_2(X'_{et})$ (7.3). If the latter inequality is strict then $H^n(X_b, j_! \pi_* \pi^* A) = 0$ for $n > d$. The assertions are now immediate from the definition of $L_b(A)$. $\qquad\square$

(7.14.1) Lemma. — *Let $I \to$ (Sch), $\lambda \mapsto X_\lambda$ be a filtering inverse system of schemes which are quasi-compact and quasi-separated, such that all transition morphisms are affine. Let $X = \varprojlim X_\lambda$. If $L_b(X_\lambda, A_\lambda) = 0$ holds for every $\lambda \in I$ and every constructible étale $\mathbb{Z}/2$-sheaf A_λ on X_λ then $L_b(X, A) = 0$ holds for every étale $\mathbb{Z}/2$-sheaf A on X.*

Proof. Let $A \in \mathrm{Mod}_{\mathbb{Z}/2}(X_{et})$. Since A is a filtering direct limit of constructible $\mathbb{Z}/2$-sheaves [SGA4 IX.2.7.2] one can assume that A is itself constructible, by (7.12.2b). Then for a suitable $\lambda \in I$ there is a constructible $\mathbb{Z}/2$-sheaf A_λ on X_λ such that $A \cong p_\lambda^* A_\lambda$ [loc.cit. 2.7.4], where $p_\lambda \colon X \to X_\lambda$ is the canonical projection. Thus the proposition follows from

(7.14.2) Lemma. — *Let $\lambda \mapsto X_\lambda$ be an inverse system as in (7.14.1). Assume that I has a final object λ_0. Let A_{λ_0} be an étale $\mathbb{Z}/2$-sheaf on X_{λ_0} and denote the preimages of A_{λ_0} on X_λ and on X by A_λ and A, respectively. Then $L_b(X, A) = \varinjlim_\lambda L_b(X_\lambda, A_\lambda)$.*

Proof. Obvious from the analogous fact for usual sheaf cohomology on the b-sites (3.4.1) together with $f_b^* j_! = j_! f_{et}^*$. $\qquad\square\square$

Note for the record that Theorem (7.7) was not used so far.

(7.15) Proposition. — *Let k be a commutative ring which contains $\frac{1}{2}$, and let X be a k-scheme of finite presentation. Then $L_b(A) = 0$ holds for every étale $\mathbb{Z}/2$-sheaf A on X.*

Proof. There is a finitely generated $\mathbb{Z}$-subalgebra k_0 of k and a k_0-scheme X_0 of finite presentation such that X is k-isomorphic to $X_0 \times_{\mathrm{spec}\, k_0} \mathrm{spec}\, k$ (compare [EGA IV, 8.8.2]). Let $\{k_i\}_{i \in I}$ be the filtering system of all finitely generated k_0-subalgebras of k, and put $X_i := X_0 \times_{\mathrm{spec}\, k_0} \mathrm{spec}\, k_i$. Each X_i is of finite type over $\mathrm{spec}\,\mathbb{Z}$, and hence $(X_i)_b$ has finite cohomological 2-dimension by Theorem (7.7). Since $X = \varprojlim X_i$ the assertion follows from (7.14.1). $\qquad\square$

(7.16) The final step will now be to generalize this to all schemes (over $\mathbb{Z}[\frac{1}{2}]$) which are quasi-compact and quasi-separated. This will be done by covering such a scheme with open affine subschemes and using the Cartan-Leray spectral sequence to pass from the local to the global.

(7.16.1) Let X be a scheme. Fix for the following a family $\mathfrak{U} = \{U_i \to X\}_{i \in I}$ in Et/X which is a covering of X in the b-topology, i.e. which is surjective and real surjective. Recall [SGA4 V.3] how the Cartan-Leray spectral sequences associated with $\mathfrak{U}$ are formed: If K is a bounded below complex in $D^+(X_b)$ let $\mathcal{H}^q(K)$ be the presheaf

$$\mathcal{H}^q(K): \quad U \longmapsto \mathbb{H}^q(U_b, K|_U)$$

on Et/X. The Cartan-Leray spectral sequence

$$E(K): \quad E_2^{pq}(K) = H^p(\mathfrak{U}, \mathcal{H}^q(K)) \implies \mathbb{H}^{p+q}(X_b, K) \tag{7}$$

converges against the hypercohomology of the complex K. Its E_2-terms are the Čech cohomology groups of the presheaves $\mathcal{H}^q(K)$ with respect to $\mathfrak{U}$. The spectral sequence $E(K)$ is functorial in K.

(7.16.2) Consider an exact sequence

$$0 \longrightarrow F' \stackrel{\alpha}{\longrightarrow} F \stackrel{\beta}{\longrightarrow} F'' \longrightarrow 0 \tag{8}$$

in $\mathrm{Ab}(X_b)$. The connecting homomorphisms $\partial: H^n(X_b, F'') \to H^{n+1}(X_b, F')$ associated with (8) can be described as follows. Let K denote the complex $0 \longrightarrow F' \stackrel{\alpha}{\longrightarrow} F \longrightarrow 0$ (with F' in degree 0), and consider the natural complex morphisms $\varphi: K \to F'$ (given by the identity in degree 0) and $\psi: K \to F''[-1]$ (given by β in degree 1). (As usual, F denotes the complex which has F in degree 0 and zero elsewhere, while $[i]$ shifts a complex by i positions to the left.) Since ψ is a quasi-isomorphism one can use it to identify $H^n(X_b, F'')$ with $\mathbb{H}^{n+1}(X_b, K)$. Under this identification ∂ is the map in (hyper-) cohomology induced by φ.

Now if $f: K \to K'$ is a quasi-isomorphism of complexes in $\mathrm{Ab}(X_b)$, the induced morphism $E(K) \to E(K')$ between the Cartan-Leray spectral sequences (7) is an isomorphism. Indeed, for every U étale over X the induced map $f|_U: K|_U \to K'|_U$ (of complexes in $\mathrm{Ab}(U_b)$) is a quasi-isomorphism; and hence $\mathcal{H}^q(K) \to \mathcal{H}^q(K')$ is an isomorphism of presheaves for all q. This shows that there is a morphism of spectral sequences $E(F''[-1]) \to E(F')$ which "realizes" the boundary maps of (8), in the sense that the map $H^p(\mathfrak{U}, \mathcal{H}^{q-1}(F'')) \longrightarrow H^p(\mathfrak{U}, \mathcal{H}^q(F'))$ (of E_2^{pq}-terms) and the map $H^{n-1}(X_b, F'') \to H^n(X_b, F')$ (of n-th limit terms) are induced by these boundaries.

(7.16.3) Now let A be an étale $\mathbb{Z}/2$-sheaf on X, and apply the remarks just made to the exact sequence

$$j_! E(A): \quad 0 \longrightarrow j_! A \longrightarrow j_! \pi_* \pi^* A \longrightarrow j_! A \longrightarrow 0.$$

One gets an infinite sequence

$$E(j_! A) \longrightarrow E(j_! A[1]) \longrightarrow E(j_! A[2]) \longrightarrow \cdots$$

of morphisms between Cartan-Leray spectral sequences. Let E denote its direct limit. The n-th limit term of E is $\lim_{i \to \infty} H^{n+i}(X_b, j_! A) = L_b(X, A)$. The E_2-terms of E are

$$E_2^{pq} = \lim_{i \to \infty} E_2^{pq}(j_! A[i]) = \lim_{i \to \infty} H^p(\mathfrak{U}, \mathcal{H}^{q+i}(j_! A)).$$

If the indexing set I of the covering $\mathfrak{U}$ is *finite* then Čech cohomology with respect to $\mathfrak{U}$ commutes with filtering direct limits of presheaves. So under this condition it follows that $E_2^{pq} = H^p(\mathfrak{U}, \mathcal{L}(A))$, where $\mathcal{L}(A)$ is the presheaf

$$\mathcal{L}(A) = \lim_{i \to \infty} \mathcal{H}^i(j_! A): \ U \ \longmapsto \ \lim_{i \to \infty} H^i(U_b, j_! A|_U) = L_b(U, A|_U)$$

on Et/X. If all multiple fibre products of the U_i over X satisfy the hypotheses of (7.15), the presheaf $\mathcal{L}(A)$ vanishes on these fibre products. This gives $E_2^{pq} = 0$ for all p, q, and hence $L_b(X, A) = 0$. In this way one gets

(7.17) Theorem. — *Let X be a scheme which is quasi-compact and quasi-separated, and on which 2 is invertible. Then*

$$L_b(X, A) = \lim_{n \to \infty} H^n(X_b, j_! A) = 0$$

for every étale $\mathbb{Z}/2$-sheaf A on X.

Proof. Choose a covering $\mathfrak{U} = \{U_i\}_{i \in I}$ of X by finitely many affine open subschemes U_i. Since X is quasi-separated, every intersection $U_{i_0} \cap \cdots \cap U_{i_p}$ is quasi-compact, and hence of finite presentation over U_{i_0}. So Proposition (7.15) applies to it. The reasoning in (7.16.3) proves the theorem. $\square$

For the following corollaries assume that X is a scheme which is quasi-compact and quasi-separated and on which 2 is invertible.

(7.18) Corollary. — $\mathrm{cd}_2(X_b)$ *is either* $\mathrm{cd}_2(X'_{et})$ *or* $1 + \mathrm{cd}_2(X'_{et})$. *If* $\mathrm{cd}_2(X_r) < \mathrm{cd}_2(X'_{et})$ *then* $\mathrm{cd}_2(X_b) = \mathrm{cd}_2(X'_{et})$.

Proof. Since $\mathrm{cd}_2(X_b) \geq \mathrm{cd}_2(X'_{et})$ is clear (7.2a) one can assume that $d := \mathrm{cd}_2(X'_{et})$ is finite. Assume that $e > d$ is an integer with $\mathrm{cd}_2(X_b) \geq e$. One have to show $e = d + 1$ and $\mathrm{cd}_2(X_r) = d$. There is $F \in \mathrm{Ab}(X_b)$ with $2F = 0$ and $H^e(X_b, F) \neq 0$ (7.1.1). From the exact sequence $0 \to j_! j^* F \to F \to i_* i^* F \to 0$ and from $H^e(X_r, i^* F) = 0$ (7.3) it follows that $H^e(X_b, j_! j^* F) \neq 0$. Now Proposition (7.13) together with the theorem imply $e = d + 1$ and $\mathrm{cd}_2(X_r) = d$. $\square$

(7.19) Corollary. — *The natural map* $\lim_{n \to \infty} H^n(X_{et}, A) \to \lim_{n \to \infty} H^n_G(X_{ret}, \nu(G)^* A)$ *is an isomorphism for every étale $\mathbb{Z}/2$-sheaf A on X. In particular,*

$$\lim_{n \to \infty} H^n(X_{et}, \mathbb{Z}/2) \xrightarrow{\sim} H^*(X_r, \mathbb{Z}/2) = \bigoplus_{q \geq 0} H^q(X_r, \mathbb{Z}/2). \qquad \square$$

(7.19.1) Write the homomorphism

$$h\colon H^n(X_{et}, \mathbb{Z}/2) \longrightarrow H^n_G(X_r, \mathbb{Z}/2) = \bigoplus_{i=0}^{n} H^i(X_r, \mathbb{Z}/2)$$

in the form

$$h(\alpha) = h_0(\alpha) + h_1(\alpha) + \cdots + h_n(\alpha),$$

with $h_i(\alpha) \in H^i(X_r, \mathbb{Z}/2)$ $(i = 0, \ldots, n)$. The 0-th component $h_0\colon H^n(X_{et}, \mathbb{Z}/2) \to H^0(X_r, \mathbb{Z}/2)$ has an easy "local" description (which is obvious from contravariant functoriality of (6.6)): If $\alpha \in H^n(X_{et}, \mathbb{Z}/2)$ and $\xi\colon x \to X$ is a real point of X, then $h_0(\alpha)$ evaluated at ξ is $\xi^* \alpha \in H^n(x_{et}, \mathbb{Z}/2) = \mathbb{Z}/2$. Write $\alpha(\xi)$ for this element of $\mathbb{Z}/2$, so $h_0(\alpha)$ is the (locally constant) map $X_r \to \mathbb{Z}/2$, $\xi \mapsto \alpha(\xi)$.

In contrast, the h_i for $i \geq 1$ are not so easily described, since clearly they do not have local character.

(7.20) **Corollary.** — *Let ℓ be any prime. If $d := \mathrm{cd}_\ell(X'_{et})$ is finite then for every ℓ-primary étale torsion sheaf A on X the natural map*

$$H^n(X_{et}, A) \longrightarrow H^n_G(X_{ret}, \nu(G)^* A) \tag{9}$$

is bijective for $n \geq d+1$. If in addition $\mathrm{cd}_\ell(X_r) < d$ then (9) is also surjective for $n = d$.

Proof. Up to injectivity for $n = d+1$ this is a direct consequence of Corollary (7.18) (for $\ell = 2$) resp. Corollary (7.4.1) (for ℓ odd) and the fundamental long exact sequence (Theorem (6.6)). To see that (9) is injective for $n = d+1$ one uses the same sequence and observes that $H^n(X_b, j_!A) \to H^n(X_{et}, A)$ factors as $H^n(X_b, j_!A) \to H^n(X_b, j_*A) \to H^n(X_{et}, A)$. Therefore the claim follows from

(7.20.1) **Lemma.** — *Assume that $d := \mathrm{cd}_\ell(X'_{et})$ is finite. Then $H^{d+1}(X_b, j_*A)$ is zero for every ℓ-primary torsion sheaf A on X_{et}. (ℓ any prime.)*

Proof. Define $B \in \mathrm{Ab}(X_{et})$ by the exact sequence $0 \to B \to \pi_* \pi^* A \xrightarrow{\mathrm{tr}} A \to 0$. Applying j_* yields an exact sequence

$$0 \longrightarrow j_*B \longrightarrow j_*\pi_*\pi^*A \longrightarrow j_*A \longrightarrow i_*\mathrm{R}^1\rho B \longrightarrow 0. \tag{10}$$

If ℓ is odd the last term is zero (3.12). The assertion follows from (10) since H^{d+1} of the second and fourth sheaf is zero (by exactness of $j_*\pi_*$ resp. by (7.3)), and since $H^{d+2}(X_b, j_*B) = 0$ by $\mathrm{cd}_\ell(X_b) \leq d+1$. $\qquad\square\square$

(7.20.2) If $\text{cd}_2(X_r)$ is equal to d then in general (9) is not surjective for $n = d$, cf. (9.8.1). In the particular case $A = \mathbb{Z}/2$, however, this is true (in other words, $H^{d+1}(X_b, j_!\mathbb{Z}/2) = 0$ holds regardless of the cohomological dimension of X_r). This follows from the commutative diagram

$$
\begin{array}{ccc}
H^d(X_{et}, \mathbb{Z}/2) & \longrightarrow & H^d_G(X_r, \mathbb{Z}/2) \\
\downarrow & & \downarrow{\scriptstyle\sim} \\
H^{d+1}(X_{et}, \mathbb{Z}/2) & \overset{\sim}{\longrightarrow} & H^{d+1}_G(X_r, \mathbb{Z}/2).
\end{array}
$$

(What is special about the sheaf $A = \mathbb{Z}/2$ is that here the boundary homomorphism $H^d_G(X_r, \nu(G)^*A) \to H^{d+1}_G(X_r, \nu(G)^*A)$ is an isomorphism.)

The next corollary answers the question raised in (7.5):

(7.21) Corollary. — *If $\text{cd}_2(X'_{et})$ is finite and the real spectrum of X is empty then also $\text{cd}_2(X_{et})$ is finite (and hence is equal to $\text{cd}_2(X'_{et})$).*

Proof. $X_{et} = X_b$ since X_r is empty. So this is a particular case of Corollary (7.18). □

(7.22) Corollary. — *If the real spectrum of X is empty then the element $s = (-1) \in H^1(X_{et}, \mathbb{Z}/2)$ is nilpotent in $H^*(X_{et}, \mathbb{Z}/2)$, i.e. $s^n = 0$ for some $n \geq 1$.*

Proof. $b = et$. □

(7.23) Remark on torsion hypotheses. As usual, results about cohomology of torsion sheaves imply some corollaries about cohomology of arbitrary sheaves. For the sake of completeness I will indicate the main points. The basic fact used for such corollaries is the following

(7.23.1) Lemma. — *Let $\mathbf{A}$ be an abelian category with sufficiently many injectives. Let $f: \mathbf{A} \to (\text{Ab})$ be a left exact functor and p a prime. Let c be an integer such that $R^n f(A) = 0$ for $n > c$ and every $A \in \mathbf{A}$ with $pA = 0$. Then for arbitrary $A \in \mathbf{A}$ one has:*
a) *There is an exact sequence $0 \to R^c f(A)/p \to R^c f(A/p) \to {}_pR^{c+1}f(A) \to 0$;*
b) *$R^{c+1}f(A)$ is p-divisible;*
c) *$R^n f(A)$ is uniquely p-divisible for $n > c+1$.*

Proof. This follows easily from the exact sequence $0 \to {}_pA \to A \overset{p}{\to} A \to A/p \to 0$. □

(7.23.2) Corollary. — *Let X be a scheme with $\frac{1}{2} \in \mathcal{O}(X)$, and let A be an arbitrary étale sheaf on X. If $n > 1 + \max\{\text{cd}(X_r), \text{cd}_2(X_b)\}$ then*
a) *the sequence $0 \to H^n(X_b, j_!A) \to H^n(X_{et}, A) \to H^n_G(X_r, \nu(G)^*A) \to 0$ is exact and splits;*
b) *$H^n(X_b, j_!A)$ is uniquely 2-divisible;*
c) *$H^n(X_{et}, A)_{2-\text{tors}} \overset{\sim}{\to} H^n_G(X_r, \nu(G)^*A)$, and both groups are annihilated by 2.*

If X is quasi-compact and quasi-separated, the corollary applies in particular if $n > \max\{1 + \dim X, 2 + \mathrm{cd}_2(X'_{et})\}$, by (7.6) and (7.18).

Proof. Here $M_{\ell-\mathrm{tors}}$ denotes the ℓ-primary torsion subgroup of an abelian group M. b) is a direct application of c) of the lemma. For any $B \in \mathrm{Ab}_G(X_r)$ the groups $H^q_G(X_r, B)$ are 2-primary torsion if $q > \mathrm{cd}(X_r)$, in particular for $q \geq n - 1$. Indeed, this follows from the spectral sequence of equivariant cohomology $E^{ij}_2 = H^i\big(G, H^j(X_r, B)\big) \Rightarrow H^{i+j}_G(X_r, B)$ since $E^{0q}_2 = 0$ and E^{ij}_2 is killed by 2 for $i \neq 0$. Combining this with b) one concludes that the connecting homomorphism $H^{q-1}_G(X_r, \nu(G)^*A) \to H^q(X_b, j_!A)$ is zero for $q \geq n$. Therefore the sequence in a) is exact. It splits since the first group is uniquely 2-divisible and the last group is 2-primary torsion. Hence one gets also the isomorphism in c). It remains to show that $H^n(X_{et}, A)_{2-\mathrm{tors}}$ is actually annihilated by 2. This follows from the sequence

$$H^n(X_{et}, A)_{2-\mathrm{tors}} \longrightarrow H^n(X'_{et}, \pi^*A)_{2-\mathrm{tors}} \xrightarrow{\mathrm{tr}} H^n(X_{et}, A)_{2-\mathrm{tors}} :$$

The composition is multiplication by 2, and the middle group is zero by (7.23.1c). $\square$

It is not hard to see that the groups figuring in c) are periodic with period 2 in the stable range, i.e. that $H^n \cong H^{n+2}$ holds (compare (3.12.1)).

When the scheme X is noetherian there is further simplification, since in high degrees all étale cohomology groups are known to be torsion:

(7.23.3) Corollary. — *Let X be a noetherian scheme of finite (Krull) dimension d.*

a) *For arbitrary sheaves $F \in \mathrm{Ab}(X_b)$, $A \in \mathrm{Ab}(X_{et})$, $B \in \mathrm{Ab}_G(X_r)$ the groups*

$$H^{n+1}(X_b, F), \quad H^n(X_{et}, A), \quad H^n_G(X_r, B)$$

are torsion for $n > d$.

b) *If A is an arbitrary étale sheaf on X, the homomorphism*

$$H^n(X_{et}, A) \longrightarrow H^n_G\big(X_r, \nu(G)^*A\big)$$

is surjective for $n = c$ and bijective for $n > c$. For $n > c$ both groups are killed by 2. Here $c := 1 + \sup\{d, \mathrm{cd}_\ell(X_b): \ell \text{ prime}\}$.

Note that by (7.18) (and (7.4.1)) the number c is equal to $1 + \sup\{d, \mathrm{cd}_\ell(X'_{et}) : \ell$ prime$\}$, unless in the case when $d = \mathrm{cd}_\ell(X_r) = \mathrm{cd}_\ell(X'_{et}) < \mathrm{cd}_\ell(X_b)$ for some prime ℓ and $\mathrm{cd}_{\ell'}(X'_{et}) \leq d$ for all primes ℓ'.

Proof. a) The third group is even 2-primary torsion: This has already been observed in the last proof (use $\mathrm{cd}(X_r) \leq d$ (7.6)). It suffices therefore to show that $H^n(X_{et}, A)$ is torsion. (The first case then follows from the exact sequence $0 \to j_!j^*F \to F \to i_*i^*F \to 0$ and from (6.6).) The fact that $H^n(X_{et}, A)$ is a torsion group for $n > d$ is certainly well known, but for lack of a reference I include a sketch of proof. The assertion holds for the spectrum of any field, and hence for $\dim X = 0$. In general, if $f_\nu : x_\nu \to X$ ($\nu = 1, \ldots, N$) are the inclusions of the generic points of the irreducible components of X, consider the homomorphism $A \longrightarrow \bigoplus_\nu f_{\nu *} f_\nu^* A$. Its kernel and cokernel have support in dimensions $< d$. Moreover, if x is the spectrum of a field, if $f : x \to X$ is any morphism and $B \in \mathrm{Ab}(x_{et})$ is arbitrary, then $H^q(X_{et}, f_* B)$ is torsion for $q \geq 1$, as is seen from the Leray spectral sequence. Thus one may conclude inductively. — b) For $n > c$ the group $H^n(X_b, j_! A)$ is torsion by a), and uniquely divisible by (7.23.1c), hence zero. Together with (7.23.2c) this implies b). $\square$

For a concrete example let R be a real closed base field and X an algebraic variety over R, of dimension d. Then $\mathrm{cd}_\ell(X_b) \leq 2d$ for all primes ℓ, since $\mathrm{cd}_\ell(X'_{et}) \leq 2d$ [Mi, Thm. VI.1.1] and $\mathrm{cd}(X_r) \leq d$. If X is affine then $\mathrm{cd}_\ell(X_b) \leq d$ (18.11); to get this one has to work, see Sect. 18. Hence b) of the last corollary applies with $c = 2d + 1$ in general, and with $c = d + 1$ if X is affine.

(7.24) Remark. The results of Sect. 5, in particular exactness of ν_*, were used only for the proof of those main results above which make assertions on cohomological dimensions. In particular, the proofs of (7.17), (7.19), (7.21) and (7.22) are independent of Sect. 5.

8. Equivariant sheaves for actions of topological groups

If a discrete group G acts on a topological space T there is a well-developed equivariant sheaf theory associated with this situation; see e.g. Chapitre V of [Gr1]. If however G is a topological group (and the action is continuous) this notion of G-sheaves is sometimes inadequate. For example, in certain situations one would like to identify the category of equivariant sheaves on T with the category of sheaves on the quotient space T/G. But the usual condition which guarantees that this can be done, namely that G operates freely and discontinuously, doesn't make sense for a topological group.

In this section a notion of equivariant sheaves will be introduced which is better adapted to actions of topological groups. A typical situation to which this material will be applied is the following: Let k be a real field with absolute Galois group $\Gamma = \mathrm{Gal}(k_s/k)$ and let $T \subset \Gamma$ be the subset of involutions. The subgroup $\Gamma' = \mathrm{Gal}(k_s/k(\sqrt{-1}))$ of Γ acts freely on T, and the quotient space $T/\Gamma' = T/\Gamma$ is the real spectrum sper k. It is desirable to identify sheaves on sper k with equivariant sheaves on T. In this way, $(\mathrm{sper}\, k)^{\sim}$ gets identified with a certain full subcategory of the category of all Γ'-equivariant sheaves on T. This is precisely the category of "continuous" Γ'-sheaves, to be introduced below.

This section has an auxiliary character: Its only purpose is to make available some techniques which are needed in Sections 9, 11, 12. Although the only applications will be to actions of profinite groups on profinite spaces, there would be no point in narrowing down the exposition so much from the beginning. On the other hand I have not aimed at completeness in any sense; rather only such material is included which will be applied later.

(8.1) First I recall briefly some notions from usual equivariant sheaf theory. Unless otherwise stated all group actions are from the left.

(8.1.1) Let X be a topological space, G a discrete group and $\sigma \colon G \times X \to X$ an action of G on X by homeomorphisms. A G-sheaf on X (cf. [Gr1 ch. V] or Example (10.10)) can be described in at least three different ways:

1. As an *espace étalé* $p \colon E \to X$ together with a G-action on E which covers the action on X:

$$\begin{array}{ccc} G \times E & \longrightarrow & E \\ {\scriptstyle 1 \times p} \downarrow & & \downarrow {\scriptstyle p} \\ G \times X & \stackrel{\sigma}{\longrightarrow} & X. \end{array} \qquad (1)$$

2. As a sheaf A on X together with an isomorphism $\phi: \mathrm{pr}_2^* A \xrightarrow{\sim} \sigma^* A$ of sheaves on $G \times X$ for which the cocycle condition

$$(\mu \times 1)^*(\phi) = (1 \times \sigma)^*(\phi) \circ \mathrm{pr}_{23}^*(\phi) \tag{2}$$

holds. Here σ, pr_2 are maps $G \times X \rightrightarrows X$, and $\mu \times 1$, $1 \times \sigma$, pr_{23} are maps $G \times G \times X \rightrightarrows G \times X$. ($\mu: G \times G \to G$ is the group multiplication.)

3. As a sheaf A on X together with a family of sheaf isomorphisms $\varphi_g: A \xrightarrow{\sim} g^* A$ ($g \in G$) which satisfy the cocycle conditions

$$\varphi_{hg} = c_{g,h} \circ g^*(\varphi_h) \circ \varphi_g \qquad (g, h \in G) \tag{3}$$

where $c_{g,h}: g^* h^* A \xrightarrow{\sim} (hg)^* A$ denotes the canonical isomorphisms.
In all three descriptions the morphisms in the category $\widetilde{X}(G)$ of G-sheaves on X are the obvious ones.

(8.1.2) The forgetful functor $\pi^*: \widetilde{X}(G) \to \widetilde{X}$ has a right adjoint π_*, and $\pi = (\pi^*, \pi_*)$ is a topos morphism from $\widetilde{X}$ to $\widetilde{X}(G)$. If G acts trivially on X then there is also the topos morphism $r = (r^*, r_*)$ which goes in the opposite direction and satisfies $r \circ \pi = \mathrm{id}$. An equivariant map $f: X \to Y$ between G-spaces induces a topos morphism $f(G) = (f(G)^*, f(G)_*)$ from $\widetilde{X}(G)$ to $\widetilde{Y}(G)$, and $f(G) \circ \pi = \pi \circ f$ holds (as topos morphisms from $\widetilde{X}$ to $\widetilde{Y}(G)$). All this is explained in greater detail (and generality) in Sect. 10.

(8.1.3) Notation. Let $p: X \to Y := X/G$ be the topological quotient, and regard Y as a trivial G-space. The composite topos morphism

$$\widetilde{X}(G) \xrightarrow{p(G)} \widetilde{Y}(G) \xrightarrow{r} \widetilde{Y}$$

will be denoted by (p_G^*, p_*^G).

(8.1.4) The notation p_*^G for the direct image functor $r_* \circ p(G)_* = \Gamma_G \circ p(G)_*$ is adapted from [Gr1]. If G acts freely and discontinuously on X then p_*^G and p_G^* are quasi-inverses of each other, see [loc.cit., 5.1]. Actually the proof shows without any conditions on the action, that the functor $p_G^*: \widetilde{Y} \to \widetilde{X}(G)$ is always fully faithful and that the adjunction map $p_G^* p_*^G A \to A$ is injective for any $A \in \widetilde{X}(G)$.

(8.2) Let now G be a topological group and $\sigma: G \times X \to X$ a continuous action on a topological space X. All topological groups are tacitly assumed to be Hausdorff. By G_δ I always denote the group G with the *discrete* topology.

(8.2.1) Definition.

a) The following two categories are naturally equivalent:

1. The category of espaces étalés $E \to X$ over X together with a *continuous* action of G on E which covers the action σ, cf. (1);

2. the category of pairs (A, ϕ) where $A \in \widetilde{X}$ and $\phi \colon \mathrm{pr}_2^* A \xrightarrow{\sim} \sigma^* A$ is an isomorphism of sheaves on $G \times X$ for which (2) holds,

the morphisms being the obvious ones in both cases. (The proof of the equivalence is straightforward and will be omitted.) Either category is called the category of G-*sheaves* on X and is denoted by $\widetilde{X}(G)$. To emphasize that G carries a topology which has to be respected I will frequently speak of the category of *continuous* G-sheaves, as opposed to the category $\widetilde{X}(G_\delta)$ of *discrete* G-sheaves, of which the former is a full subcategory.

b) The category of continuous abelian G-sheaves on X is written $\mathrm{Ab}_G(X)$. The right derived functors of $H_G^0(X, -) \colon \mathrm{Ab}_G(X) \to (\mathrm{Ab})$ are written $H_G^n(X, -)$, $n \geq 0$. If $X = *$ is the one-point space I write $(G\text{-sets})$, $(G\text{-mod})$ and $H^n(G, -)$ for $\widetilde{*}(G)$, $\mathrm{Ab}_G(*)$ and $H_G^n(*, -)$, respectively (cf. (0.5.2)).

(8.2.2) Remark. The second definition of $\widetilde{X}(G)$ in (8.2.1a) can be rephrased by saying that $\widetilde{X}(G)$ is the category of descent data of sheaves relative to the diagram of spaces

$$G \times G \times X \rightrightarrows\!\!\!\!\rightarrow G \times X \rightrightarrows X, \tag{4}$$

where $G \times G \times X \rightrightarrows\!\!\!\!\rightarrow G \times X$ are the maps pr_{23}, $\mu \times 1$ and $1 \times \sigma$, and $G \times X \rightrightarrows X$ are the maps pr_2 and σ. A more enlightening way to describe the diagram (4) is the following. Let E be the simplicial topological group

$$E = \left(\cdots G \times G \times G \rightrightarrows\!\!\!\!\rightarrow G \times G \rightrightarrows G \right) \tag{5}$$

with the obvious face and degeneration maps (deletion resp. repetition). (For simplicity only the face maps are displayed symbolically.) Then G acts "diagonally" on E, and hence also on the simplicial space $E \times X$. (X denotes here the constant simplicial space X, so $(E \times X)^n = G^{n+1} \times X$.) Write $E \times_G X := (E \times X)/G$ for the quotient simplicial space, which is componentwise the quotient of $E \times X$, with all faces and degeneracies induced from $E \times X$. Then (4) is the truncation of $E \times_G X$. Indeed, $(E \times_G X)^n$ can be identified with $G^n \times X$ by letting $(a_1, \ldots, a_n, x)$ correspond to the G-orbit of $(1, a_1, a_1 a_2, \ldots, a_1 \cdots a_n, a_1 \cdots a_n x)$ in $G^{n+1} \times X = (E \times X)^n$. Under this identification one gets the following formulas for the face maps of $E \times_G X$:

$$d_0(a_1, \ldots, a_n, x) = (a_2, \ldots, a_n, x),$$
$$d_i(a_1, \ldots, a_n, x) = (a_1, \ldots, a_i a_{i+1}, \ldots, a_n, x) \quad (0 < i < n),$$
$$d_n(a_1, \ldots, a_n, x) = (a_1, \ldots, a_{n-1}, a_n x),$$

and similar ones for the degeneracies. Under suitable hypotheses (see [Sg]) the quotient $|E| \to |E|/G$ of the geometric realization $|E|$ is a classifying space for G, and $|E \times_G X|$ is a model on which G-equivariant (singular) cohomology of X can be studied, cf. [tD].

(8.3) Remark. Let G be a topological group acting on X, and let A be a discrete G-sheaf on X. It makes sense to ask what it means for A to be a continuous G-sheaf. Let $x \in X$, and let $a \in A_x$ be an element in the stalk at x, represented by a section $s \in A(U)$ on some neighborhood U of x. If $g \in G$ is sufficiently close to the identity (namely if $g \cdot x \in U$) one can compare the elements $g \cdot a = g \cdot s_x$ and $s_{g.x}$ in the stalk $A_{g.x}$. It is independent of the choices of s and U whether or not $g \cdot s_x = s_{g.x}$ holds for all $g \in G$ in *some* neighborhood of 1. The discrete G-sheaf A is a continuous G-sheaf if and only if this always is the case.

(Using the interpretation by espaces étalés this follows readily from the following observation: Let $\sigma \colon G_\delta \times E \to E$ be an action of the discrete-made group G on a topological space E by homeomorphisms. Then σ is continuous as a map $G \times E \to E$ if and only if for each open subset $V \subset E$ and each $z \in V$ there are neighborhoods U of z (in E) and W of 1 (in G) with $\sigma(W \times U) \subset V$.)

(8.3.1) This shows that the concept of continuous G-sheaves is quite restrictive. For example, if A is a sheaf of continuous maps into some topological space Z, with G-structure given by $(g \cdot s)(x) = s(g^{-1}x)$, then for A to be a continuous G-sheaf it is necessary that all local sections of A are locally constant on G-orbits. If G is a Lie group acting on a manifold then in general the classical equivariant sheaves like those of continuous, differentiable, analytic functions fail to be continuous in the above sense.

(8.4) Example. Let G be a topological group, H a closed subgroup of G and $X = G/H$. Let A be a continuous G-sheaf on the homogeneous space X. Then for every $x = gH \in X$ the natural map $H^0(X, A)^G \longrightarrow (A_x)^{G_x}$ is bijective, where $G_x = gHg^{-1}$ is the stabilizer subgroup of x.

This example shows why the notion of continuous G-sheaves was introduced. If A were only a discrete G-sheaf the above would fail in general.

As to the proof, injectivity is clear. Regard A as an espace étalé over X, and let $a \in A_x$ be fixed by G_x. Then there is a unique global section $s \colon X \to A$ which makes

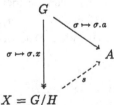

commutative; and clearly s is G-invariant.

More generally: If Z is any space with H-action and $Y = G \times_H Z$ is the G-space induced from Z (cf. [tD I.4]), there is a canonical equivalence of categories between $\widetilde{Z}(H)$ and $\widetilde{Y}(G)$. In particular, for $X = G/H$ this reduces to an equivalence $\widetilde{X}(G) \sim (H\text{-sets})$, of which the above remark is a consequence. (I'm not going to use the equivalence $\widetilde{Z}(H) \sim \widetilde{Y}(G)$.)

(8.5) Proposition. — *Let G be any topological group and X any G-space. Then $\widetilde{X}(G)$ is stable inside $\widetilde{X}(G_\delta)$ under the formation of finite inverse and arbitrary direct limits. In other words, $\widetilde{X}(G)$ is a topos, and the inclusion $\widetilde{X}(G) \subset \widetilde{X}(G_\delta)$ is the inverse image functor of a topos morphism from $\widetilde{X}(G_\delta)$ to $\widetilde{X}(G)$.*

Proof. From Giraud's criterion [SGA4 IV.1.2] it is clear that a full subcategory of a topos which is stable under finite $\varprojlim$'s and arbitrary $\varinjlim$'s is itself a topos. The stability of $\widetilde{X}(G)$ in $\widetilde{X}(G_\delta)$ under finite inverse limits is obvious. Let $I \to \widetilde{X}(G)$, $i \mapsto A_i$ be a functor and let A be the direct limit of the composite functor $I \to \widetilde{X}(G) \subset \widetilde{X}(G_\delta)$. To show that A is a continuous G-sheaf I use the criterion (8.3). Let $x \in X$ and $a \in A_x$. Since both the forgetful functor $\widetilde{X}(G_\delta) \to \widetilde{X}$ and the stalk functor $\widetilde{X} \to$ (sets) in x preserve direct limits, there are a neighborhood U of x and a section $s_i \in A_i(U)$ for some $i \in I$ such that a is represented by $\beta_i(s_i) \in A(U)$, where $\beta_i \colon A_i \to A$ is the canonical map. Since A_i is a continuous G-sheaf one has $g.(s_i)_x = (s_i)_{g.x}$ for $g \in G$ in a neighborhood of 1. Hence $g.a = g.\beta_i((s_i)_x) \underset{(*)}{=} \beta_i(g.(s_i)_x) = \beta_i((s_i)_{g.x}) = (\beta_i(s_i))_{g.x}$ for g near 1; here $(\star)$ holds since β_i preserves the G-action. This proves the proposition. $\qquad\square$

(8.6) With respect to functoriality the notion of continuous G-sheaves does not always behave as one would hope. Let $f \colon X \to Y$ be an equivariant map between G-spaces. The equivariant inverse image functor $f(G_\delta)^* \colon \widetilde{Y}(G_\delta) \to \widetilde{X}(G_\delta)$ (8.1.2) sends continuous G-sheaves to continuous G-sheaves and hence induces $f(G)^* \colon \widetilde{Y}(G) \to \widetilde{X}(G)$. From Proposition (8.5) it follows moreover that $f(G)^*$ preserves all direct and all finite inverse limits. Hence $f(G)^*$ has a right adjoint $f(G)_*$, and:

(8.6.1) Notation. The pair $\big(f(G)^*, f(G)_*\big)$ is a topos morphism from $\widetilde{X}(G)$ to $\widetilde{Y}(G)$. It will be denoted by $f(G)$.

(8.6.2) It is natural to ask whether also the direct image functor $f(G)_*$ is induced by its discrete counterpart. In other words, does also $f(G_\delta)_*$ send continuous G-sheaves to continuous G-sheaves?

In general the answer is "no". For example, if Y is reduced to a point then $\widetilde{Y}(G) = (G\text{-sets})$, the category of discrete G-sets. And $f(G_\delta)_*$ sends a G-sheaf A to the G-set $H^0(X, A)$. It is not hard to construct examples where A is a continuous G-sheaf but the G-action on the discrete set $H^0(X, A)$ is not continuous.

However under favorable conditions the answer is positive, for example if f is proper:

(8.6.3) Proposition. — *Let G be a topological group and $f \colon X \to Y$ an equivariant map between G-spaces. Assume that f is proper. Then $f(G_\delta)_*$ sends continuous G-sheaves to continuous G-sheaves, and hence $f(G)_*$ is the functor induced by $f(G_\delta)_*$.*

Proof. Let A be a continuous G-sheaf on X. One has to show that the discrete G-sheaf $f(G_\delta)_* A = \left(f_* A, G\text{-operation} \right)$ is continuous. For this let $V \subset Y$ be open and $t \in (f_* A)(V)$, and let $y \in V$. One has to show that

$$W := \left\{ g \in G: \ g.y \in V \text{ and } g.t_y = t_{g.y} \text{ in } (f_* A)_{g.y} \right\}$$

is a neighborhood of the identity in G. By proper base change the natural maps $(f_* A)_z \longrightarrow \Gamma(f^{-1}(z), A)$ are bijective for every $z \in Y$. Identify t with a section $t \in \Gamma(f^{-1}(V), A)$. Then W is the set

$$W = \left\{ g \in G: \ g.y \in V \text{ and } g.\left(t|_{f^{-1}(y)} \right) = t|_{f^{-1}(g.y)} \text{ in } \Gamma(f^{-1}(g.y), A) \right\},$$

or else

$$W = \left\{ g \in G: \ g.y \in V \text{ and } g.t_x = t_{g.x} \text{ in } A_{g.x}, \text{ for all } x \in f^{-1}(y) \right\}.$$

Since A is a continuous G-sheaf the set

$$\Omega := \left\{ (g,x) \in G \times f^{-1}(y): \ g.y \in V \text{ and } g.t_x = t_{g.x} \right\}$$

is open in $G \times f^{-1}(y)$. Since the projection $G \times f^{-1}(y) \to G$ to the first factor is proper it follows that there is a neighborhood U of 1 in G with $U \times f^{-1}(y) \subset \Omega$. Clearly $U \subset W$, which proves the assertion. $\square$

(8.6.4) For example, if the space X is compact then $H^0(X, A)$ is a continuous G-module for every continuous abelian G-sheaf A, and so there is a spectral sequence converging against $H^*_G(X, A)$ with E_2-terms $H^p(G, H^q(X, A))$. (Recall that $H^*(G, -)$ denotes cohomology of *continuous* G-modules.)

(8.7) Notation. Let X be an arbitrary G-space, and let $p: X \to Y = X/G$ be the quotient. Regard Y as a trivial G-space. The topos morphism $r: \tilde{Y}(G_\delta) \to \tilde{Y}$ (8.1.2) factors obviously through $\tilde{Y}(G_\delta) \to \tilde{Y}(G)$ (8.5); this factorization $\tilde{Y}(G) \to \tilde{Y}$ is again denoted by r. The composite topos morphism

$$\tilde{X}(G) \xrightarrow{p(G)} \tilde{Y}(G) \xrightarrow{r} \tilde{Y}$$

is denoted by (p_G^*, p_*^G), cf. (8.1.3). Observe that *both* p_G^* and p_*^G are induced by their discrete counterparts.

(8.7.1) As in the discrete case one can decompose the invariant sections functor $H^0_G(X, -)$ in different ways. This gives rise to several spectral sequences which converge against $H^*_G(X, A)$ [Gr1]:

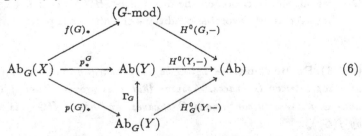

$$(6)$$

(f denotes the map $X \to *$.) The diagram (6) commutes, and the composition $\mathrm{Ab}_G(X) \to (\mathrm{Ab})$ is $H_G^0(X, -)$. If X and G are compact then $p(G)_*$ is induced by p_* and $f(G)_*$ is $H^0(X, -)$, (8.6.3).

The next proposition is one of the reasons why for topological groups the concept of continuous G-sheaves is more useful than its discrete analogue, cf. the introduction:

(8.7.2) Proposition. — *Let G be a compact group acting freely on a Hausdorff space X, and let $p: X \to Y = X/G$ be the quotient. Then the functors p_G^*, $p_*^G: \widetilde{X}(G) \leftrightarrows \widetilde{Y}$ are quasi-inverses of each other.*

Proof. By what one knows from the discrete case (8.1.4) it suffices to show that $p_G^* p_*^G A \to A$ is surjective for every continuous G-sheaf A on X. The condition can be checked stalkwise, so let $x \in X$. Then

$$\left(p_G^* p_*^G A\right)_x = \left(p_*^G A\right)_{p(x)} = \left((p_* A)_{p(x)}\right)^G \xrightarrow{\sim} H^0\left(G \cdot x, A|_{G \cdot x}\right)^G,$$

the last isomorphism by proper base change since p is a proper map. The hypotheses imply that $G \cdot x$ is isomorphic to G as a G-space. So by Example (8.4) the upper (left) arrow in

$$p_G^* p_*^G A = H^0\left(G \cdot x, A|_{G \cdot x}\right)^G \begin{array}{c} \nearrow \left(A|_{G \cdot x}\right)_x \\ \\ \searrow \quad\;\; \downarrow \sim \\ A_x \end{array}$$

is bijective, from which the assertion follows.   □

(8.8.1) The following generalization will be useful. Let Γ be a topological group acting on a Hausdorff space X, with quotient $p: X \to Y = X/\Gamma$. Assume that $N \trianglelefteq \Gamma$ is an invariant subgroup which is compact and open in Γ, such that the following conditions hold:

(a) The operation of N on X is free;

(b) the natural map $X/N \longrightarrow X/\Gamma$ is bijective.

(It is equivalent to say that, for every $x \in X$, Γ is a semidirect product $N \cdot \Gamma_x$.) Write $G := \Gamma/N$.

(8.8.2) Corollary. — *Under these hypotheses the composite functor*

$$\widetilde{X}(\Gamma) \xrightarrow{p(\Gamma)_*} \widetilde{Y}(\Gamma) \xrightarrow{\Gamma_N} \widetilde{Y}(G)$$

is an equivalence. A quasi-inverse is given by $\widetilde{Y}(G) \xrightarrow{\mathrm{can}} \widetilde{Y}(\Gamma) \xrightarrow{p(\Gamma)^*} \widetilde{X}(\Gamma)$.

Proof. Here Γ_N is the functor which takes the sheaf of N-invariants, and "can" denotes the canonical functor induced by $\Gamma \twoheadrightarrow G$. The two composite functors are clearly adjoint. The adjunction morphisms become those of p_N^* and p_*^N, after restriction from Γ to N resp. from G to $\{1\}$; and hence they are isomorphisms by (8.7.2). Compare also Remark (8.8.4) below. $\square$

(8.8.3) Remark. Since $\widetilde{Y}$ is equivalent to $\widetilde{X}(N)$ by (8.7.2), the corollary shows that $\widetilde{X}(\Gamma)$ can also be identified with the category of G-objects in $\widetilde{X}(N)$, which in turn is the same as $\widetilde{X}(N \times G)$ where $N \times G$ acts on X through $\mathrm{pr}_1 \colon N \times G \to N$. This identification goes as follows: For $x \in X$ and $a \in G$ let $s_x(a)$ be the unique element of Γ_x which projects to a under $\Gamma \twoheadrightarrow G$. The map $G \times X \to \Gamma$, $(a, x) \mapsto s_x(a)$ is continuous since $(a, x) \mapsto (s_x(a), x)$ is the inverse of the composite map

$$\{(\sigma, x) \in \Gamma \times X \colon \sigma . x = x\} \subset \Gamma \times X \longrightarrow G \times X$$

which is bijective and closed. Let $p \colon E \to X$ be the espace étalé of a Γ-sheaf on X. Then E is made an $N \times G$-sheaf by the rule $((n, a), e) \longmapsto n s_{p(e)}(a) . e$ ($e \in E$, $(n, a) \in N \times G$). This defines explicitly the equivalence of $\widetilde{X}(\Gamma)$ with $\widetilde{X}(N \times G)$.

(8.8.4) Remark. This last remark has a useful generalization in another direction, in which condition (a) above is weakened. Let a topological group Γ act on a space X. Let N be an open normal subgroup of Γ, put $G := \Gamma/N$ and assume that there is a continuous map $s \colon G \times X \to \Gamma$, $(a, x) \mapsto s_x(a)$, such that the following hold for every $x \in X$:
   (i) $s_x \colon G \to \Gamma$ is a homomorphic section of $1 \longrightarrow N \longrightarrow \Gamma \longrightarrow G \longrightarrow 1$;
   (ii) $\mathrm{im}(s_x) \subset \Gamma_x$;
   (iii) $s_{nx}(a) = n s_x(a) n^{-1}$ for $n \in N$ and $a \in G$.
Let $N \times G$ act on X through $N \times G \xrightarrow{\mathrm{pr}_1} N \subset \Gamma$. Then every map s as before gives rise to an equivalence of categories

$$\widetilde{X}(\Gamma) \sim \widetilde{X}(N \times G). \tag{7}$$

Conditions (i)–(iii) can be rephrased informally by saying that for every $x \in X$ the sequence $1 \to \Gamma_x \cap N \to \Gamma_x \to G \to 1$ is exact and splits as a *direct* product, and that s consists of a continuous family of such splittings over X which is "N-equivariant" (condition (iii)). To verify the equivalence (7) one has only to observe that for any espace étalé $p \colon E \to X$ over X the map s induces a bijection between Γ-actions and $N \times G$-actions on E over X. Indeed, one passes from a Γ-action to an $N \times G$-action by

$$(n, a) \cdot e := n s_{p(e)}(a) . e,$$

and from an $N \times G$-action to a Γ-action by

$$\sigma . e := \left(\sigma s_{p(e)}(\bar{\sigma})^{-1}, s_{p(e)}(\bar{\sigma}) \right) \cdot e$$

($\bar{\sigma} := \sigma N \in G$). These two assignments are clearly inverses of each other and are functorial in E.

In the situation of (8.8.2) there is a unique map s with (i)–(iii), which is the map of (8.8.3). So here the equivalence $\tilde{X}(\Gamma) \sim \tilde{X}(N \times G) = (\tilde{X}(N))(G)$ is *canonical*. If this equivalence is composed with that of (8.7.2) between $\tilde{X}(N)$ and $\tilde{Y}$ one gets back the equivalence of (8.8.2).

(8.9) Remark. Let now Γ be a *profinite* group and X a *profinite* Γ-space, and let $p: X \to Y = X/\Gamma$ be the quotient. The various spectral sequences for equivariant cohomology (8.7.1) degenerate since X and Y have no sheaf cohomology. For every abelian Γ-sheaf A on X one has therefore

$$H_\Gamma^n(X, A) = H^n(\Gamma, H^0(X, A)) = H^0(Y, \mathcal{H}^n(A))$$

for $n \geq 0$. Here $\mathcal{H}^n(A) := H_\Gamma^n(p(\Gamma)_* A)$, which sheaf is isomorphic to $R^n p_*^\Gamma A$ since $p(\Gamma)_*$ is exact. It is the sheaf associated to the presheaf $V \mapsto H^n(\Gamma, A(p^{-1}V))$ on Y. Note that for $x \in X$ the stalk of this sheaf at $p(x)$ is canonically identified with $H^n(\Gamma_x, A_x)$.

Assume now that the hypotheses of (8.8.4) are satisfied, and fix a map s as there. Then s makes A into an $N \times G$-sheaf, and by (7),

$$H_\Gamma^n(X, A) = H_{N \times G}^n(X, A) = H^n(N \times G, H^0(X, A)).$$

The Hochschild-Serre spectral sequence can be applied to the last group to give two spectral sequences with abutment $H_\Gamma^*(X, A)$ and E_2 terms

$$E_2^{pq} = H^p(G, H_N^q(X, A)) \quad \text{resp.} \quad E_2^{pq} = H^p\left(N, H^q(G, H^0(X, A))\right).$$

In particular, if one assumes also that N acts freely (i.e. that conditions (a) and (b) of (8.8.1) hold) then the first of them degenerates to give

$$H_\Gamma^n(X, A) = H^n(G, H_N^0(X, A)) = H^n(G, H^0(Y, p_*^\Gamma A)), \quad n \geq 0.$$

9. Cohomological dimension of X_b, II: The field case

It is the central issue of this book to relate the étale site of a scheme X to its real spectrum X_r, mainly under cohomological aspects. As we saw in Sect. 6, a key point is to determine the cohomological dimension of the site X_b. The proof of the main result (Theorem (7.17)) however has not yet been completed, the case where X is the spectrum of a field being still missing. This gap will now be closed. Ultimately the statement will be reduced to a theorem essentially due to Arason (Proposition (9.12)).

On the other hand, in order to carry out these reduction steps in the field case it is necessary to reformulate the situation entirely in terms of the absolute Galois group. This translation is an instructive exercise, but it is more than that. It is also an important step towards revealing the ultimate reasons behind the main result (7.17) which are cohomological properties of absolute Galois groups of fields. What is needed for the proof of Arason's theorem are multiplicative relations in the mod 2 cohomology ring, which — when one has in mind "general" profinite groups — are of a very special nature. This is so even after the proof has been reduced to its bones, as is discussed at the end of this section. It becomes therefore natural to try and see what happens for other profinite groups. This question forms the point of departure for a different line of study which will be taken up in Part Two of this work (Sections 11.2 and 12). There a completely different approach will show that none of these special properties of absolute Galois groups is really essential for the theorem.

Below the various comparison functors between étale site and real spectrum of a field k will be made explicit in terms of the Galois group Γ of k. It is advisable to regard sheaves on the real spectrum as equivariant sheaves on the space of real closures of k. Therefore the notions from Sect. 8 will be used freely.

(9.1) Let k be a field. Put $k' := k(\sqrt{-1})$ and fix a separable closure $k' \subset k_s$ of k'. Write $\Gamma := \mathrm{Gal}(k_s/k)$ and $\Gamma' := \mathrm{Gal}(k_s/k')$ for the absolute Galois groups of k resp. k'. I always assume in the following that -1 is not a square in k, so that $G := \Gamma/\Gamma'$ has order 2. Put $X := \mathrm{spec}\, k$, $X' := \mathrm{spec}\, k'$ and consider $\bar{x} := \mathrm{spec}\, k_s$ as an X-scheme by the fixed inclusion $k \subset k_s$. Let $\pi: X' \to X$ be the morphism induced by $k \subset k'$.

(9.1.1) Let $U \to X$ be an étale k-scheme. The group Γ acts on the set $\mathrm{Hom}_X(\bar{x}, U)$

from the left by the rule

$$(\sigma, f) \longmapsto f \circ \operatorname{spec}(\sigma)$$

($\sigma \in \Gamma$, $f \in \operatorname{Hom}_X(\bar{x}, U)$), and all stabilizer subgroups are open in Γ. It is classical (Grothendieck's formulation of Galois theory) that this sets up an equivalence between the category Et/X of étale k-schemes (which is also $\widetilde{X}_{et}$) and the category (Γ-sets) of discrete Γ-sets.

(9.1.2) Let $k \to F$ be a field extension, and let $f: (\operatorname{spec} F)\widetilde{~}_{et} \to (\operatorname{spec} k)\widetilde{~}_{et}$ be the induced morphism between the étale toposes. By (9.1.1), f corresponds to a topos morphism f from (Δ-sets) to (Γ-sets), where Γ resp. Δ is the absolute Galois group of k resp. F. (It is assumed that a compatible embedding $k_s \to F_s$ of separable closures has been chosen.) As in (0.5.3), f can be made explicit in terms of the associated homomorphism $\Delta \to \Gamma$ between the absolute Galois groups.

(9.2) Let T always denote the set of involutions in Γ (the identity being excluded). This is a closed subset of Γ on which Γ acts by conjugation. Every $t \in T$ is self-centralizing, which means that the operation of Γ' on T is free. Algebraically this is the fact that $\operatorname{Aut}(R/k) = \{1\}$ for every real closure R of k. Let $p: T \to T/\Gamma = T/\Gamma'$ be the topological quotient map. The quotient space T/Γ is the real spectrum $X_r = \operatorname{sper} k$.

(9.2.1) Let $U \in \operatorname{Et}/X$ and let $M := \operatorname{Hom}_X(\bar{x}, U)$ be the Γ-set corresponding to U (9.1.1). A pair consisting of a real closure R of k in k_s and a k-morphism $\operatorname{spec} R \to U$ corresponds to an involution $t \in T$ together with a morphism $\Gamma/\{1, t\} \to M$ of Γ-spaces; or else to an element of the space

$$T(M) := \{(t, m) \in T \times M: t.m = m\} \tag{1}$$

on which Γ acts "diagonally". This shows that there is a natural surjective map $T(M) \twoheadrightarrow U_r$, which in addition is continuous: The pair (t, m) is mapped to the element of U_r which is represented by the k-morphism $\operatorname{spec} R \to U$ corresponding to (t, m). Moreover two k-morphisms $\operatorname{spec} R \to U$, $\operatorname{spec} R' \to U$ as above represent the same point in U_r iff there is an isomorphism $\operatorname{spec} R \xrightarrow{\sim} \operatorname{spec} R'$ over U. So the fibres of $T(M) \twoheadrightarrow U_r$ are precisely the Γ-orbits in $T(M)$. This proves

(9.2.2) **Proposition.** — *Upon identifying* Et/X *with* (Γ-sets), *the functor "real spectrum"*

$$\operatorname{Et}/X \longrightarrow (\operatorname{Top})/X_r, \quad U \longmapsto U_r$$

gets identified with the functor $M \mapsto T(M)/\Gamma$, *where* $T(M)$ *is defined in* (1) *and* $T(M)/\Gamma \to T/\Gamma = X_r$ *is induced by projection to the first component.* □

Henceforth I will be less formal when stating how objects, functors etc. get identified under the indicated equivalences, and just say "is" instead of "gets identified with".

(9.2.3) Corollary. — *The real étale topology ret on (Γ-sets) is the topology under which a family $\{M_i \to M\}_{i \in I}$ is a covering if and only if $\{M_i^t \to M^t\}_{i \in I}$ is a surjective family for every $t \in T$.* □

Hence $\{M_i \to M\}_i$ is a covering for the b-topology if and only if $\{M_i^t \to M^t\}_i$ is a surjective family for every $t \in T \cup \{1\}$.

(9.3) By (8.7.2) resp. (8.8.2) there are canonical equivalences of categories

$$\widetilde{X}_r \sim \widetilde{T}(\Gamma') \quad \text{and} \quad \widetilde{X}_r(G) \sim \widetilde{T}(\Gamma) \sim (\widetilde{T}(\Gamma'))(G) = \widetilde{T}(\Gamma' \times G). \tag{2}$$

The first is given by $p_{\Gamma'}^*$ and $p_*^{\Gamma'}$ (8.7), for the second see (8.8.2). Under the last equivalence in (2), a Γ-sheaf $f: E \to T$ on T corresponds to its underlying Γ'-sheaf together with the involution $e \mapsto f(e) . e$ on it (8.8.3); i.e. in the fibre E_t the involution is translation by t. Recall from (8.9) that these equivalences give rise to the following isomorphisms, all of them canonical, for B an abelian Γ-sheaf on T and $n \geq 0$:

$$H_\Gamma^n(T, B) \cong H^n\big(\Gamma, H^0(T, B)\big) \cong H^0\big(X_r, R^n p_*^\Gamma(B)\big) \cong H^n\big(G, H_{\Gamma'}^0(T, B)\big). \tag{3}$$

To make explicit the various functors between étale and real topos (i.e. the functor ρ and the topos morphisms ν and $\nu(G)$, (6.4.1)) it is preferable to identify the objects of the real topos with equivariant sheaves on T, instead of sheaves on X_r. Under the above identifications this leads to

(9.4) Proposition. —
a) $\nu(G)^*: \widetilde{X}_{et} \to \widetilde{X}_{ret}(G)$ is the functor (Γ-sets) $\to \widetilde{T}(\Gamma)$ which sends a Γ-set M to the constant sheaf M on T on which Γ acts "diagonally"; i.e. to the espace étalé $\mathrm{pr}_1: T \times M \to T$ with Γ-action $\sigma . (t, m) = (\sigma t \sigma^{-1}, \sigma . m)$.
b) $\nu(G)_*: \widetilde{X}_{ret}(G) \to \widetilde{X}_{et}$ is the functor $\widetilde{T}(\Gamma) \to$ (Γ-sets) which sends a Γ-sheaf $p: E \to T$ on T to its set of global sections $H^0(T, E)$, on which Γ acts by $(\sigma \cdot s)(t) = \sigma . s(\sigma^{-1} t \sigma)$.
c) $\rho: \widetilde{X}_{et} \to \widetilde{X}_{ret}$ is the functor (Γ-sets) $\to \widetilde{T}(\Gamma')$ which sends M to $\mathrm{pr}_1: T(M) \to T$ (see (9.2.1); this is an espace étalé over T on which Γ, and in particular Γ', acts diagonally).
d) $\nu^*: \widetilde{X}'_{et} \to \widetilde{X}_{ret}$ is the functor (Γ'-sets) $\to \widetilde{T}(\Gamma')$ which sends N to the constant sheaf $T \times N \to T$ on T, with diagonal Γ'-action. Its right adjoint ν_* sends a Γ'-sheaf E on T to $H^0(T, E)$ with the natural Γ'-action (compare b)).

Proof. First consider c). Since for $U \in \mathrm{Et}/X$ the sheaf $\rho(h_U)$ on X_{ret} corresponds to the espace étalé $U_r \to X_r$ (2.13), the assertion is a consequence of (9.2.2).

By definition $\nu^* = \rho \pi_*$ (Sect. 5). Now $\pi_*: (\Gamma'\text{-sets}) \to (\Gamma\text{-sets})$ is the coinduction functor $N \mapsto \mathrm{coind}_{\Gamma'}^\Gamma(N) = \mathrm{Hom}_{\Gamma'}(\Gamma, N)$ (0.5.3). For $t \in T$ and $f \in$

$\mathrm{Hom}_{\Gamma'}(\Gamma, N)$ the condition $t \cdot f = f$ means $f(t) = f(1)$. This shows that $\rho \pi_* N = T\big(\mathrm{coind}_{\Gamma'}^{\Gamma}(N)\big)$ is simply the direct product espace étalé $T \times N \to T$ (and this is compatible with the Γ'-action). a) follows from this, one only has to keep track of the G-actions on (Γ'-sets) resp. on $\widetilde{T}(\Gamma')$. Finally the descriptions of $\nu(G)_*$ resp. ν_* follow by simply observing that the functors indicated in b) and d) are right adjoint to $\nu(G)^*$ resp. ν^*. $\qquad\square$

(9.4.1) Let A be a discrete Γ-module. For $n \geq 0$ I'll write $\mathcal{H}^n(A)$ for the sheaf $R^n p_*^{\Gamma} A$ on $T/\Gamma = X_r$. Recall (8.9) that this is the sheaf associated to the presheaf

$$V \longmapsto H_{\Gamma}^n\big(p^{-1}(V), A\big), \quad V \subset X_r \text{ open.}$$

Then $\mathcal{H}^n(A) = R^n \rho(A)$. The Γ'-sheaf on T corresponding to $R^n \rho(A)$ under the equivalence $\widetilde{X}_r \sim \widetilde{T}(\Gamma')$ is $\coprod_{t \in T} H^n(\langle t \rangle, A) \to T$, where the left hand set is suitably topologized as an espace étalé over T, and Γ' acts on it by conjugation.

(9.5) **Remark.** From Proposition (9.4) and (3) it follows that for every Γ-module A there are canonical isomorphisms

$$
\begin{aligned}
H_G^n\big(X_{ret}, \nu(G)^* A\big) &\cong H_{\Gamma}^n(T, A) \cong H^n\big(\Gamma, C(T, A)\big) \\
&\cong H^0\big(X_r, \mathcal{H}^n(A)\big) \cong H^n\big(G, \mathrm{Hom}_{\Gamma'}(T, A)\big),
\end{aligned}
\tag{4}
$$

$n \geq 0$. Here $C(T, A) = H^0(T, A)$ is the Γ-module of locally constant maps $T \to A$, and $\mathrm{Hom}_H(T, A)$ (for $H \subset \Gamma$ a subgroup) is its subgroup of H-equivariant such maps. The G-action on $\mathrm{Hom}_{\Gamma'}(T, A)$ is the involution

$$f \longmapsto \big(f^{\tau} : t \mapsto t . f(t)\big).\tag{5}$$

Note that the *abelian group* $\mathrm{Hom}_{\Gamma'}(T, A)$ is isomorphic to $C(T/\Gamma', A) = C(X_r, A)$. Indeed, it is known that the quotient map $T \to T/\Gamma = X_r$ has a section ([Er Thm. 2] or [Ha1 Lemma 5.2]), and a choice of such a section gives an isomorphism of Γ'-spaces $T \approx \Gamma' \times X_r$, hence the asserted isomorphism. *But for general A there is no way to express the involution on $C(X_r, A)$, other than by choice of a section.* This is why it is preferable here to think of the real spectrum topos as of the category of Γ'-sheaves on T, instead of sheaves on X_r.

(9.6) **Remark.** Let A be a Γ-module. The homomorphisms

$$H^n(X_{et}, A) \longrightarrow H_G^n(X_{ret}, \nu(G)^* A)\tag{6}$$

($n \geq 0$) induced by $\nu(G)^*$ are part of the long exact sequence for A (6.6). In the present situation (6) gets identified with a map

$$H^n(\Gamma, A) \longrightarrow H_{\Gamma}^n(T, A).\tag{7}$$

It is useful to make (7) explicit with respect to the identifications made in (4):

- Identify $H^n_\Gamma(T, A)$ with $H^n(\Gamma, C(T, A))$. Then (7) becomes the map induced by the canonical homomorphism $A \to C(T, A)$ of Γ-modules.

- Identify $H^n_\Gamma(T, A)$ with $H^0(X_r, \mathcal{H}^n(A))$. Let $t \in T$, and let $\xi := p(t)$ be the ordering of k which corresponds to t. The stalk of the sheaf $\mathcal{H}^n(A)$ at ξ is identified with $H^n(\langle t \rangle, A)$. Under this identification, (7) sends a class $\alpha \in H^n(\Gamma, A)$ to the global section of $\mathcal{H}^n(A)$ whose value in the stalk ξ is the restriction of α to $\langle t \rangle$. (Note again that this description depends on the choice of a section of p, and hence is less valuable, unless for example A is a trivial Γ-module.)

- Upon identifying $H^n_\Gamma(T, A)$ with $H^n(G, \mathrm{Hom}_{\Gamma'}(T, A))$, (7) has the same explanation as in the first case, since the Hochschild-Serre spectral sequence identifies $H^*(\Gamma, C(T, A))$ with $H^*(G, \mathrm{Hom}_{\Gamma'}(T, A))$.

(9.6.1) Example. For an example, assume that k is a number field. Let $\sigma_i \colon k \to \mathbb{R}$, $i = 1, \ldots, r$, be the real places of k at infinity. If A is a $\Gamma = \mathrm{Gal}(k_s/k)$-module, write A^{σ_i} for the $G = \mathrm{Gal}(\mathbb{C}/\mathbb{R})$-module deduced from A by σ_i. Then

$$H^n_G(k_{ret}, \nu(G)^* A) = \bigoplus_{i=1}^r H^n_{et}(\mathbb{R}, A^{\sigma_i}),$$

and the map from $H^n_{et}(k, A)$ to this group is the obvious one.

After these preparations I now turn to the completion of the proof of Theorem (7.17). I'm using the notations from Sect. 7. I remark again that we are free to use the material from (7.8)–(7.14), since Theorem (7.7) was not used for it.

(9.7) Theorem. — *If k is a field of characteristic not 2 then*

$$L_b(k, A) = \lim_{n \to \infty} H^n_b(k, j_! A) = 0$$

for every étale $\mathbb{Z}/2$-sheaf A on spec k.

Before the proof is given here are some consequences of this theorem. For brevity write $\mathrm{cd}_\ell(k_t) := \mathrm{cd}_\ell((\mathrm{spec}\, k)_t)$ if $t \in \{et, b, ret\}$. The following corollary supplies the missing link in the proof of Theorem (7.17):

(9.8) Corollary. — $\mathrm{cd}_2(k_b) = \mathrm{cd}_2\big(k(\sqrt{-1})_{et}\big)$ *unless possibly in the case when every finite extension of $k(\sqrt{-1})$ has odd degree; then* $\mathrm{cd}_2(k_b) \leq 1$.

Proof. See Proposition (7.13) and the proof of (7.18), and note that sper k has cohomological dimension 0 because it is compact and totally disconnected. The condition that $k(\sqrt{-1})$ has no extension of even degree is equivalent to $\mathrm{cd}_2\big(k(\sqrt{-1})_{et}\big) = 0$, by [Se4, ch. I §3]. □

The determination of $\mathrm{cd}_\ell(X_b)$ in the field case is completed by

(9.8.1) Proposition. — *Let k be a field and ℓ a prime. Assume that $k' = k(\sqrt{-1})$ has cohomological ℓ-dimension 0, or equivalently, that the degree of every finite extension K/k' is prime to ℓ. Assume moreover that k is real. Then $\mathrm{cd}_\ell(k_b)$ is 0 if k is real closed, and is 1 in all other cases.*

Proof. Write $X := \mathrm{spec}\, k$ and let $\Gamma = \mathrm{Gal}(k_s/k)$. The case k real closed is obvious, so let it be excluded. Fix an involution $t \in \Gamma$. There is a discrete $(\mathbb{Z}/\ell)$-module A for which $A^\Gamma \neq A^t$. (For example, take $A := (\mathbb{Z}/\ell)[\Gamma/H]$ where $H \neq \Gamma$ is an open subgroup of Γ containing t.) Let $\xi \in X_r$ be the point which corresponds to t. Since X_r is a boolean space, the stalk map $H^0_G(X_r, \nu(G)^*A) = H^0(X_r, \rho A) \to (\rho A)_\xi = A^t$ is surjective. Its composition with $A^\Gamma = H^0(X_{et}, A) \to H^0_G(X_r, \nu(G)^*A)$ is the inclusion $A^\Gamma \subset A^t$. This shows that $H^0(X_{et}, A) \to H^0_G(X_r, \nu(G)^*A)$ is not surjective. Thus $H^1(X_b, j_! A) \neq 0$ by the fundamental exact sequence (6.6). $\square$

(9.8.2) Remark. For any prime ℓ there exist fields k which satisfy the hypotheses of (9.8.1) and are not real closed. For ℓ odd this is immediate. (E.g. start from any real field k_0 which, to be safe, has at least two orderings, and take the maximal ℓ-extension k/k_0.) More interesting is the case $\ell = 2$. Let k be a real field for which $[k_s : k(\sqrt{-1})]$ is odd (infinite). Clearly such a field must be euclidean and hereditarily pythagorean. It is a particular case of theorems of Becker [Be1, III §1] that Γ' is an abelian group, actually a direct product of copies of p-adic integers for odd primes p, on which every involution $t \in \Gamma$ acts by taking the inverse. Examples of such fields can also be found in Becker's work: For instance, any field k is an example which is real closed of exact higher level n with n odd, $n > 1$ ([Be2, Satz 3.6]; read $2n$ for n in this paper).

(9.9) Corollary. — *Let Γ be the absolute Galois group of the field k (char $k \neq 2$), and let A be a discrete $(\mathbb{Z}/2)\Gamma$-module. There is a canonical isomorphism between $L_{et}(k, A) = \lim\limits_{n \to \infty} H^n(\Gamma, A)$ and the cokernel of the map*

$$\mathrm{Hom}_{\Gamma'}(T, A) \longrightarrow \mathrm{Hom}_\Gamma(T, A), \quad f \longmapsto f + f^\tau.$$

Here τ is the involution (5).

Proof. This follows from (9.7) (and the exact triangle (7.12.1)), one only has to identify $L^G_{ret}(k, A)$: From (9.5) one sees that

$$L^G_{ret}(k, A) = H^1(G, \mathrm{Hom}_{\Gamma'}(T, A)) = \ker(1 + \tau)/\mathrm{im}(1 + \tau).$$

Since $\ker(1 + \tau) = \mathrm{Hom}_\Gamma(T, A)$, the assertion follows. $\square$

Now I come to the proof of Theorem (9.7). For this one needs

(9.10) Lemma. — *Let X be a scheme with $\frac{1}{2} \in \mathcal{O}(X)$ for which $c := \mathrm{cd}_2(X_r)$ is finite. Let $f : Y \to X$ be a finite étale morphism which has odd local degree everywhere. Then for every $\mathbb{Z}/2$-sheaf $A \in \mathrm{Ab}(X_{et})$ the natural map $L_b(X, A) \to L_b(Y, f^*A)$ is injective.*

Proof. See (7.12.2c) for the map in question. The assertion follows from the trace argument: Let d be the degree of f (which can be assumed to be constant). Since the composition $A \longrightarrow f_*f^*A \stackrel{\mathrm{tr}}{\longrightarrow} A$ is multiplication by d it follows that the natural maps

$$H^n(X_b, j_*A) \longrightarrow H^n(X_b, j_*f_*f^*A) = H^n(Y_b, j_*f^*A)$$

are injective, $n \geq 0$. (Recall (3.11.1) that f_{b*} is exact.) The diagram

$$
\begin{array}{ccc}
H^n(X_b, j_!A) & \longrightarrow & H^n(Y_b, f_b^*j_!A) = H^n(Y_b, j_!f^*A) \\
\downarrow & & \downarrow \\
H^n(X_b, j_*A) & \lhook\joinrel\longrightarrow & H^n(Y_b, j_*f^*A)
\end{array}
$$

commutes for every $n \geq 0$. Since also $\mathrm{cd}_2(Y_r) \leq c$, the vertical maps are isomorphisms for $n \geq c + 2$. Hence it follows that $H^n(X_b, j_!A) \to H^n(Y_b, j_!f^*A)$ is injective for every $n \geq c + 2$, from which the lemma follows. $\square$

(9.11) Now let k be a field of characteristic not 2 and put $X := \mathrm{spec}\, k$. Let k_s be a separable closure of k and $\Gamma = \mathrm{Gal}(k_s/k)$. One has to show that $L_b(A) = \lim_{n \to \infty} H_b^n(k, j_!A)$ vanishes for every discrete $(\mathbb{Z}/2)\Gamma$-module A.

Let K/k be a maximal subextension of k_s/k of odd degree (possibly infinite), so that $\mathrm{Gal}(k_s/K)$ is a Sylow 2-subgroup of Γ. Write $K = \lim_{\longrightarrow} K_i$ as the directed union of finite extensions K_i/k. Assume that $L_b(K, A) = 0$ is known. By Lemma (7.14.2) one has $L_b(K, A) = \lim_{\longrightarrow} L_b(K_i, A)$. By (9.10) all transition maps in this direct system are injective. Hence it follows that $L_b(k, A) = 0$.

So one can assume that $\Gamma = \mathrm{Gal}(k_s/k)$ is a pro-2 group. To show $L_b(k, A) = 0$ one may assume that A is finite, by b) of (7.12.2). And the exact triangle a) of loc. cit. allows to assume that A is irreducible as a Γ-module. But Γ being a pro-2 group, the only irreducible $(\mathbb{Z}/2)\Gamma$-module is the trivial Γ-module $\mathbb{Z}/2$. So it suffices to show $L_b(k, \mathbb{Z}/2) = 0$. Now the vanishing of $L_b(k, \mathbb{Z}/2)$ turns out to be equivalent to the following theorem, the essential part of which is due to Arason:

(9.12) Proposition ([Ar, Satz 3], [AEJ, Proposition 2.4]). *Let k be a field of characteristic not 2. Then the natural map*

$$\lim_{n \to \infty} H^n(k_{et}, \mathbb{Z}/2) \longrightarrow H^0(\mathrm{sper}\, k, \mathbb{Z}/2) \tag{8}$$

is bijective.

(9.12.1) Here $H^n(k_{et}, \mathbb{Z}/2) \to H^0(\operatorname{sper} k, \mathbb{Z}/2)$ is the map which associates to a cohomology class $\alpha \in H^n(k_{et}, \mathbb{Z}/2)$ the continuous (= locally constant) map

$$\operatorname{sper} k \longrightarrow \mathbb{Z}/2, \quad \xi \longmapsto \alpha|_{k(\xi)} \in H^n(k(\xi)_{et}, \mathbb{Z}/2) = \mathbb{Z}/2, \tag{9}$$

where $k(\xi)$ is a real closure of k with respect to ξ. The transition maps for the limit on the left of (8) are cup product with (-1), the class of -1 in $k^*/k^{*2} \cong H^1(k_{et}, \mathbb{Z}/2)$; so this limit is $L_{et}(k, \mathbb{Z}/2)$ (7.10.3). The constant sheaf $\mathbb{Z}/2$ on sper k is equal to the sheaf $\mathcal{H}^n(\mathbb{Z}/2) = R^n\rho(\mathbb{Z}/2)$ for every $n \geq 0$ (9.4.1). By the description (9.6), the above map $H^n(k_{et}, \mathbb{Z}/2) \longrightarrow H^0(\operatorname{sper} k, \mathbb{Z}/2)$ is equal to the map $H^n(k_{et}, \mathbb{Z}/2) \to H^n_G(k_{ret}, \nu(G)^*\mathbb{Z}/2) = H^n_G(\operatorname{sper} k, \mathbb{Z}/2)$ induced by $\nu(G)^*$. Under the identification $H^n_G(\operatorname{sper} k, \mathbb{Z}/2) \approx H^0(\operatorname{sper} k, \mathbb{Z}/2)$ the transition maps (cup product with the generator of $H^*(G, \mathbb{Z}/2)$) become the identity, and so (8) is precisely the map $L_{et}(k, \mathbb{Z}/2) \to L^G_{ret}(k, \mathbb{Z}/2)$ of (7.12.1). From the exact triangle of loc. cit. it is clear that this map is an isomorphism if and only if $L_b(k, \mathbb{Z}/2) = 0$.

(9.13) Thus (9.12) proves Theorem (9.7), and hence also the results of Sect. 7 are completely established now. Essentially, the determination of the cohomological 2-dimension of X_b (resp. the proof of $\varinjlim H^n(X_b, j_!A) = 0$) for X quasi-compact and quasi-separated, was reduced in a series of steps to Arason's theorem above. The results of Sect. 7 can be seen as a very far reaching generalization of this theorem. It is therefore worthwhile to analyze its proof more closely and try to reveal the ultimate reasons for these facts.

(9.13.1) The essential part of (9.12) is the injectivity of the map (8). It is proved in [Ar]. The proof is reproduced in [AEJ], where also surjectivity is shown and the theorem is presented in the above formulation.

The proof proceeds as follows. If k is not real then k contains a nonreal subfield k_0 which is finitely generated over its prime field. So (-1) is nilpotent (over either field) since $\operatorname{cd}_2(k_0) < \infty$, and both groups in (8) are zero. Hence suppose that k is real. For $a \in k^*$ denote as usual by (a) the class of a in $k^*/k^{*2} = H^1(k_{et}, \mathbb{Z}/2)$ (7.10) . The image of (a) under (8) is the characteristic function of $\{\xi \in \operatorname{sper} k: a(\xi) < 0\}$. Since $H^0(\operatorname{sper} k, \mathbb{Z}/2)$ is generated by these characteristic functions as a ring, (8) is surjective since it is a ring homomorphism. To prove injectivity let $\alpha \in H^n(k_{et}, \mathbb{Z}/2)$ be an element which maps to zero; one has to show $(-1)^r \cdot \alpha = 0$ for some $r \geq 0$. Suppose false. By Zorn's Lemma one may assume that for every proper finite extension K/k the restriction $\alpha|_K$ is (-1)-primary torsion in $H^*(K_{et}, \mathbb{Z}/2)$. Thus k has clearly no finite odd extensions. One shows that $k^*/k^{*2} = \{\pm 1\}$ (which in turn implies that k is real closed and thus gives the desired contradiction) as follows: Assume that there is $d \in k^*$ with $\pm d \notin k^{*2}$. By replacing α with $(-1)^r \cdot \alpha$ for suitable r one can assume that α restricts to 0 in both $k(\sqrt{\pm d})$. By the trace (or "res-cor") exact sequence there are

$\beta, \gamma \in H^{n-1}(k_{et}, \mathbb{Z}/2)$ with $\alpha = (d) \cdot \beta = (-d) \cdot \gamma$. Hence

$$(-1) \cdot \alpha = (-1) \cdot (d) \cdot \beta = (d)^2 \cdot \beta = (d) \cdot \alpha = (d) \cdot (-d) \cdot \gamma = 0,$$

the desired contradiction.

(9.13.2) An analysis of this proof reveals that it can be modified in such a way that the only ingredient becomes the well known identity

$$(-1) \cdot x = x^2, \tag{10}$$

which holds for every $x \in H^1(k_{et}, \mathbb{Z}/2)$ and every field k, char $k \neq 2$. To make this remark more precise, let us try and see whether an analogue of Arason's theorem (9.12) remains true in a more general setting.

In looking for such a generalization, it is natural to include more general coefficient groups if possible. How should this be done properly? To find this out I will first translate the statement of Theorem (9.7) into purely group-theoretical terms. So let $\Gamma = \mathrm{Gal}(k_s/k)$. By the exact triangle (7.12.1), Theorem (9.7) is equivalent to the assertion that the map

$$\lim_{n \to \infty} H^n(k_{et}, A) \longrightarrow \lim_{n \to \infty} H^n_G(k_{ret}, \nu(G)^* A) \tag{11}$$

is an isomorphism, for every discrete $(\mathbb{Z}/2)\Gamma$-module A. By (9.6), the map (11) is the direct limit of the maps $H^n(\Gamma, A) \to H^n_\Gamma(T, A)$ for $n \to \infty$, the transition maps being cup product with (-1).

So it is clear how the desired generalization of (9.7) resp. (9.12) should read. I will now show that this generalization does indeed hold under the single hypothesis that a relation like (10) is valid. For brevity write $H^n(\Gamma) := H^n(\Gamma, \mathbb{Z}/2)$ in the following.

(9.13.3) **Theorem.** — *Let* Γ *be a profinite group with a distinguished element* $s \in H^1(\Gamma)$, *and put* $\Gamma' := \ker(s)$. *Assume that for every closed subgroup H of Γ and every $x \in H^1(H)$ the relation*

$$x^2 = s \cdot x \tag{12}$$

holds in $H^2(H)$. *Let T be the subset of involutions in* Γ.
a) $T \cap \Gamma' = \emptyset$. *In particular, T is closed in* Γ.
b) *For every discrete $(\mathbb{Z}/2)\Gamma$-module A the natural map*

$$\lim_{n \to \infty} H^n(\Gamma, A) \longrightarrow \lim_{n \to \infty} H^n_\Gamma(T, A) \tag{13}$$

is an isomorphism. Here the transition maps for both limits are cup product with s.

Observe that the virtual cohomological 2-dimension of Γ is allowed to be infinite. In case it is finite, equal to $d < \infty$, one has $\mathrm{cd}_2(\Gamma') = d$ by a) and by Serre's theorem [Se2]. In this case, therefore, the direct limits in (13) are attained already for $n > d$.

Proof. a) Let $t \in T$, let x be the non-zero element of $H^1(\langle t \rangle)$. Since $x^2 \neq 0$, relation (12) shows $\mathrm{res}^\Gamma_{\langle t \rangle}(s) \neq 0$, and hence $t \notin \Gamma'$. Now I prove b). For all n, $H^n_\Gamma(T, A) = H^n(\Gamma, C(T, A))$ (8.9), and the map (13) is induced by the canonical map $A \to C(T, A)$ of Γ-modules (which sends an element $a \in A$ to the constant function a on T). Assume $T \neq \emptyset$, and let P be the cokernel of the injection $A \hookrightarrow C(T, A)$. It is not hard to show that P satisfies

$$H^n(\langle t \rangle, P) = 0 \quad \text{for every } t \in T,\ n \geq 1. \tag{14}$$

For a proof see Lemma (12.17). (There the hypothesis is made that no two different involutions of Γ commute; in the present case this is obviously true in view of a).) Therefore it is enough to prove:

(∗) *If P is any discrete $(\mathbb{Z}/2)\Gamma$-module which satisfies property (14), then $H^*(\Gamma, P)$ is s-primary torsion, i.e.* $\lim\limits_{n \to \infty} H^n(\Gamma, P) = 0$.

Note that (∗) will also settle the case $T = \emptyset$ of the theorem, since then one can take $P := A$. The proof of (∗) copies Arason's method. Let $\alpha \in H^n(\Gamma, P)$, and assume that $s^r \cdot \alpha \neq 0$ for every $r \geq 0$. By Zorn's Lemma one can assume that the restriction of α to every proper closed subgroup of Γ is s-primary torsion. So Γ is a pro-2 group. Assume $H^1(\Gamma) = \{0, s\}$. Then Γ would be pro-cyclic; since Γ is assumed to be a counterexample to (∗), $|\Gamma| > 2$. But then $s^2 = 0$ (it suffices to check this for Γ cyclic of order 4), contradicting $s^r \cdot \alpha \neq 0$ for all r. Therefore there is $x \in H^1(\Gamma)$ with $x \notin \{0, s\}$. Since $s^r \alpha$ restricts to 0 in both $\ker(x)$ and $\ker(s+x)$ for suitable $r \geq 0$, there are $\beta, \gamma \in H^*(\Gamma, P)$ with $s^r \alpha = x \cdot \beta = (s+x) \cdot \gamma$, and one gets the same contradiction as in (9.13.1):

$$s^{r+1} \cdot \alpha = s \cdot x \cdot \beta \underset{(12)}{=} x^2 \cdot \beta = x \cdot (s+x) \cdot \gamma \underset{(12)}{=} 0. \qquad \square$$

(9.13.4) Condition (12) in Theorem (9.13.3) is only one possibility. One can find weaker conditions on the cohomology rings of Γ (and its closed subgroups!) which guarantee that the above proof goes through. However this approach remains unsatisfactory in so far as the existence of such multiplicative relations in $H^*(\Gamma, \mathbb{Z}/2)$ are somewhat unnatural conditions, if one does not restrict attention to Galois groups. They don't seem to be necessary for the theorem in question to hold.

In Sect. 12 I will follow a different approach. Although it works only if Γ has finite virtual cohomological 2-dimension (in which case the direct limit in (13) becomes constant), it is free of hypotheses on the ring structure of $H^*(\Gamma)$ and applies also otherwise to a much wider range of cases. In particular, it gives also an independent proof of Arason's theorem, and therefore of (9.12). (Indeed, by obvious limit arguments, the general case of (9.12) is reduced to the case where k is finitely generated over its prime field, and hence $\mathrm{cd}_2 k(\sqrt{-1})$ is finite.)

Part Two

10. G-toposes

The notion of G-toposes is the proper framework for the study of equivariant topos theory, which should be seen in analogy to equivariant sheaf theory on topological spaces. As usual, things become technically more complicated through the fact that toposes form a 2-category, and not just a category like topological spaces. More concretely this means that one has to study pseudo group actions on toposes, where the attribute "pseudo" indicates that the axioms for a group action are required to hold only up to a coherent system of isomorphisms.

The G-toposes fit into the much more general theory of fibered toposes, corresponding to the case where the base category of the fibred topos is the category associated with G. In the first part of this section I will mainly review from [SGA4] the basic notions from this theory, thereby specializing to the case of G-toposes. I have made the exposition rather detailed, and I hope that it will be possible to read this material without prior knowledge of fibered toposes, perhaps glancing into [SGA4] from time to time. The most important point of this section is the treatment of inverse limits of G-toposes in the second part. The inverse limit $\underleftarrow{\mathrm{Limtop}}(\mathfrak{E}/G)$ of a G-topos $\mathfrak{E}$ is the analogue of the subspace of fixpoints in a topological G-space. Inverse limits of fibered toposes are treated in [SGA4] only for left filtering base categories, which case is basically different, cf. Remark (10.17). I do not know whether inverse limits of G-toposes have been considered before.

The reason why all this highly technical material has been included in this paper is mainly of philosophical nature. From the beginning of my work I knew that the real topos of a scheme X behaves like a kind of fixtopos of the involution on $X[\sqrt{-1}]_{et}^{\sim}$; but it took me some time to realize that the language of fibered toposes is exactly what is needed to give a precise meaning to this statement (and to prove it). This proof will be given in Sect. 11.1. For a quite different case in which the inverse limit of a G-topos is determined (group extensions) see Sect. 11.2. For further discussion I refer to Sect. 14 and to the Introduction of this work.

The material and the language introduced in this section will be used in Sections 11, 13 and 14.

Let G always be a discrete group. Let BG denote the category associated with G, so BG has a single object $*$ whose monoid of endomorphisms is G. My basic source for the theory of fibered toposes is Chaper VI.7 of [SGA4]. As pointed out before, all set-theoretic questions are ignored; in particular, I will not bother to distinguish between small categories and general categories.

(10.1) Here is a reminder of some basics from the theory of fibered categories. See [SGA1 VI] for the details and [SGA4 VI.6.1] for a brief summary. Let I be a category. A functor $p: \mathfrak{F} \to I$ with target category I is also called a *category over* I. If $p: \mathfrak{F} \to I$ and $p': \mathfrak{F}' \to I$ are categories over I then $\underline{\mathrm{Hom}}_I(\mathfrak{F}, \mathfrak{F}')$ is the full subcategory of $\underline{\mathrm{Hom}}(\mathfrak{F}, \mathfrak{F}')$ consisting of the functors φ with $p' \circ \varphi = p$. These functors φ are often called the *functors over* I, or *I-functors*, from $\mathfrak{F}$ to $\mathfrak{F}'$. For $i \in I$ one writes $\mathfrak{F}_i := p^{-1}(i)$ for the fibre category over i. If $f: i \to j$ is an arrow in I and $x \in \mathfrak{F}_i$, $y \in \mathfrak{F}_j$, then $\mathrm{Hom}_f(x, y)$ is the set of $\mathfrak{F}$-arrows $\alpha: x \to y$ over f (i.e. with $p(\alpha) = f$); if $i = j$ and $f = \mathrm{id}_i$, this is $\mathrm{Hom}_{\mathfrak{F}_i}(x, y)$. An arrow $\alpha: x \to y$ in $\mathfrak{F}$ is said to be *cartesian* if for every $x' \in \mathfrak{F}_{p(x)}$ the map

$$\mathrm{Hom}_{\mathfrak{F}_{p(x)}}(x', x) \longrightarrow \mathrm{Hom}_{p(\alpha)}(x', y), \qquad \varphi \longmapsto \alpha \circ \varphi$$

is bijective. One says that $\mathfrak{F}$ is *prefibered over* I if for every arrow $f: i \to j$ in I and every $y \in \mathfrak{F}_j$ there is a cartesian arrow over f with target y. One says that $\mathfrak{F}$ is *fibered over* I if it is prefibered over I and if the composition of any two (composable) cartesian arrows in $\mathfrak{F}$ is cartesian.

(10.1.1) Suppose $\mathfrak{F}$ is fibered over I. Then one can choose for each arrow $f: i \to j$ in I a functor $f^*: \mathfrak{F}_j \to \mathfrak{F}_i$ such that, for every $y \in \mathfrak{F}_j$, f^*y represents the functor $\mathfrak{F}_i^\circ \to (\mathrm{sets})$, $x \mapsto \mathrm{Hom}_f(x, y)$. There are natural isomorphisms of functors

$$c_{f,g}: f^* g^* \xrightarrow{\sim} (gf)^* \tag{1}$$

for each diagram $\bullet \xrightarrow{f} \bullet \xrightarrow{g} \bullet$ in I, which satisfy the "cocycle condition"

$$c_{f,hg} \circ (f^* * c_{g,h}) = c_{gf,h} \circ (c_{f,g} * h^*) \tag{2}$$

for each diagram $\bullet \xrightarrow{f} \bullet \xrightarrow{g} \bullet \xrightarrow{h} \bullet$ in I. Conversely, if one is given for each $i \in I$ a category $\mathfrak{F}_i$ and for each arrow $f: i \to j$ in I a functor $f^*: \mathfrak{F}_j \to \mathfrak{F}_i$, plus isomorphisms of functors (1) for each pair of composable arrows in I such that conditions (2) are satisfied, then one can construct, in an essentially unique way, a fibered category over I, and the two procedures are inverses of each other up to equivalence. (Cf. Remark (10.3.1) for the case $I = BG$.)

(10.2) A *fibered topos* over a category I [SGA4 VI.7.1] is a fibered category $p: \mathfrak{E} \to I$ over I such that each of the fibre categories $\mathfrak{E}_i$ ($i \in I$) is a topos, and such that the inverse image functors f^* (for the structure of fibered category, f ranging over the arrows in I) preserve finite inverse and arbitrary direct limits. In other words, for each arrow $f: i \to j$ in I there has to be a topos morphism (f^*, f_*) from $\mathfrak{E}_i$ to $\mathfrak{E}_j$ and a bijection

$$\mathrm{Hom}_{\mathfrak{E}_i}(x, f^* y) \approx \mathrm{Hom}_f(x, y),$$

functorial in $x \in \mathfrak{E}_i$ and $y \in \mathfrak{E}_j$. Putting it in a different language, a fibered topos over I is essentially the same as a pseudo-functor from I to the 2-category of toposes.

(10.3) Definition. By a *G-topos* I mean a fibered topos $p: \mathfrak{E} \to BG$ over BG. The fibre category $E := \mathfrak{E}_* = p^{-1}(*)$, which is a topos, will be referred to as *the fibre* of $\mathfrak{E}$, and will frequently be denoted by the corresponding roman character.

(10.3.1) Remark. Frequently it is more convenient to think of a concrete fibered topos by using a "biscindage" (in the sense of [SGA4 VI.7.1.4]). This means, in the case of a *G*-topos $p: \mathfrak{E} \to BG$ with fibre E, that for each $g \in G$ an inverse image functor $g^*: E \to E$ (for the fibered category structure, (10.1.1)) and a right adjoint g_* of g^* have been chosen. For $g = 1$ one can always take $g^* = g_* = \mathrm{id}$. Observe moreover that there is a canonical choice for the direct image functors once the g^* are chosen, namely $g_* := (g^{-1})^*$ ($g \in G$). As in (10.1.1) one gets a family of isomorphisms $c_{g,h}: g^* h^* \xrightarrow{\sim} (hg)^*$ for $g, h \in G$ which satisfies the cocycle conditions (2). Conversely, the data consisting of the g^* and the $c_{g,h}$ gives back the *G*-topos $\mathfrak{E}$, by putting $\mathrm{ob}\,\mathfrak{E} = \mathrm{ob}\,E$ and

$$\mathrm{Hom}_{\mathfrak{E}}(y, x) = \{(g, \alpha): g \in G, \ \alpha \in \mathrm{Hom}_E(y, g^* x)\}$$

and defining composition of arrows in $\mathfrak{E}$ by the formula

$$(g, \alpha) \circ (h, \beta) := (gh, \ c_{h,g} \circ h^*(\alpha) \circ \beta),$$

the projection functor $p: \mathfrak{E} \to BG$ being obvious. Thus, a *G*-topos is essentially the same as a topos E together with a pseudo group action by G on E through topos morphisms. The word "pseudo" indicates that the axioms for a group action are required to hold only up to a system of isomorphisms $c_{g,h}$, this system having to be coherent in the sense of (2).

(10.3.2) Remark. The fact that the base category BG is a groupoid (all arrows are isomorphisms) is the reason for some simplification, compared with the general theory of fibered toposes. In fact, let $\mathfrak{E} \to I$ be a fibered category over a groupoid I. Then the cartesian arrows in $\mathfrak{E}$ are precisely the isomorphisms in $\mathfrak{E}$ [SGA1 VI.6.12]. In particular, every I-functor $\mathfrak{E} \to \mathfrak{F}$ between categories over I is automatically cartesian, i.e. carries cartesian arrows to cartesian arrows. Moreover it is obvious that $\mathfrak{E} \to I$ is a fibered topos over I if and only if every fibre category $\mathfrak{E}_i$ is a topos.

Since I do not intend to give an exposition of the general theory of fibered toposes — for this see the literature cited — I will henceforth explain most concepts directly for the case of G-toposes, without dwelling on the general case.

(10.4) Definition (cf. [SGA4 VI.7.1]). Let $\mathfrak{E}$ and $\mathfrak{F}$ be G-toposes. A *morphism of G-toposes* from $\mathfrak{E}$ to $\mathfrak{F}$ consists of a functor $\mu : \mathfrak{F} \to \mathfrak{E}$ over BG and a right adjoint $m_* : E \to F$ of the fibre $m^* : F \to E$ of μ such that m^* is left exact, i.e. such that (m^*, m_*) is a topos morphism from E to F. Since m_* is determined by μ up to canonical isomorphism, I will mostly abuse the definition and identify the given morphism of G-toposes with μ. The set $\mathrm{Homtop}_G(\mathfrak{E}, \mathfrak{F})$ of morphisms of G-toposes from $\mathfrak{E}$ to $\mathfrak{F}$ is the set of objects of the category $\underline{\mathrm{Homtop}}_G(\mathfrak{E}, \mathfrak{F})$, which by definition is a full subcategory of $\underline{\mathrm{Hom}}_{BG}(\mathfrak{F}, \mathfrak{E})^\circ$.

(10.4.1) Assume that biscindages (10.3.1) have been chosen in $\mathfrak{E}$ and $\mathfrak{F}$. Then the last definition means: A morphism of G-toposes from $\mathfrak{E}$ to $\mathfrak{F}$ is essentially the same as a topos morphism (m^*, m_*) from E to F, plus a family $\{b_g\}_{g \in G}$ of isomorphisms of functors

$$b_g : \ m^* g^* \ \xrightarrow{\sim} \ g^* m^* \tag{3}$$

such that

$$
\begin{array}{ccc}
m^* g^* h^* & \xrightarrow{\ m^* * c_{g,h}\ } & m^*(hg)^* \\
\Big\downarrow{\scriptstyle b_g * h^*} & & \searrow{\scriptstyle b_{hg}} \\
 & & (hg)^* m^* \\
 & \nearrow{\scriptstyle c_{g,h} * m^*} & \\
g^* m^* h^* & \xrightarrow{\ g^* * b_h\ } & g^* h^* m^*
\end{array}
\tag{4}
$$

commutes for $g, h \in G$. (The fact that μ is automatically cartesian (10.3.2) corresponds to the fact that the b_g are isomorphisms, and not just morphisms, of functors. The arrows b_f given in [SGA4 VI.7.1.6] seem to go in the wrong direction.)

(10.4.2) By adjunction the b_g induce isomorphisms $b_g' : m_* g_* \xrightarrow{\sim} g_* m_*$ which satisfy dual compatibility laws. I will write $a_g : m_* g^* \xrightarrow{\sim} g^* m_*$ for the isomorphism induced by $b_{g^{-1}}'$ and the canonical isomorphism $(g^{-1})_* \cong g^*$. The functor m_* in the fibre defines, together with the isomorphisms a_g ($g \in G$), a BG-functor $\mathfrak{E} \to \mathfrak{F}$ which will be denoted by μ'. Up to isomorphism this μ' depends only on μ. Using μ' instead of μ one could equally well identify $\underline{\mathrm{Homtop}}_G(\mathfrak{E}, \mathfrak{F})$ with the full subcategory of $\underline{\mathrm{Hom}}_{BG}(\mathfrak{E}, \mathfrak{F})$ consisting of the functors whose fibre has a left exact left adjoint. I will refer to μ as the *inverse image part* and to μ' as the *direct image part* of the given morphism of G-toposes.

(10.4.3) Later I will sometimes just say that a given topos morphism $m : E \to F$ is *G-equivariant*. Then this is to mean that there is a morphism $\mathfrak{E} \to \mathfrak{F}$ of G-toposes, which should be obvious from the context, whose fibre is m.

The concept of G-topos arises when one tries to extend the notion of a (topological) space with operators to the setting of toposes. The morphisms of G-toposes generalize the G-equivariant maps between G-spaces. Next one needs the generalization of G-sheaves on a G-space (for these see [Gr1 ch. V]). Later it will pay to have this done in a context more general than that of G-toposes.

(10.5) (Cf. [SGA4 VI.6].) Let $p: \mathfrak{F} \to BG$ be a fibered category over BG, and let F be its fibre. The category

$$\underline{\mathrm{Hom}}_{BG}(BG, \mathfrak{F}) =: \Gamma(\mathfrak{F}/BG)$$

of sections of p has the following description: The objects of $\Gamma(\mathfrak{F}/BG)$ are the pairs

$$\left(x, \{\varphi_g\}_{g \in G}\right)$$

where x is an object of F and $\{\varphi_g\}$ is a family of isomorphisms

$$\varphi_g : x \xrightarrow{\sim} g^*x \quad (g \in G)$$

in F which make the diagram

$$
\begin{array}{ccc}
x & \xrightarrow{\varphi_g} & g^*x \\
{\scriptstyle \varphi_{hg}} \downarrow & & \downarrow {\scriptstyle g^*(\varphi_h)} \\
(hg)^*x & \xleftarrow{c_{g,h}} & g^*h^*x
\end{array}
\tag{5}
$$

commutative for all $g, h \in G$. (Necessarily $\varphi_1 = \mathrm{id}$.) An arrow in $\Gamma(\mathfrak{F}/BG)$ from $(x', \{\varphi'_g\})$ to $(x, \{\varphi_g\})$ is an F-arrow $x' \to x$ which commutes with the φ'_g and φ_g. So one can alternatively think of $\Gamma(\mathfrak{F}/BG)$ as of the category of G-objects in F, where F carries the pseudo-operation of G given by $p: \mathfrak{F} \to BG$. For this reason I will also use the notation $F(G)$ for $\Gamma(\mathfrak{F}/BG)$, see (10.5.2) below.

(10.5.1) There is a canonical BG-functor

$$\bar{\pi} : \Gamma(\mathfrak{F}/BG) \times BG \longrightarrow \mathfrak{F} \tag{6}$$

which is the forgetful functor on objects. It has the following universal property: For any category C the functor

$$\underline{\mathrm{Hom}}\left(C, \Gamma(\mathfrak{F}/BG)\right) \longrightarrow \underline{\mathrm{Hom}}_{BG}(C \times BG, \mathfrak{F}) \tag{7}$$

induced by $\bar{\pi}$ is an equivalence of categories. In other words, $\Gamma(\mathfrak{F}/BG)$ is the inverse limit of $\mathfrak{F} \to BG$ in the sense of fibered categories [SGA4 VI.6.10]. This justifies the use of the following terminology:

(10.5.2) Definition. If $\mathfrak{F} \to BG$ is a fibered category over BG and F is its fibre, I use alternatively the notations

$$F(G) = \varprojlim(\mathfrak{F}/BG) = \Gamma(\mathfrak{F}/BG)$$

for the category $\underline{\mathrm{Hom}}_{BG}(BG, \mathfrak{F})$ described above, and call it the category of *G-objects in F*, or the *inverse limit* of $\mathfrak{F}$ over BG in the sense of fibered categories. I write $\bar{\pi} = \bar{\pi}_{\mathfrak{F}}\colon F(G) \times BG = \varprojlim(\mathfrak{F}/BG) \times BG \longrightarrow \mathfrak{F}$ for the universal BG-functor (6).

(10.5.3) The *direct* limit $\varinjlim(\mathfrak{F}/BG)$ in the sense of fibered categories [SGA4 VI.6.3] is uninteresting since it is just $\mathfrak{F}$ itself, i.e. is given by the natural BG-functor $\mathfrak{F} \to \mathfrak{F} \times BG$. This is completely trivial.

(10.5.4) Remark. Consider again a fibered category $\mathfrak{F} \to BG$ and the BG-functor $\bar{\pi}\colon F(G) \times BG \to \mathfrak{F}$ of (10.5.1). I write π^* (or π_F^*, or $\pi_{\mathfrak{F}}^*$) for the fibre $F(G) \to F$ of $\bar{\pi}$, which is just the forgetful functor $(x, \{\varphi_g\}) \mapsto x$. If products (resp. coproducts) indexed by G exist in F then π^* will have a right adjoint π_*, also written π_{F*} or $\pi_{\mathfrak{F}*}$ (resp. a left adjoint $\pi_!$, also written $\ldots$): The underlying F-objects of $\pi_* x$ resp. $\pi_! x$ are

$$\pi^*\pi_* x = \prod_{g \in G} g^* x \quad \text{resp.} \quad \pi^*\pi_! x = \coprod_{g \in G} g^* x,$$

and on these objects there are natural G-structures which make them into G-objects and thus describe the functors π_* and $\pi_!$. Observe that there are natural isomorphisms of functors

$$\beta_g \colon \pi^* \xrightarrow{\sim} g^*\pi^* \quad (g \in G) \tag{8}$$

which are made explicit using the φ_g. Assuming that π_* exists one gets from $\beta_{g^{-1}}$ an isomorphism

$$\alpha_g \colon \pi_* g^* \xrightarrow{\sim} \pi_* \quad (g \in G). \tag{9}$$

by adjunction, using that g^* is right adjoint (actually quasi-inverse) to $(g^{-1})^*$.

(10.5.5) The inverse limit $\varprojlim(\mathfrak{F}/BG) = F(G)$ is functorial in $\mathfrak{F}$. That is, if $f\colon \mathfrak{F}' \to \mathfrak{F}$ is a BG-functor between fibered categories over BG, there is a natural functor

$$\varprojlim(f/BG) = f(G)\colon F'(G) \longrightarrow F(G),$$

and the diagram of fibered categories over BG

$$
\begin{array}{ccc}
F'(G) \times BG & \xrightarrow{\bar{\pi}_{\mathfrak{F}'}} & \mathfrak{F}' \\
{\scriptstyle f(G) \times 1} \downarrow & & \downarrow {\scriptstyle f} \\
F(G) \times BG & \xrightarrow{\bar{\pi}_{\mathfrak{F}}} & \mathfrak{F}
\end{array}
$$

commutes.

(10.6.1) Now assume that the fibered category is actually a G-topos $\mathfrak{E} \to BG$, with fibre E. The coarsest topology on $\mathfrak{E}$ which makes the inclusion $\alpha_!: E \hookrightarrow \mathfrak{E}$ of the fibre continuous is called the *total topology* of $\mathfrak{E}$ [SGA4 VI.7.4]. It is not hard to see (cf. also [SGA4 VI.7.4.4]) that the category of sheaves on $\mathfrak{E}$ with respect to the total topology is just $E(G)$ (10.5.2), and that the composition $E \xrightarrow{\alpha_!} \mathfrak{E} \xrightarrow{\epsilon} \widetilde{\mathfrak{E}} = E(G)$ is the functor $\pi_!$ described in (10.5.4). I will only use the fact that $E(G)$ is a topos.

(10.6.2) Remark. Let $\mathfrak{E} \to BG$ be a G-topos. The BG-functor $\bar{\pi}: E(G) \times BG \to \mathfrak{E}$ of (10.5.1) is the inverse image part of a morphism of G-toposes from $\mathfrak{E}$ to $E(G) \times BG$, cf. (10.4), where $E(G) \times BG$ is a G-topos through the second projection. The fibre of this morphism of G-toposes is the topos morphism $\pi = \pi_E = (\pi^*, \pi_*)$ from E to $E(G)$ discussed in Remark (10.5.4). The *topos* $E(G)$ depends functorially on the G-topos $\mathfrak{E}$. That is, if a morphism of G-toposes from $\mathfrak{E}$ to $\mathfrak{F}$ is given, and if $m = (m^*, m_*)$ is its fibre, then m induces a topos morphism from $E(G)$ to $F(G)$, denoted by $m(G)$, whose inverse image component satisfies

$$m(G)^*: \quad \left(y, \{y \xrightarrow{\psi_g} g^*y\}\right) \quad \longmapsto \quad \left(m^*y, \{m^*y \xrightarrow{m^*(\psi_g)} m^*g^*y \xrightarrow{b_g} g^*m^*y\}\right);$$

and a similar formula holds for $m(G)_*$ which uses the isomorphisms a_g of (10.4.2). The fact that $(m(G)^*, m(G)_*)$ is a topos morphism, which here is easy to check directly, is proved in a much more general situation in [SGA4 VI.7.4.10]. Observe that the diagram

$$
\begin{array}{ccc}
E & \xrightarrow{\quad m \quad} & F \\
{\scriptstyle \pi_E}\downarrow & & \downarrow{\scriptstyle \pi_F} \\
E(G) & \xrightarrow{\quad m(G) \quad} & F(G)
\end{array}
\qquad (10)
$$

of topos morphisms commutes.

(10.6.3) Example. If W is any topos one can form the G-topos $W \times BG$, i.e. the fibered topos $\mathrm{pr}_2: W \times BG \to BG$. This corresponds to a space with trivial G-operation. Of course the fibre of $W \times BG$ is W, and $g^* = g_* = \mathrm{id}$ for every $g \in G$. Such a G-topos will be called *trivial*. The topos $W(G)$ of G-objects on this G-topos is identified with $\underline{\mathrm{Hom}}(BG, W)$, the category of all objects w of W with an operation of G on w (from the left); or in other words, $W(G)$ is what is called in [SGA4 IV.2.4] the classifying topos B_G of the "constant" Group G in W.

In addition to the topos morphism $\pi = \pi_W = (\pi^*, \pi_*)$ from W to $W(G)$ (10.6.2) one has, in the case of a trivial G-topos, also a topos morphism $r = r_W = (r^*, r_*)$ in the *opposite* direction. Indeed, there is a triple of adjoint functors

$$
\begin{array}{c}
W(G) \\
{\scriptstyle r_!}\downarrow \quad \uparrow{\scriptstyle r^*} \quad \downarrow{\scriptstyle r_*} \\
W
\end{array}
$$

in which r^* sends $w \in W$ to w equipped with the trivial G-operation; and $r_!$ and r_* are $\varinjlim$ and $\varprojlim$, respectively. So $r_!$ sends an object w of W on which G operates to the quotient object w/G, and r_* sends it to the fixobject w^G. Sometimes I use alternatively the notations Γ_G or $\mathbf{H}_G^0$ for the functor r_*, and write $\mathbf{H}_G^n$ for the n-th right derived functor of its abelian version ($n \geq 0$). It is obvious that the composition

$$ W \xrightarrow{\pi} W(G) \xrightarrow{r} W $$

of topos morphisms is the identity, so r may be thought of as a retraction of π. (These facts were already mentioned in (6.1).)

(10.6.4) Remark. Let $\mathfrak{E}$ be a G-topos and W a topos. The morphism $\bar{\pi}$ of G-toposes from $\mathfrak{E}$ to $E(G) \times BG$ (10.6.2) induces a functor

$$ \underline{\mathrm{Homtop}}(E(G), W) \longrightarrow \underline{\mathrm{Homtop}}_G(\mathfrak{E}, W \times BG) \tag{11} $$

for which the diagram

$$ \underline{\mathrm{Homtop}}(E(G), W) \xrightarrow{(11)} \underline{\mathrm{Homtop}}_G(\mathfrak{E}, W \times BG) $$

$$ \downarrow \qquad\qquad\qquad\qquad\qquad \downarrow $$

$$ \underline{\mathrm{Hom}}(W, E(G))^\circ \longrightarrow \underline{\mathrm{Hom}}_{BG}(W \times BG, \mathfrak{E})^\circ $$

commutes (in which the vertical arrows are the canonical fully faithful functors, and the lower horizontal functor is again induced by $\bar{\pi}$). By the universal property of $E(G) = \underline{\mathrm{Lim}}(\mathfrak{E}/BG)$ for fibered categories over BG (10.5.1), the lower horizontal functor is an equivalence of categories. Since the functor $\pi^* : E(G) \to E$ preserves all limits and is conservative, one concludes from this that also (11) is an equivalence of categories. In other words, the topos $E(G)$ has to be considered as the *direct limit* of $\mathfrak{E}$ in the sense of G-toposes, or speaking loosely, as the *quotient* of E by G in the 2-categorical sense:

(10.6.5) Definition. If $\mathfrak{E} \to BG$ is a G-topos, I add to the already introduced notations $E(G) = \underline{\mathrm{Lim}}(\mathfrak{E}/BG) = \Gamma(\mathfrak{E}/BG)$ for the topos $\Gamma(\mathfrak{E}/BG)$ the new ones $\underline{\mathrm{Limtop}}(\mathfrak{E}/G) = \underline{\mathrm{Limtop}}(E/G)$. This topos, together with the morphism

$$ \bar{\pi} : \mathfrak{E} \longrightarrow \underline{\mathrm{Limtop}}(\mathfrak{E}/G) \times BG = E(G) \times BG \tag{12} $$

of G-toposes (10.6.2), is called the *direct limit* of the G-topos $\mathfrak{E}$ (in the sense of G-toposes).

(10.6.6) As pointed out in (10.6.2), this direct limit is functorial in the sense that every morphism f of G-toposes from $\mathfrak{E}$ to $\mathfrak{F}$ induces a commutative diagram

$$ \mathfrak{E} \xrightarrow{f} \mathfrak{F} $$

$$ \pi_\mathfrak{E} \downarrow \qquad\qquad\qquad\qquad \downarrow \pi_\mathfrak{F} \tag{13} $$

$$ \underline{\mathrm{Limtop}}(\mathfrak{E}/G) \times BG \longrightarrow \underline{\mathrm{Limtop}}(\mathfrak{F}/G) \times BG $$

of morphisms of G-toposes whose fibre is the diagram (10).

(10.6.7) If W is a topos and $f : E \to W$ is a G-invariant topos morphism (i.e., is the fibre of a morphism $\mathfrak{E} \to W \times BG$ of G-toposes), the unique factorization $\bar{f} : E(G) \to W$ of f through the direct limit $\pi : E \to E(G)$ is easily made explicit. Indeed, by functoriality (10.6.6) one has the commutative square of topos morphisms (solid arrows)

$$
\begin{array}{ccc}
E & \xrightarrow{\ f\ } & W \\
\pi \downarrow & & q \downarrow \uparrow r \\
E(G) & \xrightarrow{f(G)} & W(G)
\end{array}
$$

in which $q := \pi_W$ and $r := r_W$. Since $r \circ q = \mathrm{id}_W$ it is clear that $\bar{f} = r \circ f(G)$.

(10.7) Remark. Consider again a G-topos $\mathfrak{E}$ with fibre E, and the canonical morphism $\bar{\pi} = \bar{\pi}_{\mathfrak{E}}$ of G-toposes (12). By functoriality (10.6.6), applied to $f := \bar{\pi}$, there is a commutative diagram

$$
\begin{array}{ccc}
\mathfrak{E} & \xrightarrow{\quad \bar{\pi} \quad} & \underrightarrow{\mathrm{Limtop}}(\mathfrak{E}/G) \times BG \\
\pi \downarrow & & \downarrow \\
\underrightarrow{\mathrm{Limtop}}(\mathfrak{E}/G) \times BG & \longrightarrow & \big(\underrightarrow{\mathrm{Limtop}}(\mathfrak{E}/G)\big)(G) \times BG
\end{array}
\qquad (14)
$$

of morphisms of G-toposes, where

$$
\big(\underrightarrow{\mathrm{Limtop}}(\mathfrak{E}/G)\big)(G) = \underrightarrow{\mathrm{Limtop}}\Big(\big(\underrightarrow{\mathrm{Limtop}}(\mathfrak{E}/G) \times BG\big)/G\Big) = (E(G))(G)
$$

is the category of objects with G-operation in $\underrightarrow{\mathrm{Limtop}}(\mathfrak{E}/G) = E(G)$. The fibre of (14) is the diagram

$$
\begin{array}{ccc}
E & \xrightarrow{\ \pi\ } & E(G) \\
\pi \downarrow & & q \downarrow \\
E(G) & \xrightarrow{\pi(G)} & (E(G))(G)
\end{array}
\qquad (15)
$$

of topos morphisms, in which $\pi := \pi_E$ and $q := \pi_{E(G)}$. Note that $E(G)$ is here considered as (the fibre of) a trivial G-topos. One has to be careful not to mix up the two different sorts of G-operation when working with $(E(G))(G)$!

(10.7.1) For an object of $(E(G))(G)$ with underlying $E(G)$-object z let $\mathrm{out}_g : z \to z$ be the "outer" operation by $g \in G$. So q^* is the functor which forgets this outer operation, while $\pi(G)^*$ is

$$
\pi(G)^* \big(z, \{\mathrm{out}_g\}\big) = \Big(\pi^* z,\ \pi^* z \xrightarrow{\pi^*(\mathrm{out}_g)} \pi^* z \xrightarrow{\beta_g} g^* \pi^* z\Big).
\qquad (16)
$$

So q and $\pi(G)$ are distinct. Recall (10.6.3) that q admits a retraction $r = r_{E(G)}$, a topos morphism from $(E(G))(G)$ to $E(G)$. I claim that also $r \circ \pi(G)$ is the identity of $E(G)$: Indeed, r^* puts a trivial outer operation on $z \in E(G)$, and so (16) shows that $\pi(G)^* \circ r^* = \mathrm{id}$. This has the following consequence:

(10.7.2) Proposition. — *The group G operates on the functor $\pi_*\pi^*\colon E(G) \to E(G)$, and the adjunction $\mathrm{id} \to \pi_*\pi^*$ identifies an object $z \in E(G)$ with the fixobject (= inverse limit) of this operation on $\pi_*\pi^*z$:*

$$z = (\pi_*\pi^*z)^G \quad \text{for } z \in E(G).$$

Proof. One checks immediately that $q^*\pi(G)_* = \pi_*\pi^*$, so $\pi(G)_*z$ is a G-operation on $\pi_*\pi^*z$, for $z \in E(G)$; and so $z = r_*\pi(G)_*z = \varprojlim \pi(G)_*z = (\pi_*\pi^*z)^G$. $\quad\square$

(10.7.3) I add the remark that for $g \in G$ the action of g on $\pi_*\pi^*$ is given by the composition (cf. (10.5.4))

$$\pi_*\pi^* \xrightarrow{\;\pi_**\beta_g\;} \pi_*g^*\pi^* \xrightarrow{\;\alpha_g*\pi^*\;} \pi_*\pi^*.$$

(10.8) Remark. The following interpretation of $(E(G))(G)$ reveals more clearly what is going on. First some generalities are needed about the change of group in fibered toposes. Let $\mathfrak{E} \to BG$ be a G-topos with fibre E, and let $\alpha\colon H \to G$ be a group homomorphism. Then one may form the base extension $\alpha^*\mathfrak{E}$ of $\mathfrak{E}$ with respect to α [SGA4 VI.7.1.9]. So $\alpha^*\mathfrak{E} \to BH$ is the H-topos whose fibre is E and on which H pseudo-acts through α and the pseudo-action of G. Write $E(G) = \varinjlim\mathrm{top}(\mathfrak{E}/G)$, as usual, and write $E(H) := E(H,\alpha) := \varinjlim\mathrm{top}(\alpha^*\mathfrak{E}/H)$. There is an obvious functor $\alpha^*\colon E(G) \to E(H)$ induced by α:

$$\alpha^*\colon \quad \left(x, \{x \xrightarrow{\varphi_g} g^*x\}_{g\in G}\right) \longmapsto \left(x, \{x \xrightarrow{\varphi_{\alpha(h)}} \alpha(h)^*x\}_{h\in H}\right).$$

Moreover it is clear that α^* preserves arbitrary (direct and inverse) limits. Hence α^* has a left adjoint $\alpha_!$ and a right adjoint α_*; and in particular, (α^*,α_*) is a topos morphism from $E(H)$ to $E(G)$ which will again be written α. If $\alpha'\colon H' \to H$ is a second group homomorphism then the topos morphism $E(H') \to E(G)$ induced by the composite group homomorphism $\alpha \circ \alpha'$ is up to canonical isomorphism the same as the composite topos morphism $E(H') \xrightarrow{\alpha'} E(H) \xrightarrow{\alpha} E(G)$.

Examples of this construction have been encountered before: The inclusion $\{1\} \to G$ gives, for every G-topos E, the topos morphism $\pi\colon E \to E(G)$ (10.6.2); and $G \to \{1\}$ gives, for every *trivial* G-topos W, the topos morphism $r\colon W(G) \to W$ (10.6.3). The fact that r is a retraction of π corresponds to the fact that the composite group homomorphism $\{1\} \to G \to \{1\}$ is the identity.

In the case of a general group homomorphism $\alpha\colon H \to G$, the construction of the direct image functor $\alpha_*\colon E(H) \to E(G)$ can be reduced to two special cases which resemble π and r, respectively: If α is injective then α_* can be expressed by choosing a transversal $T \subset G$ of $G \to H\backslash G = \{Hg\colon g \in G\}$; one has

$$\alpha_*\left(y, \{y \xrightarrow{\psi_h} h^*y\}_{h\in H}\right) = \left(\prod_{t\in T} t^*y, \text{ can. } G\text{-str.}\right)$$

where "can. G-str." denotes a canonical G-structure on the product which I do not make explicit. If on the other hand α is surjective with kernel N then one has

$$\alpha_*\left(y, \{y \xrightarrow{\psi_h} h^*y\}_{h \in H}\right) = \left(y^N, \text{ can. } G\text{-str.}\right)$$

where y^N is the $\varprojlim$ of the N-operation on y.

Now return to a G-topos $\mathfrak{E}$ as above, and consider the diagram of group homomorphisms

$$G \overset{i,d}{\underset{}{\rightrightarrows}} G \times G \overset{\text{pr}_1}{\longrightarrow} G \tag{17}$$

where $i(g) = (g,1)$, $d(g) = (g,g)$ and $\text{pr}_1(g,g') = g$. Pulling back $\mathfrak{E}$ to $G \times G$ via pr_1 makes E a $G \times G$-topos where the second component of $G \times G$ acts trivially. So $E(G \times G)$ is canonically the same as $(E(G))(G)$, the first (resp. second) component of $G \times G$ corresponding to the inner (resp. outer) G in $(E(G))(G)$. And the topos morphism $\text{pr}_1 : E(G \times G) \to E(G)$ is just $r = r_{E(G)}$.

Since $\text{pr}_1 \circ i = \text{pr}_1 \circ d = \text{id}$ in (17), one gets back the G-topos $\mathfrak{E}$ when one pulls back $\text{pr}_1^* \mathfrak{E}$ further along either i or d. This second pullback induces therefore topos morphisms

$$i \colon E(G) \longrightarrow E(G \times G) \quad \text{and} \quad d \colon E(G) \longrightarrow E(G \times G).$$

These latter are nothing else but $q = \pi_{E(G)}$ and $\pi(G)$, in the notation of (10.7). From this point of view it is of course immediate that $r \circ q = r \circ \pi(G) = \text{id}_{E(G)}$.

(10.9) Remark. An abelian group object of the topos $E(G)$ is the same as a G-object $B = (A, \{A \xrightarrow{\varphi_g} g^*A\})$ in E for which $A = \pi^*B$ is an abelian group object of E and the φ_g are group isomorphisms. The cohomology $H^*(E(G), B)$ is often called *equivariant cohomology* of B (or of A, if the G-structure is clear) and is frequently denoted by $H_G^*(E, B)$. One has a natural ("Hochschild-Serre") spectral sequence

$$E_2^{pq} = H^p(G, H^q(E, A)) \implies H_G^{p+q}(E, B) \tag{18}$$

for this cohomology, which can be derived as follows: Since $B \xrightarrow{\sim} (\pi_* \pi^* B)^G$ for $B \in \text{Ab}(E(G))$ by (10.7.2), the global sections functor $H^0(E(G), -) : \text{Ab}(E(G)) \to (\text{Ab})$ can be decomposed as

$$H^0(E(G), B) = \left(H^0(E(G), \pi_* \pi^* B)\right)^G = H^0(E, \pi^* B)^G = H^0(E, A)^G.$$

Since π^* has an exact left adjoint, one deduces (18).

It may be the time now for some examples:

(10.10) Example. Let G be a group acting on a topological space T. This gives in a natural way rise to a G-topos $\mathfrak{T}$ with fibre $\widetilde{T}$. The topos $\widetilde{T}(G)$, i.e. the direct limit of the G-topos $\mathfrak{T}$, is the category of G-sheaves on T as studied by Grothendieck in [Gr1, ch. V]. Let T/G be the topological quotient space of the group action. If G acts freely and discontinuously — this means that every point in T admits a neighborhood V such that $V \cap gV = \emptyset$ for every $g \neq 1$ —, then the natural topos morphism from $\widetilde{T}(G)$ to $(T/G)^{\sim}$ is an equivalence of categories (8.1.4). However this fails if G has fixpoints. In the general case the topos $\widetilde{T}(G)$ can be considered as the sheaf theoretic equivalent of the space $EG \times_G T$. The latter is considered by topologists, e.g. for the study of equivariant cohomology (see [Bd2, VII.1] or [tD III.1], for example). By definition, $EG \times_G T$ is the quotient space of $EG \times T$ by the diagonal G-action, where EG is a contractible free G-complex. The homotopy type of $EG \times T$ is that of T. But on $EG \times T$ the G-action has become free, and so $EG \times_G T$ is sort of a "free-made" quotient of T by G. Note that the projection $EG \times_G T \to BG$ to the first component (with $BG := EG/G$ here) is a fibre bundle with fibre T, so the analogy with the G-topos $\mathfrak{T} \to BG$ is quite apparent. This analogy is manifest in the fact that, under suitable hypotheses on T, the cohomology of $EG \times_G T$ and of $\widetilde{T}(G)$ with constant coefficients agree (see (13.6)). The canonical topos morphism $\pi : \widetilde{T} \to \widetilde{T}(G)$ corresponds to the inclusion $T \hookrightarrow EG \times_G T$ (which is natural after choice of a base point of EG). The projection $EG \times_G T \to T/G$ to the second factor gives, if G operates trivially, a retraction $EG \times_G T = BG \times T \to T$. Of course this parallels the retraction r of a trivial G-topos (10.6.3). In the topological case it is more natural to think of r as a projection and of π as a section of r.

In Sect. 14 a sheaf theoretic construction will be explained which corresponds to the *topological* quotient space T/G, rather than to the free-made quotient $EG \times_G T$.

(10.11) Example. This example and the next review the basic setup for this work from the point of view of G-toposes. Let $f : Y \to X$ be a morphism of schemes, and let G be a group acting on Y (from the left) by automorphisms over X. As in (4.11.1) G acts on Et/Y by $(g, V) \mapsto {}^g V = V_{g^{-1}}$. For $g \in G$ the topos morphism $g = (g^*, g_*) : \widetilde{Y}_{et} \to \widetilde{Y}_{et}$ satisfies $(g^* B)(V) = B({}^g V)$ and $(g_* B)(V) = B(V_g)$ ($B \in \widetilde{Y}_{et}$, $V \in \mathrm{Et}/Y$), hence also $(g^{-1})_* = g^*$. So $\widetilde{Y}_{et}$ together with the (g^*, g_*) is a G-topos in which even strict equalities $g^* h^* = (hg)^*$ hold. To simplify the language I will just speak of the G-topos $\widetilde{Y}_{et}$, instead of using awkward constructions like "the G-topos associated with $\widetilde{Y}_{et}$". The topos morphism $f = (f^*, f_*)$ from $\widetilde{Y}_{et}$ to $\widetilde{X}_{et}$ is "G-invariant", i.e. it is the fibre of a morphism of G-toposes from $\widetilde{Y}_{et}$ to $\widetilde{X}_{et} \times BG$ (and consequently it factors, in an essentially unique way, through $\pi : \widetilde{Y}_{et} \to \widetilde{Y}_{et}(G)$: (10.6.4)). This is seen as follows: For any X-scheme U there is a canonical Y-isomorphism $U_Y \xrightarrow{\sim} {}^g(U_Y)$, given by $1 \times g^{-1}$ in the commutative diagram

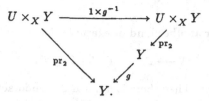

These isomorphisms induce isomorphisms for $B \in \widetilde{Y}_{et}$

$$\bar{\alpha}_g: \left(f_*g^*B\right)(U) = B\left({}^g(U_Y)\right) \overset{\sim}{\longrightarrow} B(U_Y) = (f_*B)(U)$$

$(U \in \mathrm{Et}/X)$, and hence isomorphisms $\bar{\alpha}_g: f_*g^* \overset{\sim}{\to} f_*$ of functors $\widetilde{Y}_{et} \to \widetilde{X}_{et}$. These $\bar{\alpha}_g$ satisfy the rule $\bar{\alpha}_{gh} = \bar{\alpha}_g \circ (\bar{\alpha}_h * g^*)$ $(g, h \in G)$. By adjunction the $\bar{\alpha}_{g^{-1}}$ give isomorphisms $\bar{\beta}_g: f^* \overset{\sim}{\to} g^*f^*$ which satisfy the identities (4) of Remark (10.4.1).

Explicitly, the factorization of f as $\widetilde{Y}_{et} \overset{\pi}{\longrightarrow} \widetilde{Y}_{et}(G) \overset{\bar{f}}{\longrightarrow} \widetilde{X}_{et}$ is given by $\bar{f}^*A = (f^*A, \{\bar{\beta}_g\})$ and $\bar{f}_*(B, \{\varphi_g\}) = \Gamma_G(f_*B)$, where the operation of $g \in G$ on f_*B is by

$$f_*B \overset{f_*(\varphi_g)}{\longrightarrow} f_*g^*B \overset{\bar{\alpha}_g}{\longrightarrow} f_*B$$

(see (10.6.7)). In other words, $\bar{f}_*$ is "the G-invariants" of the operation of G on $f_*\pi^*$ given by

$$f_*\pi^* \overset{f_**\beta_g}{\longrightarrow} f_*g^*\pi^* \overset{\bar{\alpha}_g*\pi^*}{\longrightarrow} f_*\pi^* \quad (g \in G).$$

(10.12) Example. Keeping the notations of the last example, assume now that $f: Y \to X$ is a finite Galois covering and that G is its Galois group. Then $\bar{f}^*$ and $\bar{f}_*$ are quasi-inverses of each other, so the toposes $\widetilde{Y}_{et}(G)$ and $\widetilde{X}_{et}$ are canonically equivalent. Indeed, from the definition of a Galois covering it follows that the category $\widetilde{Y}_{et}(G)$ of étale G-sheaves on Y is isomorphic to the category of étale sheaves on Y together with descent data with respect to f; and the canonical functor from $\widetilde{X}_{et}$ to the latter category (which is essentially the $\bar{f}^*$ of (10.11)) is an equivalence, cf. [SGA4 VIII.9.4].

I will change notation now and assume that the name of the Galois covering $Y \to X$ is π instead of f. This fits well with the former general use of π for the topos morphism $E \to E(G)$, since it was just remarked that $\widetilde{Y}_{et} \to \widetilde{X}_{et}$ can be identified with $\widetilde{Y}_{et} \to \widetilde{Y}_{et}(G)$. In accordance with the general notations, let $\widetilde{X}_{et}(G)$ denote the topos whose objects are étale sheaves on X together with an operation of G on them. Remark (10.7) shows that π induces a topos morphism $\pi(G): \widetilde{X}_{et} \to \widetilde{X}_{et}(G)$ such that, for $A \in \widetilde{X}_{et}$, the underlying sheaf of $\pi(G)_*A$ is $\pi_*\pi^*A$, and $A = \Gamma_G(\pi_*\pi^*A)$. Of course this operation of G on $\pi_*\pi^*A$ is easily made explicit: For $U \in \mathrm{Et}/X$ the operation of g on $(\pi_*\pi^*A)(U) = A(U \times_X Y)$ comes from the operation of g^{-1} on $U \times_X Y$ via the second component. And the fact that $A(U) \to (\pi_*\pi^*A)(U)^G$ is bijective comes from applying the sheaf condition to the covering $U \times_X Y \to U$. In this way one recovers the right half of diagram (3) in (6.4.1) as a particular case of diagram (15) in (10.7).

(10.13) Example. For another kind of examples let

$$1 \longrightarrow \Delta \longrightarrow \Gamma \overset{p}{\longrightarrow} G \longrightarrow 1 \tag{19}$$

be an extension of groups. Then there is a natural pseudo-action of G on the topos $(\Delta\text{-sets})$ which makes $(\Delta\text{-sets})$ a G-topos: For $g \in G$ let $g^*\colon (\Delta\text{-sets}) \to (\Delta\text{-sets})$ be the functor given by

$$g^*N = \mathrm{Hom}_\Delta\big(C(g), N\big). \tag{20}$$

Explanation: $C(g) := p^{-1}(g)$ is g, considered as a Δ-coset in Γ. This is a (left) Δ-set itself (non-canonically isomorphic to Δ), so one can form the set $\mathrm{Hom}_\Delta\big(C(g), N\big)$. This latter set becomes a Δ-set through the *right* action of Δ on $C(g)$, i.e. through the rule

$$(b \cdot f)(x) := f(xb)$$

for $f \in \mathrm{Hom}_\Delta\big(C(g), N\big)$, $b \in \Delta$, $x \in C(g)$. Although the Δ-set $C(g)$ is isomorphic to Δ, the Δ-set g^*N is not necessarily isomorphic to N. It is not hard to verify that (20) defines a pseudo-action of G on $(\Delta\text{-sets})$, i.e. that there are isomorphisms $c_{g,h}\colon g^*h^* \overset{\sim}{\to} (hg)^*$ satisfying the cocycle conditions (2). Let $\mathfrak{D}$ denote the G-topos with fibre $(\Delta\text{-sets})$ which is defined in this way. Let

$$\pi\colon (\Delta\text{-sets}) \longrightarrow (\Gamma\text{-sets})$$

be the natural topos morphism, see (0.5.3). The topos morphism π is G-invariant, i.e. it defines a morphism of G-toposes

$$\bar{\pi}\colon \mathfrak{D} \longrightarrow (\Gamma\text{-sets}) \times BG.$$

(This is reflected by the fact that, if M is a Γ-set, the Δ-set $g^*\pi^*M = \mathrm{Hom}_\Delta\big(C(g), M\big)$ is *canonically* isomorphic to the Δ-set π^*M; namely via the map $m \mapsto f_m$, $M \to \mathrm{Hom}_\Delta\big(C(g), M\big)$ where $f_m(x) = xm$ ($x \in C(g)$, $m \in M$).) I claim that π resp. $\bar{\pi}$ identifies $(\Gamma\text{-sets})$ with $\varinjlim\mathrm{top}(\mathfrak{D}/G)$, i.e. with the topos of G-objects in $(\Delta\text{-sets})$. Another way of expressing this fact is to say that to give a Γ-set is canonically the same as to give a Δ-set N plus isomorphisms $\varphi_g\colon N \overset{\sim}{\to} g^*N$ of Δ-sets ($g \in N$) which satisfy the cocycle conditions (5). Of course it is straightforward how to attack the proof this assertion, but the verification that this works is lengthy and will hence by omitted. The spectral sequence (18) of Remark (10.9) becomes here the usual Hochschild-Serre sequence for the group extension (19).

Note that the G-topos $\mathfrak{D}$ of Δ-sets is trivial in the sense of (10.6.3) if and only if the extension Γ splits as a *direct* product $\Gamma \cong \Delta \times G$.

All of the above works equally well if Γ and Δ are profinite groups and Δ is open in Γ (so that G is discrete).

(10.14) In the remainder of this section inverse limits of G-toposes will be studied. Since here and in Sect. 11 some fiddling with induced and coinduced topologies is required, it will be convenient to have shorthand notations for them.

(10.14.1) Definition. Let $f: C \to D$ be a functor between categories.

a) If σ, σ' are topologies on C then $\sigma \subset \sigma'$ means that σ' is finer than σ.

b) If σ is a topology on C then $\mathrm{coind}_f(\sigma)$ denotes the topology *coinduced* by σ on D, i.e. the coarsest topology on D which makes f continuous [SGA4 III.3.6].

c) If τ is a topology on D then $\mathrm{ind}_f(\tau)$ or $\tau|_C$ denotes the topology *induced* by τ on C, i.e. the finest topology on C which makes f continuous [SGA4 III.3.1].

d) Write $\mathrm{coind}(f) := \mathrm{coind}_f(\mathrm{can}_C)$ and $\mathrm{ind}(f) := \mathrm{ind}_f(\mathrm{can}_D)$, where "can" denotes the canonical topology on a category.

Thus, if σ is a topology on C and τ is one on D, one has

$$f: C \to D \text{ continuous} \iff \mathrm{coind}_f(\sigma) \subset \tau \iff \sigma \subset \mathrm{ind}_f(\tau).$$

The inclusions $\sigma \subset \mathrm{ind}_f \mathrm{coind}_f(\sigma)$ and $\mathrm{coind}_f \mathrm{ind}_f(\tau) \subset \tau$ hold trivially.

The following "transitivity lemma" will be needed later. I do not know whether the hypotheses made about fibre products are really essential. Fortunately they will hold in all later applications.

(10.14.2) Lemma. — *Let $C \xrightarrow{f} D \xrightarrow{g} E$ be functors between categories, let σ be a topology on C and τ a topology on E. Assume that all fibre products exist in C and in D, and that f and g preserve them. Then*

a) $\mathrm{ind}_f \mathrm{ind}_g(\tau) = \mathrm{ind}_{g \circ f}(\tau)$ *and* $\mathrm{coind}_g \mathrm{coind}_f(\sigma) = \mathrm{coind}_{g \circ f}(\sigma)$;

b) *f is continuous with respect to $(\sigma, \mathrm{ind}_g(\tau))$ if and only if $g \circ f$ is continuous with respect to (σ, τ), if and only if g is continuous with respect to $(\mathrm{coind}_f(\sigma), \tau)$.*

Proof. a) The inclusion $\mathrm{ind}_f \mathrm{ind}_g(\tau) \subset \mathrm{ind}_{g \circ f}(\tau)$ is trivial. Conversely let $\{x_i \to x\}$ be a family in C which is a covering for $\mathrm{ind}_{g \circ f}(\tau)$. Then $\{gfx_i \to gfx\}$ is a covering for τ in E, and so, by the hypotheses on D and g, $\{fx_i \to fx\}$ is a covering for $\mathrm{ind}_g(\tau)$ [SGA4 III.3.3]. By the hypotheses on C and f it follows that f is continuous with respect to $(\mathrm{ind}_{gf}(\tau), \mathrm{ind}_g(\tau))$ [SGA4 III.1.6], whence $\mathrm{ind}_{gf}(\tau) \subset \mathrm{ind}_f \mathrm{ind}_g(\tau)$. Next I show b). Using the just proved part of a) one sees that

$$gf \text{ continuous w.r.t. } (\sigma, \tau) \iff \sigma \subset \mathrm{ind}_{gf}(\tau) \underset{a)}{\iff} \mathrm{coind}_f(\sigma) \subset \mathrm{ind}_g(\tau),$$

from which b) follows. To prove the second equality in a) one has to check that g is continuous with respect to $(\mathrm{coind}_f(\sigma), \mathrm{coind}_{gf}(\sigma))$, which is obvious from b). $\square$

Now inverse limits of G-toposes are defined by the obvious universal property. The question of their existence, however, turns out to be more delicate than for direct limits.

(10.15) Definition. Let $\mathfrak{E} \to BG$ be a G-topos. An *inverse limit* of $\mathfrak{E}$ is a pair (F, f) where F is a topos and $f\colon F \times BG \to \mathfrak{E}$ is a morphism of G-toposes, such that for any topos W the functor

$$\underline{\mathrm{Homtop}}(W, F) \longrightarrow \underline{\mathrm{Homtop}}_G(W \times BG, \mathfrak{E})$$

induced by f is an equivalence of categories. If such an inverse limit F exists it will be denoted by $\underleftarrow{\mathrm{Limtop}}(\mathfrak{E}/G)$, or by $\underleftarrow{\mathrm{Limtop}}(E/G)$ where E is the fibre of $\mathfrak{E}$. I will also call F the *fixtopos* of the G-topos $\mathfrak{E}$, cf. (10.15.1).

(10.15.1) This copies from [SGA4 VI.8.1] the definition of the inverse limit of a fibered topos, specialized to the case where the base category is BG. If (F, f) is an inverse limit of $\mathfrak{E}$ as in the definition then the fibre $F \to E$ of f is a topos morphism which is universal (in the 2-categorical sense) for those topos morphisms $h\colon W \to E$ for which all compositions $g \circ h$ ($g \in G$) are "coherently" isomorphic. Thus $\underleftarrow{\mathrm{Limtop}}(E/G) \to E$ should be considered as the analogue of the subspace of fixpoints in a topological G-space.

It is obvious that every morphism $\mathfrak{E}' \to \mathfrak{E}$ of G-toposes induces a topos morphism from $F' = \underleftarrow{\mathrm{Limtop}}(\mathfrak{E}'/G)$ to $F = \underleftarrow{\mathrm{Limtop}}(\mathfrak{E}/G)$, which is essentially unique, such that the diagram

$$
\begin{array}{ccc}
F' \times BG & \longrightarrow & F \times BG \\
\downarrow & & \downarrow \\
\mathfrak{E}' & \longrightarrow & \mathfrak{E}
\end{array}
$$

of morphisms of G-toposes commutes (assuming that both inverse limit toposes exist).

(10.15.2) Remark. Let $\mathfrak{E}$ be a G-topos. It was remarked before that the natural BG-functor $\bar{\pi}\colon E(G) \times BG \to \mathfrak{E}$ identifies $E(G)$ with $\underleftarrow{\mathrm{Lim}}(\mathfrak{E}/BG)$ (10.5.1) and also with $\underrightarrow{\mathrm{Limtop}}(\mathfrak{E}/G)$ (10.6.4): The inverse limit of $\mathfrak{E}$ in the sense of fibered categories "is" the direct limit of $\mathfrak{E}$ in the sense of fibered toposes.

This becomes false when the opposite limits are considered. Indeed, the direct limit in the sense of fibered categories is $\underrightarrow{\mathrm{Lim}}(\mathfrak{E}/BG) = \mathfrak{E}$ (10.5.3); but the category $\mathfrak{E}$ is not even a topos unless $G = \{1\}$. For example, $\mathfrak{E}$ does not have a final object.

The following theorem shows that the inverse limit of a G-topos exists whenever the group G is finite. Moreover this limit is constructed explicitly:

(10.16) Theorem. — *Let G be a finite group, and let $\mathfrak{E} \to BG$ be a G-topos with fibre E. Then the inverse limit topos $\underleftarrow{\mathrm{Limtop}}(\mathfrak{E}/G)$ exists and can be constructed as follows. Let $\tau := \mathrm{coind}(\pi_*)$, so τ is the coarsest topology on $E(G)$ which makes the direct image functor $\pi_*\colon E \to E(G)$ continuous (E carrying its canonical topology). Let F be the topos of sheaves on the site $\big(E(G), \tau\big)$, and let ν^* be the composite functor*

$$\nu^*\colon E \xrightarrow{\ \pi_*\ } E(G) \xrightarrow{\ \epsilon_\tau\ } F.$$

Then ν^* is the inverse image functor of a topos morphism $\nu: F \to E$ which identifies F with $\varprojlim \mathrm{top}(\mathfrak{E}/G)$.

(10.17) Remark. When one compares this with the existence theorem for the inverse limit of a fibered topos over a *left filtering* base category [SGA4 VI.8.2.3], one is at a first glance tempted to think that the constructions are completely analogous. But actually they are basically different. To see why, let us first resume Grothendieck's construction in a few words.

Let I be a base category which is left filtering, and consider a fibered topos $p: \mathfrak{E} \to I$ over I. One is looking for an inverse limit of $\mathfrak{E}$, which means a topos F together with a cartesian morphism of fibered toposes over I from $F \times I$ to $\mathfrak{E}$ which is universal for such pairs, in a sense similar to the above definition. The construction in [loc.cit.] starts with taking the direct limit $\underline{\mathfrak{E}}$ of the fibered category $\mathfrak{E} \to I$, i.e. the localization $\underline{\mathfrak{E}}$ of $\mathfrak{E}$ with respect to the class of all cartesian arrows in $\mathfrak{E}$. Then one considers the coarsest topology on $\underline{\mathfrak{E}}$ which makes the canonical functor $\pi: \mathfrak{E} \to \underline{\mathfrak{E}}$ continuous, where $\mathfrak{E}$ carries the total topology. Let W be a topos and let $m: \mathfrak{E} \to W \times I$ be a cartesian functor over I. By the universal property of $\underline{\mathfrak{E}}$ there is a functor $\psi: \underline{\mathfrak{E}} \to W$ (which is essentially unique) such that m is isomorphic to

$$\mathfrak{E} \xrightarrow{(\pi,p)} \underline{\mathfrak{E}} \times I \xrightarrow{\psi \times 1} W \times I.$$

Now one shows (essentially this is done in [loc.cit., Lemma 8.2.2]): m is the inverse image part of a morphism of fibered toposes from $W \times I$ to $\mathfrak{E}$ if, and only if, ψ is a morphism of sites from W to $\underline{\mathfrak{E}}$. From this it is readily seen that $\widetilde{\underline{\mathfrak{E}}}$, together with $(\epsilon \circ \pi, p): \mathfrak{E} \to \widetilde{\underline{\mathfrak{E}}} \times I$, is the desired inverse limit.

What goes wrong if one drops the condition that I be left filtering is that the canonical I-functor $(\pi,p): \mathfrak{E} \to \underline{\mathfrak{E}} \times I$ will, in general, no longer be a morphism of fibered sites over I: Although for every $i \in I$ the composite functor $\mathfrak{E}_i \xrightarrow{\alpha_{i!}} \mathfrak{E} \xrightarrow{\pi} \underline{\mathfrak{E}}$ is continuous, by the very definition of the topology on $\underline{\mathfrak{E}}$, it may not be a morphism of sites. ($\alpha_{i!}: \mathfrak{E}_i \hookrightarrow \mathfrak{E}$ denotes the inclusion of the fibre over i.) Exactly this happens in the case $I = BG$. Here $\underline{\mathfrak{E}} = \mathfrak{E}$ (10.5.3), so $\widetilde{\underline{\mathfrak{E}}} = E(G)$, and

$$E \xrightarrow{\alpha_!} \mathfrak{E} = \underline{\mathfrak{E}} \xrightarrow{\epsilon} E(G)$$

is the functor $\pi_!$ of Remark (10.5.4), see (10.6.1): It is continuous, but not left exact since $\pi_!(*)$ is not a final object in $E(G)$.

The key for getting around this difficulty is Proposition (10.18.1) below. Instead of taking the direct limit $\mathfrak{E} \to \underline{\mathfrak{E}} \times BG$ of the fibered category $\mathfrak{E}$, which is universal for *all* BG-functors $\mathfrak{E} \to C \times BG$, one considers its *inverse* limit $\bar{\pi}: E(G) \times BG \to \mathfrak{E}$ (10.5.2). Of course this BG-functor goes in the "wrong" direction. But it is (the inverse image part of) a morphism of G-toposes (10.6.2), and so one can consider its direct image part $\bar{\pi}': \mathfrak{E} \to E(G) \times BG$, compare (10.4.2). It will be shown that this $\bar{\pi}'$ is universal for BG-functors $\mathfrak{E} \to C \times BG$ whose fibre is a *left exact* functor (only categories C are considered in which finite $\varprojlim$'s exist). This uses essentially the fact that the base category is a group. Of course any functor $\mathfrak{E} \to W \times BG$ which is (the inverse image part of) a morphism of G-toposes is of this sort. Then it is natural to put on $E(G)$ the coarsest topology which makes π_* continuous. This construction will do the job.

(10.18) Recall that a functor is called *left exact* if it preserves all finite inverse limits. If C, C' are categories then $\underline{\mathrm{Hom}}^{le}(C, C')$ denotes the full subcategory of $\underline{\mathrm{Hom}}(C, C')$ consisting of all left exact functors. (Gabriel's notation would be $\underline{\mathrm{Sex}}(C, C')$, cf. [Gr2 p. 50].) If $\mathfrak{F}$, $\mathfrak{F}'$ are categories over BG let $\underline{\mathrm{Hom}}^{le}_{BG}(\mathfrak{F}, \mathfrak{F}')$ be the full subcategory of $\underline{\mathrm{Hom}}_{BG}(\mathfrak{F}, \mathfrak{F}')$ of all BG-functors whose fibre functor is left exact. (For these definitions one should assume that all finite inverse limits exist in C, resp. in the fibre of $\mathfrak{F}$.) Let now $\mathfrak{E}$ be a G-topos. Recall (10.4.2) that $\bar{\pi}'$ denotes the *direct* image part of the canonical morphism $\bar{\pi}$ of G-toposes from $\mathfrak{E}$ to $E(G) \times BG$. The fibre of $\bar{\pi}'$ is the direct image functor π_*. The essential point in the proof of the theorem is

(10.18.1) **Proposition.** — *Assume that G is finite. Let W be a category in which all finite inverse limits exist. Then the functor*

$$\Phi \colon \ \underline{\mathrm{Hom}}^{le}(E(G), W) \ \longrightarrow \ \underline{\mathrm{Hom}}^{le}_{BG}(\mathfrak{E}, W \times BG)$$

induced by the BG-functor $\bar{\pi}' \colon \mathfrak{E} \longrightarrow E(G) \times BG$ is an equivalence of categories.

Proof. Since π_*, the fibre of $\bar{\pi}'$, is a left exact functor, the functor Φ is well defined. The proposition will be proved by explicitly constructing a functor which is quasi-inverse to Φ.

The inverse limit of the fibered category $\mathrm{pr}_2 \colon W \times BG \to BG$ is the category $W(G)$ of G-objects in W. As in Example (10.6.3) there are two pairs of adjoint functors

$$
\begin{array}{ccc}
W & & W(G) \\[2pt]
\scriptstyle \pi_W^* \uparrow \ \downarrow \scriptstyle \pi_{W*} & \text{and} & \scriptstyle r_W^* \uparrow \ \downarrow \scriptstyle r_{W*} \\[2pt]
W(G) & & W
\end{array}
$$

since finite inverse limits exist in W.

Let $f \colon \mathfrak{E} \longrightarrow W \times BG$ be a functor over BG whose fibre is left exact. The functor $f(G) \colon E(G) \to W(G)$ induced by f (10.5.5) is again left exact, and hence so is the composite functor

$$\Psi f \colon \ E(G) \ \xrightarrow{\ f(G)\ } \ W(G) \ \xrightarrow{\ r_{W*}\ } \ W.$$

It is clear that Ψf depends functorially on f. Therefore a functor

$$\Psi: \underline{\mathrm{Hom}}^{le}_{BG}(\mathfrak{E}, W \times BG) \longrightarrow \underline{\mathrm{Hom}}^{le}(E(G), W)$$

has been defined. I claim that Ψ is quasi-inverse to Φ.

Let $\eta: E(G) \to W$ be a left exact functor. Then $\Phi\eta$ is the composite BG-functor

$$\mathfrak{E} \xrightarrow{\bar{\pi}'} E(G) \times BG \xrightarrow{\eta \times 1} W \times BG.$$

So $(\Psi \circ \Phi)(\eta)$ is the composition $r_{W*} \circ \eta(G) \circ \pi(G)_*$ in the diagram

$$(21)$$

The left triangle in (21) commutes (10.7.1). Since η is left exact it is clear that also the right square commutes (up to isomorphism, always). So $(\Psi \circ \Phi)(\eta) \cong \eta$, and therefore $\Psi \circ \Phi$ is isomorphic to the identity. Conversely let $f: \mathfrak{E} \to W \times BG$ be a functor over BG whose fibre f_0 is left exact. Then $(\Phi \circ \Psi)(f)$ is the composition $(r_{W*} \times 1) \circ (f(G) \times 1) \circ \bar{\pi}'$ in the diagram

$$(22)$$

Since $r_{W*} \circ \pi_{W*}$ is the identity it suffices to show: The natural morphism of functors

$$(f(G) \times 1) \circ \bar{\pi}' \longrightarrow (\pi_{W*} \times 1) \circ f \tag{23}$$

(indicated by the dotted arc in (22)) is an isomorphism. It suffices to check this in the fibre, which is the upper square in the diagram

$$(24)$$

Since π_W^* is conservative and the lower square in (24) commutes, it suffices that the outer square in (24) commutes. But there the morphism of functors (23) corresponds to the morphism $f_0(\prod_G x) \to \prod_G f_0(x)$ $(x \in E)$, which is an isomorphism by the hypothesis on f_0. This proves the proposition. $\square$

From the proof one sees that the proposition is actually true when one replaces the categories $\underline{\mathrm{Hom}}^{le}$ by the respective categories $\underline{\mathrm{Hom}}^{fdp}$ of functors preserving finite direct products.

(10.18.2) To complete the proof of the theorem let now W be a topos. Since π_* is left exact, the BG-functor $\bar{\pi}': \mathfrak{E} \longrightarrow E(G) \times BG$ is a morphism of G-sites from $E(G) \times BG$ to $\mathfrak{E}$ (i.e. the fibre is a morphism of sites), where $E(G)$ is given the topology τ of Theorem (10.16). Hence every morphism of sites from W to $E(G)_\tau :=$ $(E(G), \tau)$ gives, by composition with $\bar{\pi}'$, a morphism of G-sites (= morphism of G-toposes) from $W \times BG$ to $\mathfrak{E}$. This explains the left vertical arrow in the commutative diagram

$$
\begin{array}{ccccc}
\underline{\mathrm{Morsite}}(W, E(G)_\tau) & \longhookrightarrow & \underline{\mathrm{Hom}}^{le}(E(G), W)^\circ & \longhookrightarrow & \underline{\mathrm{Hom}}(E(G), W)^\circ \\
\downarrow & & \Phi\downarrow & & \downarrow \\
\underline{\mathrm{Homtop}}_G(W \times BG, \mathfrak{E}) & \longrightarrow & \underline{\mathrm{Hom}}^{le}_{BG}(\mathfrak{E}, W \times BG)^\circ & \longhookrightarrow & \underline{\mathrm{Hom}}_{BG}(\mathfrak{E}, W \times BG)^\circ.
\end{array}
\tag{25}
$$

Let F be the category of sheaves on $E(G)_\tau$, as in the theorem. Since $\underline{\mathrm{Homtop}}(W, F)$ $\to \underline{\mathrm{Morsite}}(W, E(G)_\tau)$ is an equivalence of categories [SGA4 IV.4.9.4], the assertion of the theorem is that the left vertical arrow in (25) is an equivalence. The horizontal functors are fully faithful, and Φ is an equivalence of categories by the proposition just proved. So one is left to show the following: Let $\eta: E(G) \to W$ be a left exact functor for which $\Phi(\eta)$ is a morphism of G-toposes. Prove that η is a morphism of sites (from W to $E(G)_\tau$)! But this is easy: Since all fibre products exist in $E(G)$ and in W, and since η preserves them, it suffices to show that η is continuous. But $\eta \circ \pi_*$ is continuous by hypothesis, so η is continuous by Lemma (10.14.2b).

This completes the proof of the theorem. $\qquad\qquad\Box$

(10.19) Let G be a finite group and let $\mathfrak{E} \to BG$ be a G-topos with fibre E. Let $F := \underleftarrow{\mathrm{Limtop}}(\mathfrak{E}/G)$ be the fixtopos and $\nu: F \to E$ the canonical topos morphism. I am going to display now the various topos morphisms between E, F and $E(G)$ in a systematic way. The sequence of topos morphisms

$$
F \xrightarrow{\nu} E \xrightarrow{\pi} E(G)
\tag{26}
$$

is the fibre of the canonical sequence of morphisms of G-toposes

$$
\underleftarrow{\mathrm{Limtop}}(\mathfrak{E}/G) \times BG \longrightarrow \mathfrak{E} \longrightarrow \underleftarrow{\mathrm{Limtop}}(\mathfrak{E}/G) \times BG.
\tag{27}
$$

From (27) one gets (10.6.2) the following diagram of topos morphisms (solid arrows)

$$
\begin{array}{ccccc}
F & \xrightarrow{\nu} & E & \xrightarrow{\pi} & E(G) \\
{\scriptstyle r_F}\uparrow\downarrow{\scriptstyle \pi_F} & & \downarrow{\scriptstyle \pi} & & {\scriptstyle \pi_{E(G)}}\uparrow\downarrow{\scriptstyle r_{E(G)}} \\
F(G) & \xrightarrow{\nu(G)} & E(G) & \xrightarrow{\pi(G)} & (E(G))(G).
\end{array}
\tag{28}
$$

Since F resp. $E(G)$ are (the fibres of) trivial G-toposes, they come along with the retractions r (10.6.3) which are indicated by broken arrows.

The basic identities between the functors associated with (28) are summarized in

(10.19.1) Proposition. — *The solid arrows part of* (28) *commutes. Moreover the following hold (always up to canonical isomorphism):*
  a) $r_F \circ \pi_F = \mathrm{id}_F$ *and* $r_{E(G)} \circ \pi_{E(G)} = r_{E(G)} \circ \pi(G) = \mathrm{id}_{E(G)}$ *(as topos morphisms).*
  b) $\pi^* \nu(G)_* = \nu_* \pi_F^*$ *and* $\nu(G)^* \pi_* = \pi_{F*} \nu^*$.
  c) $\pi_{E(G)}^* \pi(G)_* = \pi_* \pi^* = \pi(G)^* \pi_{E(G)*}$.
  d) $\nu^* = r_{F*} \nu(G)^* \pi_*$ *and* $\nu_* = \pi^* \nu(G)_* r_F^*$.
If one regards F as the topos of sheaves on the site $(E(G), \tau)$, $\tau := \mathrm{coind}(\pi_*)$, as in Theorem (10.16), then the functor $\epsilon_\tau : E(G) \to F$ is isomorphic to the functor $r_{F*} \nu(G)^*$.

Proof. a) has already been observed in (10.6.3) and (10.7.1), b) and c) follow from Remark (10.6.2). d) follows from a) and b):

$$r_{F*} \nu(G)^* \pi_* \underset{b)}{=} r_{F*} \pi_{F*} \nu^* \underset{a)}{=} \nu^* \quad \text{and} \quad \pi^* \nu(G)_* r_F^* \underset{b)}{=} \nu_* \pi_F^* r_F^* \underset{a)}{=} \nu_*.$$

The assertion about ϵ_τ follows since $\nu^* = \epsilon_\tau \circ \pi_*$ by Theorem (10.16) and the factorization of ν^* through π_* is unique up to isomorphism. $\qquad\square$

At this point, the reader should feel reminded of diagram (3) in (6.4.1) and the discussion there. In the following Sect. 11.1 it will be explained why (6.4.1) is indeed a particular case of what was done here in generality.

11. Inverse limits of G-toposes: Two examples

In this section the fixtopos (*alias* inverse limit) is determined for the two G-toposes which are at the center of our interest: The G-topos $\widetilde{X}'_{et}$ ($G = \mathbb{Z}/2$) and the G-topos arising from an extension of discrete or profinite groups. The first case fulfills a promise made in Sect. 5: It is shown that the real spectrum of a scheme X (more precisely, its associated topos of sheaves) is the fixtopos of the G-action on $\widetilde{X}'_{et}$, via the topos morphism ν. (For this one needs $\frac{1}{2} \in \mathcal{O}(X)$.)

An example of a more basic nature will be studied in (13.2), namely equivariant sheaves on a topological space with operators. The topological and the topos-theoretical notions of fix-object are shown to agree, and therefore the topological situation may serve as a source of inspiration for more abstract cases.

From a practical point of view these identifications of fixtoposes may be of little value. But I think that these facts are "philosophically" interesting enough to be included in this treatise. In addition, although they do not seem to *prove* results like those of Sections 5, 7 or 12, they tell that one should *expect* such results. It is conceivable that there are further interesting examples of G-toposes, not yet treated, where this philosophy would lead to the prediction of analogous theorems. See the end of Sect. 14 for a further discussion of this point.

11.1 Example 1: The real topos of a scheme

Let X be a scheme, put $X' = X \times_{\mathrm{spec}\, \mathbb{Z}} \mathrm{spec}\, \mathbb{Z}[\sqrt{-1}]$ and denote the projection $X' \to X$ by π. The action of the group $G = \{1, \sigma\}$ turns $\widetilde{X}'_{et}$ into (the fibre of) a G-topos, as explained in Example (10.11). Write $\mathfrak{X}'$ for this G-topos.

(11.1.1) Theorem. — *Assume that X has no points of residue characteristic 2. Then the topos morphism ν from $\widetilde{X}_{ret}$ to $\widetilde{X}'_{et}$ identifies $\widetilde{X}_{ret}$ with the inverse limit $\varprojlim\mathrm{top}(\widetilde{X}'_{et}/G)$ of the G-topos $\mathfrak{X}'$.*

(11.1.2) Remark. There is a canonical morphism of G-toposes

$$\mathfrak{X}' \longrightarrow \widetilde{X}_{et} \times BG \tag{1}$$

whose fibre is the topos morphism $\pi: \widetilde{X}'_{et} \to \widetilde{X}_{et}$ (Example (10.11)). Let $\bar{\pi}'$ denote the direct image part of (1), cf. (10.4.2). The fibre of the composite BG-functor

$$\mathfrak{X}' \xrightarrow{\bar{\pi}'} \widetilde{X}_{et} \times BG \xrightarrow{\rho \times 1} \widetilde{X}_{ret} \times BG \tag{2}$$

is $\rho\pi_* = \nu^*$, which shows that the topos morphism $\nu : \widetilde{X}_{ret} \to \widetilde{X}'_{et}$ is "G-equivariant" and hence factors, essentially uniquely, through a topos morphism from $\widetilde{X}_{ret}$ to $\underleftarrow{\text{Limtop}}(\widetilde{X}'_{et}/G)$. This is true without any hypothesis on X. But the theorem breaks down as soon as X has a point of characteristic 2: Any geometric point of X' of characteristic 2 gives a G-invariant topos morphism (sets) $\to \widetilde{X}'_{et}$, which of course does not factor through the real spectrum.

(11.1.3) Since 2 is invertible on X the covering π is Galois, and hence $\widetilde{X}'_{et}(G)$ is canonically equivalent to $\ddot{X}_{et}$ (Example (10.12)). By the *ret*-topology on $\ddot{X}_{et}$, or on any of its subcategories, I mean the topology on $\widetilde{X}_{et}$ induced by the functor $\rho : \widetilde{X}_{et} \to \widetilde{X}_{ret}$, where $\widetilde{X}_{ret}$ is given its canonical topology. So a family $\{A_i \to A\}$ in $\widetilde{X}_{et}$ is a covering family for *ret* if and only if $\{\rho A_i \to \rho A\}$ is epi in $\widetilde{X}_{ret}$ [SGA4 III.3.3]. By applying the Comparison Lemma [SGA4 III.4.1] to the fully faithful inclusion $\mathrm{Et}/X \subset \widetilde{X}_{et}$, where both categories carry the real étale topology, one sees that $\widetilde{X}_{ret}$ is identified with the category of sheaves on the site $(\widetilde{X}_{et}, ret)$, and that ρ is isomorphic to the corresponding associated sheaf functor $\epsilon_{ret} : \widetilde{X}_{et} \to \widetilde{X}_{ret}$. As we saw in Sect. 5, the *ret*-topology on $\widetilde{X}_{et}$ makes $\pi_* : \widetilde{X}'_{et} \to \widetilde{X}'_{et}(G) \sim \widetilde{X}_{et}$ continuous. So Theorem (10.16) says that, in order to prove the above theorem, one must show that every topology on $\widetilde{X}_{et}$ which makes $\pi_* : \widetilde{X}'_{et} \to \widetilde{X}_{et}$ continuous is finer than the *ret*-topology.

(11.1.3.1) Let τ denote the topology on $\widetilde{X}_{et}$ which is coinduced by π_* (10.14.1). So τ is coarser than *ret*, and one has to show that both are equal. The plan of the proof is to verify that the topos of sheaves on the site $(\widetilde{X}_{et}, \tau)$ has sufficiently many points. By Deligne's criterion it suffices for this to show that this topos is locally coherent. Once this is known, it suffices to verify that $\nu : \widetilde{X}_{ret} \to \widetilde{X}'_{et}$ has the universal property of the fixtopos with respect to *points*, i.e. to morphisms of G-toposes from (sets) $\times BG$ to $\mathfrak{X}'$. But here one is immediately done since the automorphism groups of geometric points of a scheme, and in particular, the involutions in these groups, are well known.

Observe that for every epi family $\{F_i \to F\}$ in $\widetilde{X}'_{et}$ the family $\{\pi_* F_i \to \pi_* F\}$ in $\widetilde{X}_{et}$ is a covering family for τ, by the very definition of τ.

(11.1.4) Actually I will not work with $\widetilde{X}_{et}$ and τ themselves. Rather one has to make a detour and use a similarly defined topology on a suitable subcategory of Et/X. The proof given below depends on the fact that there are full subcategories C of Et/X and C' of Et/X' which satisfy the following list of properties:

(1) C (resp. C') is closed under fibre products in Et/X (resp. in Et/X');

(2) for every $U \to X$ in C (resp. every $V \to X'$ in C') the scheme U (resp. V) is quasi-compact;

(3) every $V \in \mathrm{Et}/X'$ has a covering family $\{V_i \to V\}$ for the étale topology by objects $V_i \in C'$ (in other words, C' is generating for the étale topology);

(4) every $U \in \mathrm{Et}/X$ has a *Zariski* open covering $\{U_i \subset U\}$ by objects $U_i \in C$;

(5) the functors $\pi^*: \widetilde{X}_{et} \to \widetilde{X}'_{et}$ and $\pi_*: \widetilde{X}'_{et} \to \widetilde{X}_{et}$ restrict to functors between
C and C'; i.e. $\pi^*(C) \subset C'$ and $\pi_*(C') \subset C$ hold.

For example, one may take as the objects of C all étale morphisms $f: U \to X$ for
which U is an affine scheme and $f(U)$ is contained in an open affine subscheme of
X; and as the objects of C' all étale morphisms $g: V \to X'$ for which V is an affine
scheme and $\pi \circ g(V)$ is contained in an open affine subscheme of X. Properties
(1)–(4) and the first part of (5) are obvious, while the second part of (5) follows
from (4.4) and (4.6.1).

Assume now that C and C' have been chosen according to (1)–(5) above. Via
the Yoneda functor h, the category C resp. C' is regarded as a full subcategory of
$\widetilde{X}_{et}$ resp. $\widetilde{X}'_{et}$; I will often suppress the mentioning of h. The restriction of π_* to
a functor $C' \to C$ will be denoted by Res: $C' \to C$.

(11.1.5) Lemma. — *Every epi family $\{A_i \rightarrowtail A\}$ in $\widetilde{X}_{et}$ which consists of mono-
morphisms is a covering family for τ. In particular, every Zariski open covering in
Et/X is a covering family for τ.*

Proof. This uses the following simple fact: For any monomorphism $B \rightarrowtail A$ in $\widetilde{X}_{et}$
the diagram

$$
\begin{array}{ccc}
B & \longrightarrow & \pi_*\pi^*B \\
\downarrow & & \downarrow \\
A & \longrightarrow & \pi_*\pi^*A
\end{array}
$$

is cartesian in $\widetilde{X}_{et}$. Here the horizontal arrows are the adjunction maps. This is
easily seen by looking at the stalks. So if $\{A_i \to A\}$ is a family as in the lemma, it
is obtained by pulling back a τ-covering, namely $\{\pi_*\pi^*A_i \to \pi_*\pi^*A\}$, and hence is
a τ-covering by itself. $\square$

(11.1.6) Lemma. — *The full subcategory C of $\widetilde{X}_{et}$ is generating for the site
$(\widetilde{X}_{et}, \tau)$.*

Proof. The term "generating" is used in the sense of *génératrice*, as in [SGA4 II.3].
So one has to show that every $A \in \widetilde{X}_{et}$ admits a τ-covering $\{U_i \to A\}_{i \in I}$ by objects
$U_i \in C$. Since C' is generating for $\widetilde{X}'_{et}$ (3), one finds an epi family $\{V_i \to \pi^*A\}_{i \in I}$
in $\widetilde{X}'_{et}$ with $V_i \in C'$. The family

$$
\{\text{Res } V_i = \pi_* V_i \longrightarrow \pi_*\pi^*A\}_{i \in I}
$$

is a covering for τ, by the definition of τ. For each $i \in I$ form the pullback

$$
\begin{array}{ccc}
U_i & \longrightarrow & \text{Res } V_i \\
\downarrow & & \downarrow \\
A & \longrightarrow & \pi_*\pi^*A
\end{array}
$$

in $\widetilde{X}_{et}$. The family $\{U_i \to A\}_{i \in I}$ is a τ-covering. Since the adjunction morphism $A \to \pi_*\pi^*A$ is mono, so are its base extensions $U_i \to \mathrm{Res}(V_i)$. Hence U_i is a subsheaf of the étale sheaf $\mathrm{Res}(V_i) = \pi_*V_i$ on X. But every subsheaf of a representable étale sheaf is itself representable [SGA4 VIII.6.1], so U_i is an open subscheme of $\mathrm{Res}(V_i)$. Use (11.1.5) and (4) to cover U_i, with respect to τ, by objects in C. $\square$

By the real étale topology ret on C I mean the topology on C induced by the topology ret on Et/X (or on $\widetilde{X}_{et}$).

(11.1.7) Lemma. — *To prove the theorem it suffices to show that the topology coinduced on C by $\mathrm{Res}\colon C' \to C$ is the real étale topology. (Here C' carries the étale topology.)*

Proof. Consider the commutative square of categories and functors

$$
\begin{array}{ccc}
C' & \overset{h'}{\lhook\joinrel\longrightarrow} & \widetilde{X}'_{et} \\[4pt]
{\scriptstyle \mathrm{Res}}\big\downarrow & & \big\downarrow{\scriptstyle \pi_*} \\[4pt]
C & \overset{h}{\lhook\joinrel\longrightarrow} & \widetilde{X}_{et}
\end{array}
$$

in which h' and h are the canonical fully faithful (Yoneda) embeddings. From Lemma (10.14.2) one gets the inclusion

$$
\mathrm{coind}_{\mathrm{Res}}(et) = \mathrm{coind}_{\mathrm{Res}}\,\mathrm{ind}(h') \subset \mathrm{ind}_h\,\mathrm{coind}(\pi_*) = \tau|_C
$$

of topologies on C. Indeed, one has to show e.g. that $(C',et) \xrightarrow{\mathrm{Res}} (C,\tau|_C)$ is continuous, which by (10.14.2b) is equivalent to $(C',et) \xrightarrow{h\circ\mathrm{Res}} (\widetilde{X}_{et},\tau)$ being continuous, and the latter continuity is obvious writing $h \circ \mathrm{Res} = \pi_* \circ h'$. Therefore, if one knows $\mathrm{coind}_{\mathrm{Res}}(et) \supset ret$, one gets the following chain of inclusions of topologies on C:

$$
ret \subset \mathrm{coind}_{\mathrm{Res}}(et) \subset \tau|_C \subset ret|_C = ret,
$$

from which $\tau|_C = ret$ follows. So in the diagram

$$
\begin{array}{ccc}
(C,\tau|_C) & \overset{h}{\longrightarrow} & (\widetilde{X}_{et},\tau) \\[4pt]
\big\| & & \big\downarrow{\scriptstyle \mathrm{id}} \\[4pt]
(C,ret) & \overset{h}{\longrightarrow} & (\widetilde{X}_{et},ret)
\end{array}
$$

(of continuous functors between sites) all three arrows other than "id" are known to induce equivalences between the respective categories of sheaves. (For the arrows h this follows from the Comparison Lemma, using (11.1.6) for the upper h and (11.1.3) for the lower h.) Hence also "id" induces an equivalence between the associated toposes, which shows $\tau = ret$ on $\widetilde{X}_{et}$. $\square$

So put $\sigma := \text{coind}_{\text{Res}}(et)$, the coarsest topology on C which makes Res: $C' \to C$ continuous. Clearly $\sigma \subset ret$, and one has to show the reverse inclusion.

(11.1.8) Lemma. — *Every object of C is quasi-compact with respect to σ.*

Proof. Let $\tilde{\sigma}$ be the topology on C associated with the pretopology whose covering families are those families $\{U_i \to U\}_{i \in I}$ in C for which there is a *finite* subset J of I such that $\{U_j \to U\}_{j \in J}$ is a covering for σ. Every object of C is quasi-compact with respect to $\tilde{\sigma}$. Since every object of C' is quasi-compact (with respect to the étale topology) and Res preserves fibre products it is clear that Res: $C' \to C$ is continuous also with respect to $\tilde{\sigma}$. But σ is the coarsest topology which makes Res continuous, and $\tilde{\sigma} \subset \sigma$, so $\tilde{\sigma} = \sigma$. $\quad\square$

(11.1.8.1) Corollary. — *The topos of sheaves on the site (C, σ) has sufficiently many points.*

Proof. This topos is locally coherent by Lemma (11.1.8) (compare [SGA4 VI.2]), and hence Deligne's theorem [SGA4 VI.9] applies. $\quad\square$

(11.1.9) By [SGA4 IV.6.5], a topology on a category C whose associated topos of sheaves has sufficiently many points is entirely determined by these points, i.e. by the fibre functors $C \to$ (sets). Applying this remark to the topologies σ and ret on C, one sees that it suffices to show now that the (fully faithful) functor $\underline{\text{pt}}(C, ret) \to \underline{\text{pt}}(C, \sigma)$ is an equivalence of categories, or merely, that it is essentially full. The category $\underline{\text{pt}}(C, \sigma)$ is identified with the category of fibre functors on (C, σ), i.e. of functors $f: C \to$ (sets) which are left exact and continuous with respect to σ [SGA4 IV.6.3]. By (10.14.2b), f is continuous w.r.t. σ if and only if $f \circ \text{Res}: C' \to$ (sets) is continuous (w.r.t. the étale topology). What remains to be shown is therefore:

(11.1.9.1) Lemma. — *Let $f: C \to$ (sets) be a left exact functor such that $f \circ \text{Res}: C' \to$ (sets) is a fibre functor for (C', et). Then f is isomorphic to the stalk functor at a real point of X.*

Proof. Write $\chi := f \circ \text{Res}: C' \to$ (sets). The fibre functors of (C', et), i.e. the points of $\tilde{X}'_{et}$, are known, they all arise from geometric points of X' [SGA4 VIII.7.9]. So there is a geometric point $\bar{x}' \to X'$, centered at $x' \in X'$, such that the functors χ and $V \mapsto V_{\bar{x}'}$ on C' are isomorphic. Recall (10.7.2) that G operates on $\pi_* \pi^*$, and that the adjunction morphism $\text{id} \to \pi_* \pi^*$ induces an isomorphism $A \xrightarrow{\sim} \Gamma_G(\pi_* \pi^* A)$ for each $A \in \tilde{X}_{et}$. This G-operation induces one on the functor $f\pi_* \pi^* = \chi \circ \pi^*: C \to$ (sets). Note that $\chi \circ \pi^*$ is a fibre functor of (C, et). Since f is left exact, the canonical map

$$f(U) \longrightarrow \left(\chi \circ \pi^*(U)\right)^G = \left(\chi(U_{X'})\right)^G \qquad (3)$$

is bijective for every $U \in C$. The G-action on $\chi \circ \pi^*$ cannot be the trivial action. For if it were, then $f \cong \chi \circ \pi^*$ and hence $\chi = f \circ \pi_* \cong \chi \circ \pi^* \pi_*$. But this is absurd, since $\pi^* \pi_* V \cong V \times_{X'} V_\sigma$ (10.5.4) would imply $\chi(V) \cong \chi(V) \times \chi(V_\sigma)$ for $V \in C'$. So there is a proper involution acting on the fibre functor $\chi \circ \pi^*$ of (C, et), such that f is isomorphic to $U \mapsto \left(\chi \circ \pi^*(U) \right)^G$. The automorphism group of $\chi \circ \pi^*$ is isomorphic to the absolute Galois group of the residue field $\kappa(x)$, where $x = \pi(x')$; and the involutions in this group correspond to the real closures of $\kappa(x)$ inside a fixed algebraic closure. If such an involution is chosen, the fibre functor of the corresponding real point of X is precisely given by (the right hand side of) (3), see (3.9.2). So f is indeed (isomorphic to) the fibre functor of a real point of X. This completes the proof of the theorem. $\quad\square\square$

(11.1.10) In the case where $X = \operatorname{spec} k$ is the spectrum of a field of characteristic $\neq 2$, an independent proof of Theorem (11.1.1) will arise from the example of group extensions treated in Sect. 11.2 below.

(11.1.11) **Remark.** Theorem (11.1.1) has the following generalization. Let $\pi: Y \to X$ be a finite Galois covering of schemes with Galois group $G \neq \{1\}$. If $|G| > 2$ then $\varprojlim \operatorname{top}(\widetilde{Y}_{et}/G)$ is the "empty" topos. If G has order 2 then $\varprojlim \operatorname{top}(\widetilde{Y}_{et}/G)$ is (equivalent to) the topos of sheaves on the space $Z := X_r - \pi_r(Y_r)$. Note that Z, the space of all orderings on X which do not extend to Y, is a clopen subspace of X_r. The proof can be given using the same methods (compare also Remark (1.17)). I will skip over it here since the result won't be used in the sequel.

11.2 Example 2: Group extensions

Consider an extension of discrete or profinite groups

$$1 \longrightarrow \Delta \longrightarrow \Gamma \longrightarrow G \longrightarrow 1, \tag{4}$$

with G finite in the latter case. As mentioned in Example (10.13), this makes $(\Delta\text{-sets})$ into a G-topos $\mathfrak{D}$, and the natural topos morphism from $(\Delta\text{-sets})$ to $(\Gamma\text{-sets})$ is identified with the fibre of the quotient morphism $\mathfrak{D} \longrightarrow \varinjlim \operatorname{top}(\mathfrak{D}/G) \times BG$; or expressed in a less formal way, it identifies $(\Gamma\text{-sets})$ with the category of G-objects in $(\Delta\text{-sets})$, with respect to the given (pseudo) action of G.

   Here the inverse limit of the G-topos $\mathfrak{D}$ is studied, in the case where G is finite. The main result says that this fixtopos of $\mathfrak{D}$ is equivalent to the category of Δ-equivariant sheaves on the space of splittings of the sequence (4). The profinite case of this theorem generalizes the field case of Theorem (11.1.1), as follows. If k is a field of characteristic not 2, with absolute Galois group Γ, and if Γ' is the subgroup which fixes $\sqrt{-1}$, then $1 \to \Gamma' \to \Gamma \to G \to 1$ is a sequence (4) of profinite groups (suppose $\sqrt{-1} \notin k$). Put $X := \operatorname{spec} k$. The pseudo actions by G on $\tilde{X}'_{et}$ and on (Γ'-sets) being equivalent, the result of this section says that the fixtopos $\underleftarrow{\operatorname{Limtop}}(\tilde{X}'_{et}/G)$ is equivalent to the category of Γ'-sheaves on the space of involutions of Γ. But this category is equivalent to the category of sheaves on $\operatorname{sper} k$, as was shown in Sect. 9. So the result below is well in accordance with Sect. 11.1.

(11.2.1) Definition. Let G be a locally compact group. I write $\operatorname{Subg}(G)$ for the space of closed subgroups of G. A subbasis of open sets for the topology of $\operatorname{Subg}(G)$ is given by the family of all subsets

$$D(U) := \{H \in \operatorname{Subg}(G): H \cap U \neq \emptyset\}$$

and

$$D'(K) := \{H \in \operatorname{Subg}(G): H \cap K = \emptyset\},$$

where U resp. K ranges over all open resp. all compact subsets of G. The space $\operatorname{Subg}(G)$ is covariantly functorial for homomorphisms with compact kernel which are open onto their image, and contravariantly functorial for open homomorphisms.

(11.2.1.1) Remark. The space $\operatorname{Subg}(G)$ is always compact. For lack of a convenient reference here is a short proof of this fact. On the set $\mathfrak{F}G$ of all closed subsets of G consider the analogous topology which is generated by the subsets

$$\tilde{D}(U) := \{F \in \mathfrak{F}G: F \cap U \neq \emptyset\} \quad \text{and} \quad \tilde{D}'(K) := \{F \in \mathfrak{F}G: F \cap K = \emptyset\}$$

($U \subset G$ open, $K \subset G$ compact). Clearly $\mathfrak{F}G$ is a Hausdorff space, and $\operatorname{Subg}(G)$ is a closed subspace of $\mathfrak{F}G$. So it suffices to show that $\mathfrak{F}G$ is quasi-compact. Consider a covering

$$\mathfrak{F}G = \bigcup_\lambda \tilde{D}(U_\lambda) \cup \bigcup_\mu \tilde{D}'(K_\mu) \tag{5}$$

by subbasic open sets. By the Alexander Subbase Theorem it suffices that (5) admits a finite subcovering. But (5) is equivalent to the condition that there is an index μ with $K_\mu \subset \bigcup_\lambda U_\lambda$. So the assertion is obvious.

Note also that if G is totally disconnected then so is $\operatorname{Subg}(G)$ (and hence is a boolean space). If G is a profinite group and $\{G_\alpha\}$ is the system of its finite (discrete) quotient groups then $\operatorname{Subg}(G) = \underleftarrow{\lim} \operatorname{Subg}(G_\alpha)$.

(11.2.2) Fix an exact sequence (4). I treat simultaneously the cases where the groups in (4) are discrete resp. where they are profinite. It is always assumed that G is a finite group. Let $\mathfrak{K}$ be the space of splittings of (4), i.e. the set of subgroups K of Γ with $K \cap \Delta = \{1\}$ and $K\Delta = \Gamma$, equipped with the subspace topology from $\operatorname{Subg}(\Gamma)$. If Γ is discrete then also $\mathfrak{K}$ is discrete. If Γ is profinite then $\mathfrak{K}$ is closed in $\operatorname{Subg}(\Gamma)$, and hence is a boolean space by itself. The group Γ acts continuously on $\mathfrak{K}$ by conjugation. As in Sect. 8, $\widetilde{\mathfrak{K}}(\Delta)$ denotes the category of (continuous) Δ-sheaves on $\mathfrak{K}$; in the discrete case continuity is an empty condition, of course.

(11.2.3) Let $\pi = (\pi^*, \pi_*)$ denote the topos morphism from (Δ-sets) to (Γ-sets) induced by $\Delta \subset \Gamma$ (0.5.3). In particular, $\pi_* : (\Delta\text{-sets}) \to (\Gamma\text{-sets})$ is the coinduction functor. By (10.16), the inverse limit $\underleftarrow{\mathrm{Lim}}\mathrm{top}(\mathfrak{D}/G)$ is the topos of sheaves on the site $((\Gamma\text{-sets}), \tau)$, where τ is the coarsest topology on (Γ-sets) which makes π_* continuous. The first step is to identify this topology τ:

(11.2.3.1) Proposition. — *A family of maps $\{M_i \to M\}_{i \in I}$ in (Γ-sets) is a covering for τ if and only if for every $K \in \mathfrak{K}$ the family $\{M_i^K \to M^K\}_{i \in I}$ is surjective.*

Proof. The families $\{M_i \to M\}$ for which $\{M_i^K \to M^K\}$ is surjective for every $K \in \mathfrak{K}$ form obviously a pretopology on (Γ-sets). Call the corresponding topology $\tilde{\tau}$. First one shows that π_* is continuous with respect to $\tilde{\tau}$. Since fibre products exist in both categories and are preserved by π_*, it suffices to see that π_* sends surjective families to $\tilde{\tau}$-coverings. Let $\{N_i \to N\}$ be a surjective family in (Δ-sets). For $f \in \pi_*(N) = \mathrm{Hom}_\Delta(\Gamma, N)$, to be fixed under $K \in \mathfrak{K}$ means that $f(xt) = f(x)$ for all $x \in \Gamma$ and $t \in K$, which is equivalent to $f(t) = f(1)$ for every $t \in K$. Given $f \in \mathrm{Hom}_\Delta(\Gamma, N)^K$, lift the element $n := f(1)$ to some $n_i \in N_i$ and let $f_i \in \mathrm{Hom}_\Delta(\Gamma, N_i)^K$ be the unique element satisfying $f_i(t) = n_i$ for all $t \in K$. Then f_i is a preimage of f in $\pi_*(N_i)^K$.

So $\tilde{\tau}$ is finer than τ. To show that also the opposite inclusion holds, the strategy will be to start out from a $\tilde{\tau}$-covering $\mathfrak{U} = \{M_\alpha \to M\}_{\alpha \in I}$ of a Γ-set M and to show that there is a family $\{M'_\beta \to M\}_\beta$ which is derived from the π_*'s of surjective families by pullbacks and refinements, with the property that each $M'_\beta \to M$ factors through some $M_\alpha \to M$. Once such a family $\{M'_\beta \to M\}$ has been found, it follows that $\mathfrak{U}$ is a τ-covering.

Let M be a Γ-set, and let $\{M_i\}_i$ be the family of Γ-orbits in M. The same argument as for (11.1.5) shows that $\{M_i \subset M\}_i$ is a covering for τ. Therefore one can assume that M is connected, say $M = \Gamma/H := \{xH : x \in \Gamma\}$ with H an open subgroup of Γ. Replacing a given $\tilde{\tau}$-covering $\mathfrak{U} = \{M_\alpha \to M\}_{\alpha \in I}$ by a finer one does not affect the proof. So one can assume that each of the M_α is itself connected. Altogether this means that there are two cases to be considered for the $\tilde{\tau}$-covering $\mathfrak{U} = \{M_\alpha \to M\}_{\alpha \in I}$:

Case I: $\mathfrak{U} = \{\Gamma/P_\alpha \to \Gamma/H\}_{\alpha \in I}$ with open subgroups P_α of H and $\Gamma/P_\alpha \to \Gamma/H$ the canonical maps ($\alpha \in I$);

Case II: $\mathfrak{U} = \{\emptyset \to \Gamma/H\}$.

Write $\mathfrak{K}[H] := \{K \in \mathfrak{K}: K \subset H\}$. If P is any subgroup of Γ and $K \in \mathfrak{K}$, then $(\Gamma/P)^K$ consists of those cosets xP for which $K \subset xPx^{-1}$. So the condition that $\mathfrak{U}$ is a covering for $\tilde{\tau}$ implies certainly the following condition $(*)$:

Case I: For each $K \in \mathfrak{K}[H]$ there are $\alpha \in I$ and $h \in H$ with

$(*)$ $K \subset hP_\alpha h^{-1}$;

Case II: $\mathfrak{K}[H] = \emptyset$;

and conversely, one verifies without trouble that $(*)$ implies that $\mathfrak{U}$ is a $\tilde{\tau}$-covering.

First consider the case where $H\Delta \neq \Gamma$. Here the Δ-set $\pi^*M = \Gamma/H$ is disconnected, i.e. can be written $\pi^*M = N_1 \amalg N_2$ with non-empty Δ-sets N_1, N_2. For $i = 1$, 2 the left square in the diagram

$$
\begin{array}{ccc}
\emptyset & \longrightarrow & \pi_*N_i = \mathrm{Hom}_\Delta(\Gamma, N_i) \\
\downarrow & & \downarrow \qquad\qquad \downarrow \\
M & \longrightarrow & \pi_*\pi^*M = \mathrm{Hom}_\Delta(\Gamma, N_1 \amalg N_2)
\end{array}
$$

is cartesian. Hence $\{\emptyset \to M\}$ is a τ-covering, and *a fortiori* $\mathfrak{U}$ is one.

The remaining case is when $H\Delta = \Gamma$, so that the Δ-set π^*M is connected. A "fine" covering of π^*M is one of the form $\Delta/Q \to \Gamma/H$ where Q is a small open subgroup of $H \cap \Delta$ and the map is induced by $\Delta \subset \Gamma$. (In the discrete case one can directly use $Q = \{1\}$ here.) So consider open subgroups Q of $H \cap \Delta$, and define M_Q to be the fibre product of Γ-sets

$$
\begin{array}{ccc}
M_Q & \longrightarrow & \pi_*(\Delta/Q) \\
\downarrow & & \downarrow \\
\Gamma/H & \longrightarrow & \pi_*\pi^*(\Gamma/H)
\end{array}
$$

in which the right hand vertical arrow is π_* applied to the canonical map $\Delta/Q \to \Gamma/H$. The map $M_Q \to M = \Gamma/H$ is a τ-covering. I will show: If Q is chosen small enough then

($**$)

> Case I: for any connected component ($=$ orbit) M' of M_Q, either Δ is not transitive on M', or the restriction of $M_Q \to \Gamma/H$ to M' factors through one of the Γ/P_α;
>
> Case II: Δ is not transitive on any connected component of M_Q.

By using the case $H\Delta \neq \Gamma$ treated above this will imply in both cases I and II that $\mathfrak{U}$ is a τ-covering.

(11.2.3.2) Lemma. — *There is an open subgroup W of Δ with the following property: In case I, whenever K is a (closed) subgroup of H with $K\Delta = \Gamma$ and $K \cap \Delta \subset W$, there are $\alpha \in I$ and $h \in H$ such that $K \subset hP_\alpha h^{-1}$. In case II, there is no $K \subset H$ with $K\Delta = \Gamma$ and $K \cap \Delta \subset W$.*

Proof. In the discrete case this follows from $(*)$, taking $W = \{1\}$. So assume that Γ is profinite. For W an open subgroup of Δ put

$$
\mathfrak{K}_W := \{K \in \mathrm{Subg}(\Gamma) \colon K \subset H,\ K\Delta = \Gamma \text{ and } K \cap \Delta \subset W\}.
$$

Clearly $\mathfrak{K}_W$ is closed (even clopen) in $\mathrm{Subg}(\Gamma)$, and $\bigcap_W \mathfrak{K}_W = \mathfrak{K}[H]$. On the other hand the set

$$
\mathfrak{P} := \{K \in \mathrm{Subg}(\Gamma) \colon \text{there are } \alpha \in I \text{ and } h \in H \text{ with } K \subset hP_\alpha h^{-1}\}
$$

(resp. $\mathfrak{P} = \emptyset$ in case II) is open in $\mathrm{Subg}(\Gamma)$. By $(*)$ one knows $\mathfrak{K}[H] \subset \mathfrak{P}$. By compactness of $\mathrm{Subg}(\Gamma)$ there is an open subgroup W of Δ with $\mathfrak{K}_W \subset \mathfrak{P}$, as desired. □

For the following fix W as in the lemma. Let Q be an open subgroup of $H \cap \Delta$ contained in W which is normal in Γ. I claim that Q satisfies $(**)$. To see this one has to study the stabilizers of points in M_Q. For $f \in \pi_*(\Delta/Q) = \mathrm{Hom}_\Delta(\Gamma, \Delta/Q)$ the stabilizer Γ_f of f satisfies $\Gamma_f \cap \Delta = Q$. Indeed, for $y \in \Delta$ and $x \in \Gamma$ one has $(y \cdot f)(x) = f(xy) = xyx^{-1} \cdot f(x)$, which is equal to $f(x)$ if and only if $xyx^{-1} \in Q$. Hence the stabilizer Γ_m of any $m = (1.H, f) \in M_Q$ satisfies $\Gamma_m \cap \Delta = Q$, and of course $\Gamma_m \subset H$. Now either $\Gamma_m \Delta \neq \Gamma$, in which case the orbit $\Gamma.m$ can be discarded for the topology τ; or else one is in case I and there are $\alpha \in I$ and $h \in H$ with $\Gamma_m \subset h P_\alpha h^{-1}$, by Lemma (11.2.3.2). This shows that $(**)$ is indeed satisfied, and so the proof of the proposition is complete. □

So in principle the fixtopos $\underleftarrow{\mathrm{Limtop}}((\Delta\text{-sets})/G)$ has been identified, as the topos of sheaves on the site $((\Gamma\text{-sets}), \tau)$ where τ is the topology described in Proposition (11.2.3.1). A more explicit description is, however, possible; this is the main result of this section:

(11.2.4) Theorem. — *Let $1 \to \Delta \to \Gamma \to G \to 1$ be an exact sequence of discrete or profinite groups, with G finite. Let $\mathfrak{K}$ be the space of splittings of this sequence. Then the fixtopos of the G-topos $(\Delta\text{-sets})$ is equivalent to the topos $\widetilde{\mathfrak{K}}(\Delta)$ of (continuous) Δ-sheaves on $\mathfrak{K}$, via the following topos morphism $\nu = (\nu^*, \nu_*)$ from $\widetilde{\mathfrak{K}}(\Delta)$ to $(\Delta\text{-sets})$: The inverse image functor ν^* sends a Δ-set N to the espace étalé $\mathfrak{K} \times N \xrightarrow{\mathrm{pr}_1} \mathfrak{K}$ over $\mathfrak{K}$ (with the "diagonal" Δ-action); and ν_* sends a Δ-sheaf $B \to \mathfrak{K}$ on $\mathfrak{K}$ to $H^0(\mathfrak{K}, B)$, its set of global sections.*

The Δ-action on $H^0(\mathfrak{K}, B)$ is by $(y \cdot s)(K) = y . s(y^{-1} K y)$.

(11.2.4.1) Let F denote the topos of sheaves on the site $((\Gamma\text{-sets}), \tau)$ where τ is the topology described in (11.2.3.1). Given a Γ-set M, put

$$\mathfrak{K}(M) := \{(K, m) \in \mathfrak{K} \times M \colon m \in M^K\}.$$

Via the first projection this is an espace étalé over $\mathfrak{K}$, and Γ acts diagonally on $\mathfrak{K}(M)$, so $\mathfrak{K}(M)$ is a Γ-sheaf on $\mathfrak{K}$. In particular it can be considered as a Δ-sheaf.

The functor $M \mapsto \mathfrak{K}(M)$ from $(\Gamma\text{-sets})$ to $\widetilde{\mathfrak{K}}(\Delta)$ preserves fibre products and sends τ-coverings to coverings in $\widetilde{\mathfrak{K}}(\Delta)$. Hence it is a site morphism from $\widetilde{\mathfrak{K}}(\Delta)$ to $((\Gamma\text{-sets}), \tau)$. Let $\phi = (\phi^*, \phi_*)$ be the induced topos morphism from $\widetilde{\mathfrak{K}}(\Delta)$ to F. In the following it will be proved that ϕ^* is an equivalence of categories. This will imply the assertions of the theorem. In particular, the identification of ν there follows from the observation that for every Δ-set N there is a canonical isomorphism $\nu^* N = \mathfrak{K}(\pi_* N) \cong \mathfrak{K} \times N$ in $\widetilde{\mathfrak{K}}(\Delta)$, and that the functor $B \mapsto H^0(\mathfrak{K}, B)$ is clearly right adjoint to ν^*.

Let $\epsilon \colon (\Gamma\text{-sets}) \to F$ be the functor which sends M to the τ-sheaf associated with the presheaf $h_M = \mathrm{Hom}_\Gamma(-, M)$. The main point in the proof is to show

(11.2.4.2) Lemma. — *For any two Γ-sets M, N the inverse image functor ϕ^* induces a bijection*

$$(\epsilon M)(N) = \mathrm{Hom}_F(\epsilon N, \epsilon M) \xrightarrow{\;\approx\;} \mathrm{Hom}_{\widetilde{\mathfrak{K}}(\Delta)}\big(\mathfrak{K}(N), \mathfrak{K}(M)\big).$$

Proof. The proof is not hard but quite technical. For $M \in (\Gamma\text{-sets})$ let M also denote the presheaf h_M on $(\Gamma\text{-sets})$. Then $\epsilon M = L \circ L(M)$ where L is the functor of [SGA4 II.3] with respect to the topology τ. Recall that, for any presheaf P on $(\Gamma\text{-sets})$,

$$(LP)(N) = \varinjlim_{\mathfrak{U}} H^0(\mathfrak{U}, P)$$

where

$$H^0(\mathfrak{U}, P) = \ker\left(\prod_\alpha P(N_\alpha) \rightrightarrows \prod_{\alpha,\beta} P(N_\alpha \times_N N_\beta) \right) \tag{6}$$

and the limit is taken over the τ-coverings $\mathfrak{U} = \{N_\alpha \to N\}_\alpha$ of N. Let M be a Γ-set. In order to calculate the presheaf LM observe first that $(LM)(\coprod_\alpha N_\alpha) = \prod_\alpha (LM)(N_\alpha)$ for any family $\{N_\alpha\}$ of Γ-sets, so one can restrict to connected Γ-sets $N = \Gamma/H$. Write again $\mathfrak{K}[H] := \{K \in \mathfrak{K} \colon K \subset H\}$. If $\mathfrak{K}[H] = \emptyset$ then $\{\emptyset \to \Gamma/H\}$ is a τ-covering. Otherwise every τ-covering of Γ/H is also a covering with respect to the canonical topology. This shows

$$(LM)(\Gamma/H) = \begin{cases} M^H & \text{if } \mathfrak{K}[H] \neq \emptyset, \\ * & \text{if } \mathfrak{K}[H] = \emptyset, \end{cases} \tag{7}$$

for $H \subset \Gamma$ any open subgroup. Let now $\mathfrak{U} = \{\Gamma/P_\alpha \to \Gamma/H\}_{\alpha \in I}$ be a τ-covering, with $P_\alpha \subset H \subset \Gamma$ open subgroups and $\Gamma/P_\alpha \to \Gamma/H$ the canonical maps ($\alpha \in I$, $I \neq \emptyset$). From (6) and (7) one reads off the following description of $H^0(\mathfrak{U}, LM)$: An element of this set is a family

$$\mathbf{m} = (m_\alpha)_\alpha \in \prod_{\substack{\alpha \in I \text{ with} \\ \mathfrak{K}[P_\alpha] \neq \emptyset}} M^{P_\alpha} \tag{8}$$

subject to the condition that

$$x \cdot m_\alpha = y \cdot m_\beta \tag{9}$$

holds for all pairs α, β in I and x, y in Γ for which $xH = yH$ and there exists $C \in \mathfrak{K}$ with $xP_\alpha \in (\Gamma/P_\alpha)^C$ and $yP_\beta \in (\Gamma/P_\beta)^C$.

There is a canonical map $\mathbf{m} \mapsto f_{\mathbf{m}}$ from $H^0(\mathfrak{U}, LM)$ to $\mathrm{Hom}_{\widetilde{\mathfrak{K}(\Delta)}}(\mathfrak{K}(\Gamma/H), \mathfrak{K}(M))$, as follows. Given $\mathbf{m} = (m_\alpha)_\alpha$ as above and given $(K, xH) \in \mathfrak{K}(\Gamma/H)$, put

$$f_{\mathbf{m}}(K, xH) := (K, x' \cdot m_\alpha) \in \mathfrak{K}(M) \tag{10}$$

where $x' \in \Gamma$ and $\alpha \in I$ are chosen such that $Kx'P_\alpha = x'P_\alpha$ and $x'H = xH$, which is possible since $\mathfrak{U}$ is a τ-covering. It follows directly from (9) that the map $f_{\mathbf{m}}$ is well-defined. Moreover it is immediate that $f_{\mathbf{m}}$ is Γ-equivariant. Finally the map $\mathbf{m} \mapsto f_{\mathbf{m}}$ from $H^0(\mathfrak{U}, LM)$ to $\mathrm{Hom}_{\widetilde{\mathfrak{K}(\Delta)}}(\mathfrak{K}(\Gamma/H), \mathfrak{K}(M))$ is injective since for $\alpha \in I$ and $K \in \mathfrak{K}[P_\alpha]$ the element $m_\alpha \in M^{P_\alpha}$ can be recovered from $f_{\mathbf{m}}$ by $f_{\mathbf{m}}(K, 1H) = (K, m_\alpha)$.

Since the maps $H^0(\mathfrak{U}, LM) \longrightarrow \mathrm{Hom}_{\widetilde{\mathfrak{K}(\Delta)}}(\mathfrak{K}(\Gamma/H), \mathfrak{K}(M))$ are compatible with refinement of $\mathfrak{U}$ it follows that there is a natural injective map

$$(\epsilon M)(\Gamma/H) = \varinjlim_{\mathfrak{U}} H^0(\mathfrak{U}, LM) \longrightarrow \mathrm{Hom}_{\widetilde{\mathfrak{K}(\Delta)}}(\mathfrak{K}(\Gamma/H), \mathfrak{K}(M)) \tag{11}$$

which by construction is the map induced by ϕ^*. It remains to check that (11) is also surjective.

First note that for any two Γ-sets N, M, any map $\mathfrak{K}(N) \to \mathfrak{K}(M)$ in $\widetilde{\mathfrak{K}}(\Delta)$ is actually Γ-invariant, since $\Delta K = \Gamma$ and elements in the fibres over $K \in \mathfrak{K}$ are stabilized by K. Let $f: \mathfrak{K}(\Gamma/H) \to \mathfrak{K}(M)$ be a map in $\widetilde{\mathfrak{K}}(\Delta)$. For $K \in \mathfrak{K}(H)$ write $f(K, 1.H) =: (K, m_K)$, with $m_K \in M^K$. It is possible to choose for every $K \in \mathfrak{K}(H)$ an open subgroup P_K of H such that $K \subset P_K$ and P_K stabilizes m_K, and $m_{K'} = m_K$ holds for every $K' \in \mathfrak{K}$ with $K' \subset P_K$.

The family $\mathfrak{U} = \{\Gamma/P_K \to \Gamma/H\}_{K \in \mathfrak{K}[H]}$ is a τ-covering, the arrows being the canonical ones. I claim that $\mathbf{m} := (m_K)_{K \in \mathfrak{K}[H]}$ is an element of $H^0(\mathfrak{U}, LM)$, compare (8) and (9). To check this let K, $L \in \mathfrak{K}[H]$ and x, $y \in \Gamma$ with $xH = yH$, and suppose that there is $C \in \mathfrak{K}$ with $CxP_K = xP_K$ and $CyP_L = yP_L$. One has to show (9), i.e. $x \cdot m_K = y \cdot m_L$. Now $K' := x^{-1}Cx \subset P_K$ and $L' := y^{-1}Cy \subset P_L$, so $m_{K'} = m_K$ and $m_{L'} = m_L$. Thus

$$(C, xm_K) = x \cdot (K', m_{K'}) = f(C, xH) = f(C, yH) = y \cdot (L', m_{L'}) = (C, ym_L)$$

which proves the claim.

So $\mathbf{m} \in H^0(\mathfrak{U}, LM)$, and one checks immediately that the corresponding map $f_{\mathbf{m}}$ defined by (10) is just the f one started with. The lemma is proved. $\qquad\square$

(11.2.4.3) Now the proof of the theorem is completed as follows. Let $C \subset F$ be the full subcategory whose objects are the sheaves ϵM, for M a Γ-set. The lemma says that the restriction of ϕ^* to C is fully faithful, so C is also a full subcategory of $\widetilde{\mathfrak{K}}(\Delta)$ via ϕ^*. Moreover the topology on C induced from $\widetilde{\mathfrak{K}}(\Delta)$ is the same as the topology induced from F. This follows since a family $\{M_i \to M\}$ in $(\Gamma$-sets$)$ is a τ-covering iff the family $\{\mathfrak{K}(M_i) \to \mathfrak{K}(M)\}$ is epi in $\widetilde{\mathfrak{K}}(\Delta)$.

So by [SGA4 IV.1.2.1] the assertion of the theorem will follow once it is shown that C is generating for $\widetilde{\mathfrak{K}}(\Delta)$, i.e. that every Δ-sheaf on $\mathfrak{K}$ can be covered by a family of $\mathfrak{K}(M_\alpha)$'s. For this let $B \in \widetilde{\mathfrak{K}}(\Delta)$, considered as an espace étalé $B \to \mathfrak{K}$ with Δ-action. Let $K \in \mathfrak{K}$, let W be an open neighborhood of K in $\mathfrak{K}$ and let b be a section of B over W. Let $b_L \in B_L$ denote the "value" of b in the stalk over L, for $L \in W$. Since B is a *continuous* Δ-sheaf one can find an open neighborhood V of the identity in Δ such that $b_{vLv^{-1}} = v \cdot b_L$ holds for all $v \in V$ and all L in a small neighborhood $W_1 \subset W$ of K in $\mathfrak{K}$ (which in particular is so small that $vLv^{-1} \in W$ for $L \in W_1$ and $v \in V$). Let H be an open normal subgroup of Γ which is contained in V, and write $P := HK$, an open subgroup of Γ with $P \cap \Delta = H$. Assume that H has been chosen so small that $\mathfrak{K}[P]$ is contained in W_1. This is possible since the intersection of all $\mathfrak{K}[HK]$, with H running over the open normal subgroups of Γ, is $\{K\}$. Define a map $f: \mathfrak{K}(\Gamma/P) \to B$ over $\mathfrak{K}$ as follows: Any element in $\mathfrak{K}(\Gamma/P)$ can be written (L, xP) with $L \in \mathfrak{K}$ and $x \in \Delta$. Put $f(L, xP) := x \cdot b_{x^{-1}Lx}$. Note that $x^{-1}Lx \in \mathfrak{K}[P] \subset W_1$, so $b_{x^{-1}Lx}$ is defined. The map f is well-defined: If $xP = yP$ with $y \in \Delta$ then $h := y^{-1}x \in P \cap \Delta = H$ is contained in V, and so $h \cdot b_{x^{-1}Lx} = b_{hx^{-1}Lxh^{-1}} = b_{y^{-1}Ly}$, i.e. $xb_{x^{-1}Lx} = yb_{y^{-1}Ly}$. Moreover f is obviously Δ-equivariant. And f is continuous since $\mathfrak{K}[P] \ni L \mapsto f(L, 1.P) = b_L$ is continuous. In summary, for a given b_K in an arbitrary stalk B_K of B one has found a Γ-set M and a $\widetilde{\mathfrak{K}}(\Delta)$-map $\mathfrak{K}(M) \to B$ which has b_K in its image. So C is generating for $\widetilde{\mathfrak{K}}(\Delta)$. The proof of the theorem is complete. $\square$

(11.2.5) Remark. In the case of discrete groups there is some simplification in the proofs of this section, due to the fact that Γ itself is a discrete Γ-set. An alternative approach to the profinite case would be to obtain it by passing the discrete case to the limit. However this would not only require to consider filtering inverse limits of toposes, rather also such limits of G-toposes, and moreover to show that formation of these limits is compatible with formation of $\underleftarrow{\mathrm{Lim}}\mathrm{top}(-/G)$, etc. Although there do not seem to be any principal difficulties in this approach, it would be highly technical and quite unpleasant to read. Therefore I have preferred to give a direct treatment. A similar situation occurred in Sect. 8 where Γ-sheaves on profinite spaces (with Γ profinite) could also have been introduced via such limit techniques, instead of explicitly speaking of "continuous" Γ-sheaves.

(11.2.6) Further Complements.

(11.2.6.1) The pseudo action of G on (Δ-sets) induced by (4) should be thought of as "having fixpoints" (under all of G) if and only if the sequence (4) splits, since the existence of a splitting is necessary and sufficient for the fixtopos not to be the "empty" topos. Hence the pseudo action of G on (Δ-sets) is to be considered as free if and only if Γ has no subgroups which intersect Δ trivially (other than $\{1\}$), or equivalently, iff all elements of prime order in Γ are contained in Δ. Note also that if (4) splits then there is for every $K \in \mathfrak{K}$ a natural bijection $\mathfrak{K}/\Delta \approx H^1(K, \Delta)$ (non-abelian cohomology).

(11.2.6.2) Write $F := \underleftarrow{\mathrm{Lim}}\mathrm{top}((\Delta\text{-sets})/G) \sim \widetilde{\mathfrak{K}}(\Delta)$. Let $\mathfrak{K}_0$ be a system of representatives of the conjugacy classes in $\mathfrak{K}$ (under Δ or under Γ, this is the same). For $K \in \mathfrak{K}$ denote by $N_\Delta(K)$ its normalizer ($=$ centralizer) in Δ. If Γ is discrete there is an obvious equivalence of categories

$$F \sim \widetilde{\mathfrak{K}}(\Delta) \sim \prod_{K \in \mathfrak{K}_0} (N_\Delta(K)\text{-sets}),$$

and $\nu^*\colon (\Delta\text{-sets}) \to F$ gets identified with the functor which sends a Δ-set N to the family $\left(\mathrm{res}^{\Delta}_{N_\Delta(K)} N\right)_{K \in \mathfrak{K}_0}$. In the profinite case the situation is hardly very different, at least when $\mathfrak{K}_0$ can be chosen as a closed subset of $\mathfrak{K}$; only a condition is added which says that the objects in the product category "vary continuously" over the base space.

(11.2.6.3) Consider the topos morphism $\nu(G)\colon F(G) \to (\Delta\text{-sets})(G) \sim (\Gamma\text{-sets})$ (10.19). Since there is a canonical map $s\colon G \times \mathfrak{K} \to \Gamma$ as in Remark (8.8.4), it follows from [loc. cit.] that $F(G) = (\widetilde{\mathfrak{K}}(\Delta))(G)$ is canonically equivalent to $\widetilde{\mathfrak{K}}(\Gamma)$. Under this identification the inverse image functor $\nu(G)^*$ sends a Γ-set M to the constant sheaf $\mathfrak{K} \times M \to \mathfrak{K}$ with the diagonal Γ-action. If Γ is discrete and $\mathfrak{K}_0$ is a system of representatives as above, this becomes

$$F(G) \sim \widetilde{\mathfrak{K}}(\Gamma) \sim \prod_{K \in \mathfrak{K}_0} (N_\Gamma(K)\text{-sets}),$$

and $\nu(G)^*$ sends a Γ-set M to the family $\left(\mathrm{res}^{\Gamma}_{N_\Gamma(K)} M\right)_{K \in \mathfrak{K}_0}$. In particular, if A is a Γ-module one has isomorphisms

$$H^n_\Gamma(\mathfrak{K}, A) \cong \prod_{K \in \mathfrak{K}_0} H^n(N_\Gamma(K), A), \quad n \geq 0.$$

Assume now that G is cyclic of prime order p. Let A be a Γ-module and consider the canonical maps (pullback by $\nu(G)^*$)

$$H^n(\Gamma, A) \longrightarrow H^n\left(F(G), \nu(G)^*A\right) = H^n_\Gamma(\mathfrak{K}, A), \quad n \geq 0. \tag{12}$$

Suppose that Γ is discrete and that Δ has finite cohomological p-dimension $d < \infty$. Then (12) is an isomorphism for every $n > d$ and every p-primary torsion Γ-module A. This follows from results about *Farrell cohomology* $\widehat{H}^*(\Gamma, A)$ (see [Br ch. X]): On the one hand $H^n(H, A) \to \widehat{H}^n(H, A)$ is an isomorphism for $n > d$ and all subgroups H of Γ; and on the other one has

$$\widehat{H}^*(\Gamma, A) \xrightarrow{\ \sim\ } \widehat{H}^*_\Gamma(\mathfrak{K}, A) \cong \prod_{K \in \mathfrak{K}_0} \widehat{H}^*(N_\Gamma(K), A),$$

see [loc.cit., Cor. X.7.4].

Note how this is completely analogous to the main result of Sect. 7 (Corollary (7.20))! In the field case the latter result translates into an analogue of the just mentioned facts for *profinite* groups, namely absolute Galois groups of fields. This was discussed extensively in Sect. 9. The question was raised there (9.13) whether the proof given for such Galois groups could be extended to more general profinite groups. The remark just made shows that this result generalizes indeed to *discrete* groups, a fact which strongly supports the idea that it should also be true for general profinite groups. The proof that this is actually so will be the subject of Sect. 12. For a complementary discussion see Sect. 14, in particular Remark (14.9).

(11.2.6.4) The results of this section illustrate the fact that, in general, the inverse limit of a G-topos is a much more delicate object than its direct limit. This becomes apparent even in the case of a trivial G-topos: If E is any topos there is a canonical topos morphism

$$E \longrightarrow \underleftarrow{\mathrm{Lim}}\mathrm{top}(E \times BG/G), \tag{13}$$

by the universal property of the inverse limit. The idea of the $\underleftarrow{\mathrm{Lim}}\mathrm{top}$ as a kind of "G-fixtopos" would suggest that (13) should be an equivalence. But this is too naïve. Consider for example an exact sequence (4) and suppose that it splits as a direct product $\Gamma \cong \Delta \times G$, which means that the G-topos (Δ-sets) is trivial. Then $\mathfrak{K} = \mathrm{Hom}(G, \Delta)$, the set of group homomorphisms on which Δ acts by conjugation. If there is only the trivial homomorphism $G \to \Delta$ one gets indeed (Δ-sets) as the $\underleftarrow{\mathrm{Lim}}\mathrm{top}$. But in all other cases the $\underleftarrow{\mathrm{Lim}}\mathrm{top}$ takes care of the fact that there are topos morphisms with target (Δ-sets) on which G acts in a non-trivial way. This is a feature of the 2-categorical nature of toposes which has no counterpart for usual topological spaces.

12. High degree cohomology of profinite groups

Let p be a prime, Γ a profinite group and A a discrete Γ-module which is p-primary torsion. Assume that Γ has finite virtual cohomological p-dimension d. Assume moreover that Γ contains no subgroups isomorphic to $\mathbb{Z}/p \times \mathbb{Z}/p$. Under these assumptions, the cohomology groups $H^n(\Gamma, A)$ are for $n > d$ canonically isomorphic to the Γ-equivariant sheaf cohomology groups $H^n_\Gamma(\mathfrak{X}, A)$, where $\mathfrak{X}$ is the space of subgroups of order p of Γ. This is the content of the main result of this section (Theorem (12.13)).

This theorem fits into the framework of this paper under two aspects. Firstly, it generalizes the field case of the main result of Part One (Corollary (7.19)) from absolute Galois groups to much more general profinite groups, and primes different from 2; see (12.20.1). Thereby it resolves the question of how special the latter result is to Galois groups of fields, cf. the discussion in (9.13). Secondly, consider the particular case when Γ has an open normal subgroup Δ of index p which is p-torsionfree. Put $G = \Gamma/\Delta$. The fixtopos of the G-topos (Δ-sets) was found in Sect. 11.2 to be the topos $\widetilde{\mathfrak{X}}(\Delta)$ of Δ-sheaves on $\mathfrak{X}$. By the observations in (8.9), the groups $H^n_\Gamma(\mathfrak{X}, A)$ are nothing but G-equivariant cohomology of the G-object $\nu(G)^* A$ in this fixtopos. So Theorem (12.13) follows exactly the same pattern as the main result of Part One (7.19). I refer to Sect. 14 for a further discussion of such common aspects.

There are some known results for *discrete* groups which are closely related to result here. Indeed, there is a theorem of K.S. Brown which, for any discrete group Γ with $\mathrm{vcd}(\Gamma) < \infty$, identifies high degree cohomology of p-primary Γ-modules with their Γ-equivariant cohomology on a certain simplicial complex. This complex is formed by the elementary abelian p-subgroups $\neq \{1\}$ of Γ. In particular, if Γ happens to have no rank 2 such subgroups, this theorem is in complete analogy to (12.13); but Brown doesn't need such kind of hypothesis for his result. So Theorem (12.13) below must be seen as a profinite analogue of the "p-rank ≤ 1" case of Brown's theorem. As far as I know there is no such profinite analogue in the literature so far.

It is natural to try and see if the method of proof used by Brown can be transferred from the discrete to the profinite case. A key point in his proof is the use of the theorem of Eilenberg and Ganea [Br, p. 205]. This theorem states in particular that a discrete group of finite cohomological dimension acts freely on a suitable *finite-dimensional* CW-complex. Unfortunately, no profinite analogue of this theorem seems to be in sight.

Therefore a different approach had to be found for profinite groups. It lies at the basis of the proof below that one can calculate profinite group cohomology via *projective* resolutions, instead of injective ones. For this one has to consider also profinite Γ-modules, instead of only discrete ones. I found the details of this calculus in D. Haran's paper [Ha2], but it goes back (at least) to a paper by A. Brumer [Bm], who was in fact working in much greater generality. In Haran's paper a remark of Serre is amplified to give a proof of Serre's profinite theorem using projective resolutions, thereby avoiding cup products and Steenrod powers. For the convenience of the reader I have included below what is needed about profinite Γ-modules; some proofs are omitted since they consist of straightforward verifications or can be found in [Ha2].

The method used here is limited to the case of p-rank ≤ 1. However its ideas have recently been generalized by the author to give a proof valid without this restriction. Thus, analogues of Brown's theorems for profinite groups exist in full generality. For this, and for a more complete treatment, see the forthcoming paper [Sch2].

Throughout this section let Γ be a profinite group and p a fixed prime. For brevity write $\mathbb{F}_p$ for the field $\mathbb{Z}/p\mathbb{Z}$ in the following.

(12.1) Definition. A *profinite $\mathbb{F}_p\Gamma$-module* is a topological left Γ-module A whose underlying abelian group is profinite and annihilated by p; or equivalently, an inverse limit of finite discrete left Γ-modules which are annihilated by p. The group of continuous Γ-homomorphisms between two topological Γ-modules A, B is denoted $\mathrm{Hom}_\Gamma(A, B)$. The category $\mathrm{Mod}_{\mathrm{prof}}(\mathbb{F}_p\Gamma)$ of profinite $\mathbb{F}_p\Gamma$-modules and continuous Γ-homomorphisms is an abelian category. If $\Gamma = \{1\}$ I speak of the category $\mathrm{Mod}_{\mathrm{prof}}(\mathbb{F}_p)$ of *profinite $\mathbb{F}_p$-modules.* A "discrete $\mathbb{F}_p\Gamma$-module" will as usual mean a discrete (left) Γ-module (with open stabilizer subgroups!), not necessarily finite, which is annihilated by p.

(12.2.1) Examples of profinite $\mathbb{F}_p\Gamma$-modules arise as follows. Let X be a boolean topological space on which Γ acts continuously. There is a filtering inverse system $\{X_\alpha\}$ of discrete quotient Γ-sets of X such that $X = \varprojlim X_\alpha$. Therefore such a Γ-space will be called a *profinite Γ-space.* Let $\mathbb{F}_p[X_\alpha]$ be the permutation Γ-module of X_α, i.e. the $\mathbb{F}_p$-vector space with basis X_α on which Γ acts by permutation of the basis, and put $\mathbb{F}_p[X] := \varprojlim \mathbb{F}_p[X_\alpha]$. This is a profinite $\mathbb{F}_p\Gamma$-module. It contains X as a topological subspace which is Γ-invariant and on which the Γ-action is the given one. The case $X = \Gamma$ gives the *completed group algebra* $\mathbb{F}_p\Gamma :=$ $\varprojlim \mathbb{F}_p[\Gamma/N]$ of Γ, where N is running through the open normal subgroups of Γ. The profinite $\mathbb{F}_p\Gamma$-modules in the sense of (12.1) are precisely the topological left modules over this completed group algebra $\mathbb{F}_p\Gamma$ which are profinite (as abelian group, or as module); this justifies the terminology. For any profinite $\mathbb{F}_p\Gamma$-module

A the natural (restriction) map

$$\mathrm{Hom}_{\Gamma}\big(\mathbb{F}_p[X], A\big) \longrightarrow \mathrm{Hom}_{\Gamma}(X, A) \tag{1}$$

is bijective (the right hand side denotes the continuous Γ-maps $X \to A$).

(12.2.2) In particular, let X be a boolean topological space, and let Γ act on $\Gamma \times X$ through the first component. The profinite $\mathbb{F}_p\Gamma$-module $\mathbb{F}_p[\Gamma \times X]$ is called the *free* profinite $\mathbb{F}_p\Gamma$-module on X, and is denoted by $F_{\Gamma}(X)$. The terminology is justified by the fact that, for any profinite $\mathbb{F}_p\Gamma$-module A, the natural map

$$\mathrm{Hom}_{\Gamma}\big(F_{\Gamma}(X), A\big) \longrightarrow C(X, A) = \{\text{continuous maps } X \to A\} \tag{2}$$

is bijective. Every free profinite $\mathbb{F}_p\Gamma$-module is a projective object in the abelian category $\mathrm{Mod}_{\mathrm{prof}}(\mathbb{F}_p\Gamma)$. Indeed, for $P \in \mathrm{Mod}_{\mathrm{prof}}(\mathbb{F}_p\Gamma)$ to be projective it is (necessary and) sufficient that every lifting problem

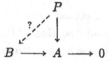

with *finite* $\mathbb{F}_p\Gamma$-modules A, B is solvable [Ha2, Lemma 3.2]; and for $P = F_{\Gamma}(X)$ this lifting property is immediate from the bijectivity of (2).

(12.2.3) This shows that the abelian category $\mathrm{Mod}_{\mathrm{prof}}(\mathbb{F}_p\Gamma)$ has enough projective objects (which is not true for the category of discrete $\mathbb{F}_p\Gamma$-modules, unless Γ is finite!): For any profinite $\mathbb{F}_p\Gamma$-module A the canonical map $F_{\Gamma}(A) \to A$ (determined by id_A via (2)) is clearly surjective. The projective objects are exactly the direct summands of free profinite $\mathbb{F}_p\Gamma$-modules.

(12.2.4) If M is a profinite $\mathbb{F}_p\Gamma$-module and A is a *discrete* $\mathbb{F}_p\Gamma$-module then $\mathrm{Hom}(M, A)$ (the group of locally constant group homomorphisms $M \to A$) is again a *discrete* $\mathbb{F}_p\Gamma$-module: $\mathrm{Hom}(M, A) = \varinjlim_{\alpha} \mathrm{Hom}(M_\alpha, A)$ where M is written as a filtering inverse limit $M = \varprojlim_{\alpha} M_\alpha$ of finite $\mathbb{F}_p\Gamma$-modules M_α. The Γ-action on $\mathrm{Hom}(M, A)$ is by $(x \cdot f)(m) = x \cdot f(x^{-1}m)$. Note that $\mathrm{Hom}(M, A)$ is exact as a functor of A. If X is a profinite Γ-space then $\mathrm{Hom}\big(\mathbb{F}_p[X], A\big)$ is the Γ-module of all locally constant maps $X \to A$; this module will usually be denoted by $H^0(X, A)$ or $C(X, A)$. More generally I use this notation to denote the discrete Γ-module of locally constant maps $X \to A$ for *any* discrete Γ-module A (not necessarily over $\mathbb{F}_p$).

(12.3) **Remark.** By Pontryagin duality, the abelian category of profinite (left) $\mathbb{F}_p\Gamma$-modules is dual to the category of discrete (right) $\mathbb{F}_p\Gamma$-modules: One passes from either category to the other by forming the dual module $M^* := \mathrm{Hom}(M, \mathbb{F}_p)$. In this way one may transfer results from one category to the other. For example, the existence of projective resolutions in $\mathrm{Mod}_{\mathrm{prof}}(\mathbb{F}_p\Gamma)$ (12.2.3) follows immediately from the existence of injective resolutions in the dual category of discrete $\mathbb{F}_p\Gamma$-modules.

One can compute the cohomology of discrete $\mathbb{F}_p\Gamma$-modules via projective resolutions by profinite $\mathbb{F}_p\Gamma$-modules. This could also be proved using duality, but I'll give a direct proof here:

(12.4) Proposition. — *Let A be a discrete $\mathbb{F}_p\Gamma$-module, not necessarily finite, and let*

$$\cdots \longrightarrow P_2 \longrightarrow P_1 \longrightarrow P_0 \longrightarrow \mathbb{Z}/p \longrightarrow 0$$

be a projective resolution of $\mathbb{Z}/p$ in $\mathrm{Mod}_{\mathrm{prof}}(\mathbb{F}_p\Gamma)$. Then $H^*(\Gamma, A)$ is canonically isomorphic to the cohomology of the complex

$$0 \longrightarrow \mathrm{Hom}_\Gamma(P_0, A) \longrightarrow \mathrm{Hom}_\Gamma(P_1, A) \longrightarrow \cdots. \tag{3}$$

Proof. Write $A = \varinjlim_\alpha A_\alpha$ as the directed union of its finite submodules. Then, for any complex P of profinite $\mathbb{F}_p$-modules, $\mathrm{Hom}_\Gamma(P, A)$ is the filtering direct limit of the complexes $\mathrm{Hom}_\Gamma(P, A_\alpha)$. So one may assume A finite. If P and $\tilde{P}$ are any two projective resolutions of $\mathbb{Z}/p$ then the cohomology groups of $\mathrm{Hom}_\Gamma(P, A)$ and $\mathrm{Hom}_\Gamma(\tilde{P}, A)$ are canonically isomorphic. Hence one can choose a special resolution P, for example the profinite bar resolution of $\mathbb{Z}/p$. It can be obtained as the inverse limit of the bar resolutions of the finite quotients Γ_α of Γ. The limit complex P is exact since $\varprojlim$ is exact for systems of finite abelian groups. Since $\mathrm{Hom}_\Gamma(-, A)$ turns the $\varprojlim$ into a $\varinjlim$ this shows that the cohomology of (3) is $\varinjlim_\alpha H^*(\Gamma_\alpha, A)$, limit over the finite quotients Γ_α of Γ which act on A, and thus is $H^*(\Gamma, A)$. $\quad\square$

(12.5) Proposition [Ha2, Cor. 3.7]. — *The cohomological p-dimension $\mathrm{cd}_p(\Gamma)$ (for discrete Γ-modules) is finite if and only if $\mathbb{Z}/p$ has a projective resolution of finite length (by profinite $\mathbb{F}_p\Gamma$-modules).*

Proof. $\mathrm{cd}_p(\Gamma)$ is the supremum of all $n \geq 0$ for which there is a finite $\mathbb{F}_p\Gamma$-module A with $H^n(\Gamma, A) \neq 0$. One implication is immediate from (12.4). For the other assume that $d := \mathrm{cd}_p(\Gamma) < \infty$. Let $\cdots \to P_1 \to P_0 \to \mathbb{Z}/p \to 0$ be any (infinite) projective resolution in $\mathrm{Mod}_{\mathrm{prof}}(\mathbb{F}_p\Gamma)$. Then $\mathrm{im}(P_{d+1} \to P_d) =: Z_{d+1}$ is easily seen to be projective, and so $0 \to Z_{d+1} \to P_d \to \cdots \to P_0 \to \mathbb{Z}/p \to 0$ is a projective resolution (of finite length $d+1$). $\quad\square$

It can be shown that $\mathbb{Z}/p$ has even a projective resolution of length $d = \mathrm{cd}_p(\Gamma)$ [Bm, §3]. This fact won't be used, however.

(12.6) Definition. Let A, B be profinite $\mathbb{F}_p$-modules. A profinite $\mathbb{F}_p$-module C together with a bilinear continuous map $\beta \colon A \times B \to C$ will be called a *completed* *tensor product* of A and B if the obvious universal property holds: For any $D \in \text{Mod}_{\text{prof}}(\mathbb{F}_p)$ the map

$$\text{Hom}(C, D) \longrightarrow \{\text{continuous bilinear maps } A \times B \to D\}$$

induced by β is bijective. The notation for C will be $A \widehat{\otimes} B$. The completed tensor product exists always and can be constructed as the inverse limit $\varprojlim A_\alpha \otimes B_\beta$, where $\{A_\alpha\}$, $\{B_\beta\}$ are the inverse systems of finite quotients of A and B. Note that $\widehat{\otimes}$ is an exact functor in both arguments. The canonical map $\beta \colon A \times B \to A \widehat{\otimes} B$ will be written $(a, b) \mapsto a \widehat{\otimes} b$.

(12.6.1) If A and B are profinite $\mathbb{F}_p \Gamma$-modules then $A \widehat{\otimes} B$ is again a profinite $\mathbb{F}_p \Gamma$-module via the rule $\sigma . (a \widehat{\otimes} b) = (\sigma . a) \widehat{\otimes} (\sigma . b)$. This is clear since it is true for the finite approximations $A_\alpha \otimes B_\beta$.

(12.6.2) For example, if X, Y are profinite Γ-spaces, then $\mathbb{F}_p[X] \widehat{\otimes} \mathbb{F}_p[Y]$ is naturally isomorphic to $\mathbb{F}_p[X \times Y]$ (as profinite $\mathbb{F}_p \Gamma$-modules). This is clear by finite approximation.

(12.7) If C, D are complexes of profinite $\mathbb{F}_p \Gamma$-modules (with differentials of degree -1) one forms the completed tensor product complex $C \widehat{\otimes} D$. It has the components

$$\left(C \widehat{\otimes} D\right)_n = \bigoplus_{p+q=n} C_p \widehat{\otimes} D_q,$$

and differential $\partial(x \widehat{\otimes} y) = \partial(x) \widehat{\otimes} y + (-1)^{\deg(x)} x \widehat{\otimes} \partial(y)$ for homogeneous elements x, y. Since $\widehat{\otimes}$ is exact the usual proof of the Künneth formula shows that

$$H_*(C) \widehat{\otimes} H_*(D) \xrightarrow{\sim} H_*(C \widehat{\otimes} D)$$

as graded topological $\mathbb{F}_p \Gamma$-modules. (Under suitable boundedness conditions on C and D the components of $C \widehat{\otimes} D$ will again be profinite.)

(12.8) I need the concept of tensor induction for profinite modules. It completely parallels the case of discrete groups, which can be found e.g. in [BnI, 3.15] and [BnII, 4.1]. To arrive at smoother formulas I use however slightly different notations and conventions, compare [Pa, sect. 10.3].

(12.8.1) Let Γ be a profinite group, and fix an open subgroup Δ of Γ of index $n = [\Gamma : \Delta]$. Let Σ_n be the symmetric group on n letters. I assume that Σ_n acts from the *right* on $\{1, \ldots, n\}$, in order to get nicer formulas. Recall the construction of the *wreath product* $\Delta \wr \Sigma_n$: This is the split extension

$$1 \longrightarrow \Delta^n \longrightarrow \Delta \wr \Sigma_n \longrightarrow \Sigma_n \longrightarrow 1$$

of profinite groups, where the operation of Σ_n on $\Delta^n = \Delta \times \cdots \times \Delta$ (n times) is by permutation of the entries. So $\Delta \wr \Sigma_n$ is $\Delta^n \times \Sigma_n$ as a set, with multiplication rule

$$(y_1, \ldots, y_n; \pi) \cdot (y_1', \ldots, y_n'; \pi') = (y_1 y_{1\pi}', \ldots, y_n y_{n\pi}'; \pi\pi'). \tag{4}$$

A choice of representatives $\sigma_1, \ldots, \sigma_n \in \Gamma$ for the cosets $\Delta\sigma$ ($\sigma \in \Gamma$) gives rise to an open embedding $\Gamma \hookrightarrow \Delta \wr \Sigma_n$, namely by mapping $x \in \Gamma$ to the tuple $(y_1(x), \ldots, y_n(x); \pi_x)$, where $y_i(x) \in \Delta$ and $\pi_x \in \Sigma_n$ are determined by the equations

$$\sigma_i x = y_i(x) \sigma_{i\pi_x}, \quad i = 1, \ldots, n. \tag{5}$$

A different choice of representatives σ_i gives rise to a conjugate embedding of Γ. For the following fix a particular transversal $\sigma_1, \ldots, \sigma_n$ and the corresponding embedding.

Let B be a profinite $\mathbb{F}_p\Delta$-module. The n-fold completed tensor product $B^{\widehat{\otimes} n} := B \widehat{\otimes} \cdots \widehat{\otimes} B$ (n times) is a profinite $\mathbb{F}_p(\Delta \wr \Sigma_n)$-module in a natural way, by the formula

$$(y_1, \ldots, y_n; \pi) \cdot (b_1 \widehat{\otimes} \cdots \widehat{\otimes} b_n) = y_1 b_{1\pi} \widehat{\otimes} \cdots \widehat{\otimes} y_n b_{n\pi}.$$

By definition the (completed) *tensor induced module* $\operatorname{ten}_\Delta^\Gamma(B)$ (the notation of [BnI] is $B^{\widehat{\otimes}\Gamma}$) is the restriction of this $(\Delta \wr \Sigma_n)$-module $B^{\widehat{\otimes} n}$ to Γ, with $\Gamma \hookrightarrow \Delta \wr \Sigma_n$ embedded as above. So the action of Γ is given by

$$x \cdot (b_1 \widehat{\otimes} \cdots \widehat{\otimes} b_n) = y_1(x) b_{1\pi_x} \widehat{\otimes} \cdots \widehat{\otimes} y_n(x) b_{n\pi_x}. \tag{6}$$

(12.8.2) For the generalization to complexes of modules one has to be careful about signs: see [Sw p. 608]. Let D be a complex of profinite $\mathbb{F}_p\Delta$-modules (differentials always of degree -1). The n-fold tensor power $D^{\widehat{\otimes} n} = D \widehat{\otimes} \cdots \widehat{\otimes} D$ is a complex of $\Delta \wr \Sigma_n$-modules by the rule

$$(y_1, \ldots, y_n; \pi) \cdot (d_1 \widehat{\otimes} \cdots \widehat{\otimes} d_n) = (-1)^\nu \cdot y_1 d_{1\pi} \widehat{\otimes} \cdots \widehat{\otimes} y_n d_{n\pi} \tag{7}$$

where $d_i \in D$ are homogeneous elements and ν is given by

$$\nu = \sum_{\substack{i < j \\ i\pi > j\pi}} \deg(d_{i\pi}) \deg(d_{j\pi}). \tag{8}$$

A tedious verification [Pa] shows that this rule defines indeed a group operation which is compatible with the boundary maps. The restriction of the $(\Delta \wr \Sigma_n)$-complex $D^{\widehat{\otimes} n}$ to Γ is called the *tensor induction* $\operatorname{ten}_\Delta^\Gamma(D)$ of the complex D to Γ. It is a complex of topological $\mathbb{F}_p\Gamma$-modules, which are profinite under suitable conditions on D (e.g. if $D_i = 0$ for $i \ll 0$).

(12.9) Remark. Let G be a discrete group and H a subgroup of finite index n. It is explained in [BnI, p. 89] how the concept of tensor induction is similar to the usual induction functor $B \mapsto \mathbb{Z}G \otimes_{\mathbb{Z}H} B$ from H-modules to G-modules. However it seems preferable to view this as a similarity to the *coinduction* functor $B \mapsto \text{Hom}_{\mathbb{Z}H}(\mathbb{Z}G, B)$, rather than to induction. Although induction and coinduction are isomorphic on modules (since $[G : H]$ is finite) they are different on sets with group action. If V is an H-set and $\mathbb{Z}[V]$ is the associated permutation H-module, the tensor induced module $\text{ten}_H^G(\mathbb{Z}[V])$ bears no relation to the induced G-set $G \times_H V$. Rather it is the permutation G-module on the G-set *coinduced* by V:

$$\text{ten}_H^G(\mathbb{Z}[V]) = \mathbb{Z}[\text{coind}_H^G(V)].$$

This identification is completely canonical: Identifying $\text{coind}_H^G(V) = \text{Hom}_H(G, V)$ with V^n by means of the transversal $\sigma_1, \ldots, \sigma_n$ one finds from (5) the formula $x \cdot (v_1, \ldots, v_n) = (y_1(x)v_{1\pi_x}, \ldots, y_n(x)v_{n\pi_x})$ for the G-action, which is in accordance with (6).

The same holds *mutatis mutandis* in the profinite category: Let V be a profinite Δ-space and $\mathbb{F}_p[V]$ the associated profinite $\mathbb{F}_p\Delta$-module (12.2.1). Then there is a canonical isomorphism

$$\text{ten}_\Delta^\Gamma(\mathbb{F}_p[V]) \cong \mathbb{F}_p[U] \tag{9}$$

of profinite $\mathbb{F}_p\Gamma$-modules, where $U := \text{coind}_\Delta^\Gamma(V) = \text{Hom}_\Delta(\Gamma, V)$ is a profinite Γ-space by the usual action. (The topology on U is the product topology on V^n after making the identification $U \approx V^n$ by means of the transversal $\sigma_1, \ldots, \sigma_n$; this topology does not depend on the choice of the transversal.) The proof of the isomorphism (9) is clear by finite approximation.

(12.10) Remark. More generally let Q be a non-negative complex of profinite $\mathbb{F}_p\Delta$-modules, and assume that there are profinite Δ-spaces V_r with $Q_r \cong \mathbb{F}_p[V_r]$, $r \geq 0$. Let $P := \text{ten}_\Delta^\Gamma(Q)$ be the tensor induction of the complex Q (12.8.2). The relation of the components P_r to Γ-spaces coinduced from the V_i is more complicated here. This is mainly due to the $\pm$ sign in (7). For simplicity make the following hypothesis, which fortunately can be assumed to hold in later applications: Either $p = 2$, or $[\Gamma : K]$ is odd where $K = \bigcap_{x \in \Gamma} x\Delta x^{-1}$. Let $r \geq 0$. Then

$$P_r = \bigoplus_{i_1 + \cdots + i_n = r} Q_{i_1} \widehat{\otimes} \cdots \widehat{\otimes} Q_{i_n} = \bigoplus_{i_1 + \cdots + i_n = r} \mathbb{F}_p[V_{i_1} \times \cdots \times V_{i_n}]$$

(cf. (12.6.2)) is free as a profinite $\mathbb{F}_p$-module on its subspace

$$B := \coprod_{i_1 + \cdots + i_n = r} V_{i_1} \times \cdots \times V_{i_n}.$$

Let U_r denote this space B together with the Γ-action

$$x \cdot (v_1, \ldots, v_n) = (y_1(x)v_{1\pi_x}, \ldots, y_n(x)v_{n\pi_x}) \tag{10}$$

$(x \in \Gamma, v_j \in V_{i_j}, i_1 + \cdots + i_n = r)$. (Note that (10) coincides with the Γ-action on P_r only up to a sign!) In terms of coinduction, U_r is the sub-Γ-space of

$$U := \text{coind}_\Delta^\Gamma \Big(\coprod_{i \geq 0} V_i \Big) = \text{Hom}_\Delta \Big(\Gamma, \coprod_{i \geq 0} V_i \Big)$$

consisting of the "weight r" elements. (The weight of an element $f \in U$ is by definition $i_1 + \cdots + i_n$ where i_j are such that $f(\Delta \sigma_j) \subset V_{i_j}$.) However, the Γ-action on P_r does not in general preserve the subspace B of P_r, it does so only up to signs $\pm$. But from the hypothesis made before it follows that, in the case when p is odd, there are no $x \in \Gamma$ and $b \in B$ with $x \cdot b = -b$ (in P_r). Indeed, this would contradict $[\Gamma : K] = \text{odd}$ since π_x would have to have even order in Σ_n. From this it follows that there is a locally constant map $\epsilon: B \to \{\pm 1\}$ such that the subset $\tilde{B} := \{\epsilon(b)b : b \in B\}$ of P_r is stable under Γ. (Indeed, if p is odd then B can be covered by clopen subsets B_λ of B such that $-x \cdot b \notin B_\lambda$ for $x \in \Gamma$ and $b \in B_\lambda$; from such patches one builds a map ϵ as desired.) Since $\tilde{B} \cong U_r$ as a Γ-space one has thus found an isomorphism $P_r \cong \mathbb{F}_p[U_r]$ of profinite $\mathbb{F}_p\Gamma$-modules.

(A more conceptual formulation of the last argument would go like this. The canonical map $B \cup (-B) \to U_r$ is a $2:1$-covering of profinite Γ-spaces (assume p odd). What was shown above is a particular case of the following more general fact: An n-fold covering $\pi: Y \to X$ of profinite Γ-spaces is trivial (i.e. Γ-isomorphic to the sum of n copies of X) if and only if for every $x \in X$ the isotropy subgroup Γ_x acts trivially on $\pi^{-1}(x)$.)

(12.11) Remark. Serre's theorem for discrete groups ([Sw Thm. 9.2], [Se3 p. 96]) states: If Γ is a torsion-free group and Δ is a subgroup of finite index then $\text{cd}(\Gamma) = \text{cd}(\Delta)$. The essential part of the proof consists in showing that $\text{cd}(\Delta) < \infty$ implies $\text{cd}(\Gamma) < \infty$. In [Se3] Serre gives a topological proof: From a theorem of Eilenberg-Ganea one knows that $\text{cd}(\Delta) < \infty$ implies that there is a finite-dimensional $K(\Delta, 1)$ space. Its universal covering space T is hence a finite-dimensional contractible cell complex with a free cellular Δ-action. The proof consists of constructing from T a finite-dimensional contractible free Γ-complex, namely one takes the coinduced Γ-space $Z = \text{Hom}_\Delta(\Gamma, T)$. To show that Γ acts freely on Z one observes that all isotropy groups of the action on Z intersect Δ trivially. So they must be finite, hence trivial by hypothesis.

Serre himself explains how to translate this proof into purely algebraic terms ([Se3 p. 98]; this algebraic proof had been published previously in [Sw pp. 607-608]): Instead of a free contractible Δ-space of finite dimension one takes a projective resolution Q of $\mathbb{Z}$ over Δ which has finite length. The formation of the coinduced Γ-space is replaced by taking the tensor induced complex $P := \text{ten}_\Delta^\Gamma(Q)$: compare Remark (12.9). The complex P is a resolution of finite length of $\mathbb{Z}$ by Γ-modules which are projective as Δ-modules; and from the hypothesis that Γ has no torsion one can conclude that the P_n are projective also as Γ-modules.

As Serre points out in [Se3], one advantage of this algebraic proof is that it is capable of being transferred to profinite groups, and that in this way a more elementary proof can be given for the profinite analogue of the above theorem. This profinite version states, for a profinite group Γ without p-torsion, that $cd_p(\Gamma) = cd_p(\Delta)$ holds for any open subgroup Δ. Serre proved it first in [Se2] using the multiplicative structure of cohomology rings and Steenrod powers. The paper [Ha2] elaborates at length on Serre's remark, it contains the details of the profinite version of the algebraic proof. Most of this material has already been summarized in this section.

(12.12) Let Γ be a profinite group. If Δ is an open subgroup of Γ then $cd_p(\Delta) \neq cd_p(\Gamma)$ can happen only if $cd_p(\Gamma) = \infty$ and $cd_p(\Delta) < \infty$; in this case, Γ contains an element of order p, by Serre's theorem. The *virtual cohomological p-dimension* of Γ is by definition

$$vcd_p(\Gamma) := \inf\{cd_p(\Delta)\colon \Delta \text{ is an open subgroup of } \Gamma\}.$$

Assume that $vcd_p(\Gamma) < \infty$. If Γ contains elements of order p then $H^n(\Gamma, A) \neq 0$ for arbitrarily large n and suitable $\mathbb{F}_p\Gamma$-modules A. The following theorem expresses this cohomology in high degrees in terms of the subgroups of order p of Γ, provided that Γ contains no $\mathbb{Z}/p \times \mathbb{Z}/p$. Let $\mathfrak{X}$ be the space of all subgroups T of Γ of order p, considered as a subset of $\mathrm{Subg}(\Gamma)$ (cf. (11.2.1)) with the subspace topology. Clearly $\mathfrak{X}$ is closed in $\mathrm{Subg}(\Gamma)$, and hence is a boolean space, on which Γ acts (continuously) by conjugation.

(12.13) Theorem. — *Assume that $d := vcd_p(\Gamma)$ is finite, and that Γ contains no subgroup isomorphic to $\mathbb{Z}/p \times \mathbb{Z}/p$. Let A be any discrete p-primary Γ-module. Then the canonical homomorphism*

$$H^n(\Gamma, A) \longrightarrow H^n_\Gamma(\mathfrak{X}, A) = H^0\big(\mathfrak{X}/\Gamma, \mathcal{H}^n(A)\big) \tag{11}$$

is an isomorphism for every $n > d$, where $\mathfrak{X}$ is the space of subgroups of Γ of order p.

(12.13.1) Here $H^n_\Gamma(\mathfrak{X}, A)$ is the Γ-equivariant cohomology of the constant sheaf A on $\mathfrak{X}$ (which is a Γ-sheaf by the diagonal action on the espace étalé $\mathfrak{X} \times A$). By (8.9) this group is canonically isomorphic to $H^0\big(\mathfrak{X}/\Gamma, \mathcal{H}^n(A)\big)$, where $\mathcal{H}^n(A) := R^n p_*^\Gamma A$ is a sheaf on the quotient space $\mathfrak{X}/\Gamma$ whose stalk at the orbit of $T \in \mathfrak{X}$ is $H^n(T, A)$. Recall also that $H^n_\Gamma(\mathfrak{X}, A) = H^n\big(\Gamma, C(\mathfrak{X}, A)\big)$, and under this identification the homomorphism (11) is induced by the obvious map $A \to C(\mathfrak{X}, A)$ of Γ-modules (which sends $a \in A$ to the constant map a on $\mathfrak{X}$).

Note that Serre's profinite theorem [Se2] is contained in Theorem (12.13) as a particular case. The proof of (12.13) can be considered a generalization of the "elementary" proof of Serre's profinite theorem, which is outlined in [Se3] (see also [Sw]) and carried out in [Ha2]; see Remark (12.11). Serre's theorem is however not used in the proof below.

(12.13.2) If one considers arbitrary discrete Γ-modules A (not necessarily p-primary torsion) then Theorem (12.13) shows that the map $H^n(\Gamma, A)_{p-\text{tors}} \to H^n_\Gamma(\mathfrak{T}, A)_{p-\text{tors}}$ is an isomorphism for $n > d + 1$. See (12.16.3) for the easy argument.

(12.13.3) The plan of the proof is as follows. First one shows that it suffices to prove that (11) is an isomorphism for large n (with a bound which is independent of A). Then it is proved that the map $A \to C(\mathfrak{T}, A)$ induces an isomorphism of cohomology in *all* positive degrees when it is considered as a map of T-modules, for any $T \in \mathfrak{T}$. This allows one to reduce the proof of the theorem to the following statement: There is an integer $e \geq 0$ such that $H^q(\Gamma, A)$ vanishes for $q > e$ whenever A is a discrete $\mathbb{F}_p\Gamma$-module for which $H^n(T, A) = 0$ for all $T \in \mathfrak{T}$ and $n \geq 1$. To prove this one takes an open subgroup Δ of Γ with $\text{cd}_p(\Delta) < \infty$ and a projective resolution Q of $\mathbb{Z}/p$ over Δ. One forms the tensor induced complex $P = \text{ten}^\Gamma_\Delta(Q)$ (which is a resolution of $\mathbb{Z}/p$ over Γ), regards $H^*(\Gamma, A)$ as the hyper-ext $\text{Ext}_{\mathbb{F}_p\Gamma}(P, A)$ and studies one of the associated spectral sequences.

   In the actual proof below the arguments have been arranged in a slightly different order, to allow a smoother exposition.

(12.14) Lemma. — *Let Γ be a profinite group and H a closed subgroup of Γ. For every discrete Γ-module A there are canonical isomorphisms*

$$H^n(\Gamma, C(\Gamma/H, A)) \cong H^n(H, A), \quad n \geq 0.$$

Proof. Assume that H is open in Γ. The functor $A \mapsto A^H$ from discrete Γ-modules to abelian groups can be decomposed in two ways:

$$A^H = H^0(\Gamma, C(\Gamma/H, A)) = H^0(H, \text{res}^\Gamma_H(A)).$$

Both functors $A \mapsto C(\Gamma/H, A)$ and $A \mapsto \text{res}^\Gamma_H(A)$ are exact and have an exact left adjoint (namely $M \mapsto M \otimes \mathbb{Z}[\Gamma/H]$ in the first case and ind^Γ_H in the second). Hence the derived functors of $A \mapsto A^H$ can be expressed as

$$H^n(\Gamma, C(\Gamma/H, A)) \cong H^n(H, A).$$

If H is not necessarily open one writes H as a filtering intersection of open subgroups and passes the above isomorphism to the limit. $\square$

(12.15) From now on assume that $d := \text{vcd}_p(\Gamma)$ is finite. Fix an open subgroup Δ of Γ of finite cohomological p-dimension, so $\text{cd}_p(\Delta) = d$.

(12.15.1) Lemma. — *Let X be a profinite Γ space and let B be a discrete $\mathbb{F}_p\Delta$-module. Then $H^n\big(\Gamma, C(X, \text{coind}^\Gamma_\Delta B)\big) = 0$ for $n > d$.*

Proof. Put $A := \text{coind}_\Delta^\Gamma B$. Writing $X = \varprojlim_\alpha X_\alpha$ as a filtering inverse limit of finite Γ-spaces one has $C(X, A) = \varinjlim_\alpha C(X_\alpha, A)$, so it suffices to assume that X is finite. Then one can even reduce to the case $X = \Gamma/H$ with H an open subgroup of Γ. By (12.14)

$$H^*(\Gamma, C(\Gamma/H, A)) = H^*(H, \text{res}_H^\Gamma A).$$

The double coset formula (e.g. [Br p. 69]) expresses $\text{res}_H^\Gamma A$ as a finite direct sum of H-modules which are (co-)induced from subgroups $H \cap \sigma \Delta \sigma^{-1}$ ($\sigma \in \Gamma$). These latter subgroups have $\text{cd}_p \leq d$, whence the assertion. ⊔

(12.16) If A is any discrete Γ-module write $M(A) := C(\mathfrak{X}, A)$, and denote the cokernel of $A \to M(A)$ by $P(A)$. Assume $\mathfrak{X} \neq \emptyset$ for the following. One has the short exact sequence

$$0 \longrightarrow A \longrightarrow M(A) \longrightarrow P(A) \longrightarrow 0 \tag{12}$$

of discrete Γ-modules. Clearly M and P are additive functors of A, and also (12) is functorial in A. Since M is obviously an exact functor the same is true for P, by the snake lemma. Similarly M preserves filtering direct limits, and hence so does P.

(12.16.1) Still assume $\mathfrak{X} \neq \emptyset$. If the theorem is true then certainly

$$(*) \qquad H^n(\Gamma, P(A)) = 0 \text{ for } n > d \text{ and every } p\text{-primary discrete } \Gamma\text{-module } A$$

will hold, cf. (12.13.1). Actually this condition $(*)$ is equivalent to the theorem. The point is only to show that $(*)$ implies that $H^{d+1}(\Gamma, A) \to H^{d+1}(\Gamma, M(A))$ is injective for every such A. To see why this is true, put $B := \text{res}_\Delta^\Gamma(A)$ and consider the exact sequence

$$0 \longrightarrow C \longrightarrow \text{coind}_\Delta^\Gamma(B) \xrightarrow{\text{tr}} A \longrightarrow 0 \tag{13}$$

in which C is defined to be the kernel of the trace. Applying M and P to (13) yields a commutative 3×3-square of discrete Γ-modules with exact rows and columns. From this one gets in particular the commutative square

$$
\begin{array}{ccc}
H^{d+1}(\Gamma, A) & \longrightarrow & H^{d+2}(\Gamma, C) \\
\Big\downarrow{\scriptstyle a} & & \Big\downarrow{\scriptstyle c} \\
H^{d+1}(\Gamma, M(A)) & \longrightarrow & H^{d+2}(\Gamma, M(C))
\end{array}
\tag{14}
$$

where the vertical arrows come from (12) (for A resp. C) and the horizontal ones are the boundary maps coming from (13), resp. from M applied to (13). Condition $(*)$ implies that the map c in (14) is an isomorphism, and the upper horizontal arrow is one since $\text{cd}_p(\Delta) = d$. Hence a is injective.

(12.16.2) Since the functor P preserves filtering $\varinjlim$'s and is exact, it is clear that $(*)$ is equivalent to

(**) $H^n(\Gamma, P(A)) = 0$ *for* $n > d$ *and every finite* $\mathbb{F}_p\Gamma$-*module* A.

So what has been shown so far is that in case $\mathfrak{T} \neq \emptyset$ it suffices to prove $(**)$, in order to get the theorem.

(12.16.3) For A an arbitrary discrete Γ-module one has the exact sequence

$$0 \longrightarrow P(_pA) \longrightarrow P(A) \overset{p}{\longrightarrow} P(A) \longrightarrow P(A/p) \longrightarrow 0.$$

It follows (7.23.1) that $H^n(\Gamma, P(A))$ is p-divisible for $n > d$ and p-torsionfree for $n > d+1$. From this one deduces (12.13.2). (To show that the map in question is injective in degree $n = d+2$, modify the argument of (12.16.1) in an obvious way.)

The next lemma says that the modules $P(A)$ have trivial cohomology when they are restricted to any subgroup $T \in \mathfrak{T}$:

(12.17) **Lemma.** — *Let* Γ *be a profinite group which is virtually* p-*torsion free, and let* $\mathfrak{T}$ *be the space of subgroups of order* p *of* Γ. *Suppose that* Γ *contains no subgroup isomorphic to* $\mathbb{Z}/p \times \mathbb{Z}/p$. *Let* T *be a subgroup of order* p *in* Γ. *Then the map*

$$H^n(T, A) \longrightarrow H^n(T, C(\mathfrak{T}, A)) = H_T^n(\mathfrak{T}, A)$$

is an isomorphism for every $n \geq 1$ *and every discrete* Γ-*module* A.

Heuristically this is plausible since $\mathfrak{T}$ considered as a T-space has only one fixpoint T, outside of which the action is free.

Proof. If I is an injective discrete Γ-module then $H_T^n(\mathfrak{T}, I) = 0$ for $n \geq 1$, since $H_T^n(\mathfrak{T}, I) = H^0(\mathfrak{T}/T, \mathcal{H}^n(I))$ and $\mathcal{H}^n(I)$ has stalks $H^n(S, I)$, $S \in \mathfrak{T}$. Therefore a dimension-shifting argument shows that it suffices to prove the lemma for $n = 1$.

Again write $M(A) := C(\mathfrak{T}, A)$. The evaluation map $M(A) \to A$, $f \mapsto f(T)$ in T is a homomorphism of T-modules and is a splitting of the sequence (12) of T-modules. Hence it is clear that $H^n(T, A) \to H^n(T, M(A))$ is injective. Let t be a generator of T. To prove surjectivity one first makes the following observation. If $f \in M(A)$ with $f(T) = 0$ is given, there is a clopen neighborhood K of T in $\mathfrak{T}$ such that

$$\bigcup_{i=0}^{p-1} t^i K t^{-i} = \mathfrak{T},$$

and such that f vanishes on $t^i K t^{-i} \cap t^j K t^{-j}$ for any two integers i, j with $i-j \not\equiv 0$ (mod p).

To see this choose a clopen neighborhood L of T in $\mathfrak{T}$ with $tLt^{-1} = L$ on which f vanishes. (E.g. start with any clopen L_1 with $f|L_1 \equiv 0$ and put $L := \bigcap_i t^i L_1 t^{-i}$.)

The action of T on $\mathfrak{T} - L$ by conjugation is free since Γ contains no $\mathbb{Z}/p \times \mathbb{Z}/p$. Hence there is a clopen subset K' of $\mathfrak{T} - L$ such that $\mathfrak{T} - L$ is the disjoint union of the $t^i K' t^{-i}$, $i = 0, \ldots, p-1$. Putting $K := K' \cup L$ gives the desired neighborhood K of T.

Write $N := 1 + t + \cdots + t^{p-1}$. Since $H^1(T, M(A))$ is the cohomology of the complex

$$M(A) \xrightarrow{1-t} M(A) \xrightarrow{N} M(A),$$

one can represent a given element of $H^1(T, M(A))$ by $f \in M(A)$ with $N \cdot f = 0$. Since also $N \cdot f(T) = 0$ (in A), one can replace f with $f - f(T)$ and therefore assume $f(T) = 0$. Let K be a clopen neighborhood of T as above. Define a continuous map $h \colon \mathfrak{T} \to A$ by

$$h\big|_K \equiv 0, \quad h(t^{-i} S t^i) = -\sum_{j=1}^{i} t^{-j} f(t^{j-i} S t^{i-j}) \quad \text{for } S \in K \text{ and } i = 1, \ldots, p-1.$$

This is well-defined in view of the properties of K. We show $f = (1-t) \cdot h$, which will prove the lemma. Let $S \in K$. Then

$$((1-t) \cdot h)(S) = h(S) - t \cdot h(t^{-1} S t) = 0 - t \cdot (-t^{-1}) \cdot f(S) = f(S),$$

and for $1 \le i \le p-2$

$$\begin{aligned}
((1-t) \cdot h)(t^{-i} S t^i) &= h(t^{-i} S t^i) - t \cdot h(t^{-(i+1)} S t^{i+1}) \\
&= -\sum_{j=1}^{i} t^{-j} \cdot f(t^{j-i} S t^{i-j}) + t \cdot \sum_{j=1}^{i+1} t^{-j} \cdot f(t^{j-i-1} S t^{i+1-j}).
\end{aligned}$$

Reindexing the second sum one sees that this is equal to $f(t^{-i} S t^i)$. For $i = p-1$ the equality $((1-t) \cdot h)(t S t^{-1}) = f(t S t^{-1})$ follows from $N \cdot f = 0$. $\qquad \square$

I claim that the theorem will now follow from

(12.18) Proposition. — *Let Γ be a profinite group with $\mathrm{vcd}_p(\Gamma) < \infty$. Then there is an integer $e \ge 0$ with the following property: Whenever A is a discrete $\mathbb{F}_p \Gamma$-module such that $H^n(T, A) = 0$ for all $n \ge 1$ and all finite elementary abelian p-subgroups T of Γ, one has $H^q(\Gamma, A) = 0$ for $q > e$.*

(12.18.1) Let us see how (12.18) implies Theorem (12.13). If (in the theorem) Γ is p-torsion free this is the conclusion "$\mathrm{cd}_p(\Gamma) < \infty \Rightarrow \mathrm{cd}_p(\Gamma) = \mathrm{cd}_p(\Delta)$" which follows from a standard argument involving the trace sequence (13). So assume $\mathfrak{T} \ne \emptyset$. It suffices to verify (**). Let A be a discrete $\mathbb{F}_p \Gamma$-module. Form the exact sequence (13), and apply the functor P to it to get another exact sequence, call it (†). From the exact sequence (12) for $\mathrm{coind}_\Delta^\Gamma(B)$ and from (12.15.1) it follows that $H^n(\Gamma, P(\text{coind}_\Delta^\Gamma B)) = 0$ for $n > d$. By (12.17) the modules $P(C)$ and $P(A)$ satisfy the hypotheses of Proposition (12.18). So this proposition, together with a dimension shifting argument applied to (†), shows $H^n(\Gamma, P(A)) = 0$ for $n > d$, i.e. (∗∗).

It remains to prove Proposition (12.18). I first need the following fact which is due to Chouinard. I am grateful to J.-P. Serre for pointing out this reference to me. Since the proof is not long, I will include it here:

(12.18.2) Lemma [Ch]. — *Let p be a prime, G a finite group and M an $\mathbb{F}_p G$-module. Assume that $H^1(T, M) = 0$ for every elementary abelian p-subgroup T of G. Then $H^n(G, M) = 0$ for every $n \geq 1$.*

Proof. One may assume that G is a p-group. Then it is well known that the vanishing of $H^n(G, M)$ for one $n \geq 1$ implies that M is actually free as an $\mathbb{F}_p G$-module, and consequently $H^n(G, M) = 0$ for all $n \geq 1$ [Br, ch. VI, Thm. 8.5]. The proof is an adaption of Serre's proof in [Se2]. Proceeding inductively, one can assume $\widehat{H}^*(U, M) = 0$ for every proper subgroup U of G, and that G is not elementary abelian. Fix $z \in H^1(G, \mathbb{Z}/p)$, $z \neq 0$, let $U := \ker(z)$, and let $\beta: H^1(G, \mathbb{Z}/p) \to H^2(G, \mathbb{Z}/p)$ be the Bockstein homomorphism associated to the exact sequence $0 \to \mathbb{Z}/p \to \mathbb{Z}/p^2 \to \mathbb{Z}/p \to 0$ of G-modules. As in [Se2, §4] one sees that cup product with $\beta(z)$ is an isomorphism $H^q(G, M) \xrightarrow{\sim} H^{q+2}(G, M)$ for all $q \geq 1$, since $\widehat{H}^*(U, M) = 0$. As shown by Serre [loc. cit., Prop. (4)], there are finitely many elements $z_1, \ldots, z_n \in H^1(G, \mathbb{Z}/p)$, $z_i \neq 0$, such that $\beta(z_1) \cdots \beta(z_n) = 0$ in $H^{2n}(G, \mathbb{Z}/p)$. It follows that $H^q(G, M) = 0$, $q \geq 1$. □

(12.18.3) Back to the situation of Proposition (12.18). If S is a Sylow p-subgroup of Γ then the restriction map $H^*(\Gamma, A) \to H^*(S, A)$ is injective. Hence one can replace Γ by S, or in other words, assume that Γ is a pro-p group. The only purpose of this reduction step is to get around the difficulties with the signs in tensor induced complexes, see Remark (12.10). Let Δ be an open normal subgroup of Γ with $\text{cd}_p(\Delta) < \infty$. Let $Q \to \mathbb{Z}/p$ be a projective resolution of finite length l by profinite $\mathbb{F}_p \Delta$-modules (12.5). Let $P = \text{ten}_\Delta^\Gamma(Q)$ be the tensor induced complex, so that $P \to \mathbb{Z}/p$ is a quasi-isomorphism (12.7) and P has finite length $e := l \cdot [\Gamma : \Delta]$. Let A be a discrete $\mathbb{F}_p \Gamma$-module. It follows that $A \to \text{Hom}(P, A)$ is a quasi-isomorphism of complexes of discrete Γ-modules, and so $H^*(\Gamma, A)$ is the Γ-hypercohomology of $\text{Hom}(P, A)$. The first hypercohomology spectral sequence for this complex has

$$E_1^{rs} = H^s(\Gamma, \text{Hom}(P_r, A))$$

and converges against $H^*(\Gamma, A)$. I'm going to show now that one has $E_1^{rs} = 0$ for $s \neq 0$. This will imply $H^q(\Gamma, A) = 0$ for $q > e = l \cdot [\Gamma : \Delta]$, and therefore prove (12.18).

(12.18.4) Since the Q_r are projective over Δ, there are profinite $\mathbb{F}_p\Delta$-modules N_r such that $F_r := N_r \oplus Q_r$ is free over Δ, say $F_r = \mathbb{F}_p[V_r]$ with free profinite Δ-spaces V_r ($r \geq 0$). Let N be the complex consisting of the N_r, with zero differentials, and let $F := N \oplus Q$. Then $P = \text{ten}_\Delta^\Gamma(Q)$ is a direct summand of $\text{ten}_\Delta^\Gamma(F)$. From Remark (12.10) it follows that $\left(\text{ten}_\Delta^\Gamma(F)\right)_r \cong \mathbb{F}_p[U_r]$ where

$$U_r = \left\{ f \in \text{Hom}_\Delta\left(\Gamma, \coprod_j V_j\right) : \text{weight } f = r \right\},$$

a profinite Γ-space.

Let $f \in U_r$ and let Γ_f be the stabilizer subgroup of f in Γ. Then $\Gamma_f \cap \Delta = \{1\}$ since Δ acts freely on each V_i; and so Γ_f is finite.

Now $\text{Hom}(P_r, A)$ is (as a Γ-module) a direct summand of $\text{Hom}\left((\text{ten}_\Delta^\Gamma F)_r, A\right) = C(U_r, A)$. Hence the same is true for cohomology, i.e. $E_1^{rs} = H^s(\Gamma, \text{Hom}(P_r, A))$ is a direct summand of $H^s(\Gamma, C(U_r, A))$. By (8.9), the latter group is identified with $H^0(U_r/\Gamma, \mathcal{H}^s(A))$. The stalk of the sheaf $\mathcal{H}^s(A)$ at the orbit of $f \in U_r$ is $H^s(\Gamma_f, A)$. Since Γ_f is a finite group, the hypotheses of (12.18) imply $H^s(\Gamma_f, A) = 0$ for $s \neq 0$, by Chouinard's theorem (12.18.2). Therefore $E_1^{rs} = 0$ for $s \neq 0$, which completes the proof of the proposition and of Theorem (12.13). $\square\square$

(12.19) Corollary. — *Let Γ be a profinite group which contains no subgroup $\mathbb{Z}/p \times \mathbb{Z}/p$ and for which $d := \text{vcd}_p(\Gamma) < \infty$. For any discrete p-primary Γ-module A the natural map*

$$H^n(\Gamma, A) \longrightarrow \prod_{\substack{T \text{ in } \mathfrak{T} \\ \text{mod conjugation}}} H^n(N_\Gamma(T), A)$$

is a dense embedding for $n > d$. Here $\mathfrak{T}$ is the space of subgroups of Γ of order p.

Proof. That is, the map is injective and its image is dense with respect to the product topology (where the factors of the direct product are given the discrete topology). The corollary follows directly from (12.13) since $\mathfrak{T}/\Gamma$ is a boolean space. $\square$

(12.20) Remarks.

(12.20.1) Theorem (12.13) is a far generalization of the main result (7.19) of Part One in the *field* case: Given a field k of characteristic not 2, with $d := \text{cd}_2 k(\sqrt{-1}) = \text{vcd}_2(k)$ finite, this theorem was seen in Sect. 9 to be equivalent to the assertion that the map

$$H^n(\Gamma, A) \longrightarrow H_\Gamma^n(T, A)$$

is bijective for $n > d$ ($\Gamma = \text{Gal}(k_s/k)$ the absolute Galois group of k, A any discrete 2-primary Γ-module and T the profinite Γ-space of all real closures of k in k_s).

Recall that in Sect. 9 the proof of this field case was reduced quite easily to the case $A = \mathbb{Z}/2$, which corresponds to Arason's theorem. So the latter theorem lies at the point of intersection of two quite different theories, each of which can be considered as a very far reaching generalization of it: On the one hand the study of the relationship between étale site and real spectrum of a scheme, and on the other the result of this section (12.13) about profinite group cohomology in high degrees. I want to point out again (see (9.13)) that Arason's theorem has an easy proof which makes use of specific relations in the cohomology *ring* $H^*(k, \mathbb{Z}/2)$ of a field. For more general profinite groups, even in the case $p = 2$, there are no such relations, and there does not seem to be a proof based on cup products. Rather one has to use profinite projective resolutions. (Observe however that there is another noteworthy case in which the multiplicative structure can be used, namely Serre's theorem; cf. the initial proof in [Se2].)

(12.20.2) In general, the map $H^n(\Gamma, A) \to H^n_\Gamma(\mathfrak{T}, A)$ in (12.13) is *not* surjective for $n = d$. This was already observed in (9.8.1), in a different language (compare also the preceding remark): Essentially, the argument given there showed that if a profinite group $\Gamma \neq \mathbb{Z}/2$ has an open subgroup Δ of index 2 such that Δ is of *odd* profinite order (hence $d := \text{vcd}_2(\Gamma) = 0$), there is a finite $\mathbb{F}_2\Gamma$-module A for which (11) is not surjective for $n = d = 0$. A similar construction works for other primes $p \neq 2$.

However, $H^d(\Gamma, A) \to H^d_\Gamma(\mathfrak{T}, A)$ is indeed surjective under the hypotheses that $d \geq 1$ and every $T \in \mathfrak{T}$ is self-normalizing. (Note how this generalizes (9.8.1).) To see this one shows $H^i_\Gamma(\mathfrak{T}, \text{coind}^\Gamma_\Delta B) = 0$ for $i \geq 1$ and every p-primary Δ-module B, which implies $H^d(\Gamma, P(A)) = 0$ for every p-primary Γ-module A, cf. (12.16).

(12.20.3) For discrete groups, K.S. Brown has proved the following theorem [Br, X.7.3]: Suppose Γ is a group with $d := \text{vcd}(\Gamma) < \infty$. Then Farrell cohomology $\hat{H}^*(\Gamma, -)$ is defined. Let $\mathfrak{T}$ be the simplicial complex of all non-trivial elementary abelian p-subgroups of Γ. Then there is an isomorphism (in all degrees)

$$\hat{H}^*(\Gamma, A) \xrightarrow{\sim} \hat{H}^*_\Gamma(\mathfrak{T}, A), \tag{15}$$

for any p-primary Γ-module A. The right hand term is equivariant Farrell cohomology.

Since the natural maps $H^n(\Gamma) \to \hat{H}^n(\Gamma)$ are isomorphisms in degrees $n > d$ and are surjective for $n = d$ (any coefficients), this gives in particular isomorphisms

$$H^n(\Gamma, A) \xrightarrow{\sim} H^n_\Gamma(\mathfrak{T}, A) \tag{16}$$

for $n \geq d + e$, by one of the spectral sequences for equivariant cohomology, where p^e is the maximal order of an elementary abelian p-subgroup of Γ. If in particular Γ contains no $\mathbb{Z}/p \times \mathbb{Z}/p$ then $\mathfrak{T}$ is (at most) 0-dimensional, and (16) becomes an isomorphism

$$H^n(\Gamma, A) \xrightarrow{\sim} \prod_{\substack{T \text{ in } \mathfrak{T} \\ \text{mod conjugation}}} H^n(N_\Gamma(T), A) \tag{17}$$

for $n > d$, cf. [Br, Cor. X.7.4]. The results of this section are the profinite counterpart of (16) resp. (17). As mentioned before, Brown's results have meanwhile been generalized to profinite groups in full generality [Sch2].

(12.20.4) Of particular interest is the following case. Let Γ be a profinite group which has an open normal subgroup Δ of index p with $d := \mathrm{cd}_p(\Delta) < \infty$. Then (12.13) applies. For every $T \in \mathfrak{T}$ one has $N_\Gamma(T) = C_\Gamma(T)$, and this group is the direct product $C_\Delta(T) \times T$. From (8.9) one gets a natural spectral sequence

$$E_2^{pq} = H^p\Big(G, H_\Delta^q(\mathfrak{T}, A)\Big) \implies H_\Gamma^{p+q}(\mathfrak{T}, A),$$

with $G = \Gamma/\Delta \cong \mathbb{Z}/p$, for every discrete Γ-module A. Note that the abutment is $H^*(\Gamma, A)$ in degrees $> d$ if A is p-primary torsion (12.13).

Now assume in addition that every $T \in \mathfrak{T}$ is self-normalizing. Then Δ acts freely on $\mathfrak{T}$, and so

$$H_\Gamma^n(\mathfrak{T}, A) = H^n(G, \mathrm{Hom}_\Delta(\mathfrak{T}, A))$$

for all $n \geq 0$ and every discrete Γ-module A. Since G is cyclic it is obvious that this cohomology is periodic with period two; and hence, after a generator of G has been chosen, cup product with the image of $(\Gamma \to G \cong \mathbb{Z}/p) \in H^1(\Gamma, \mathbb{Z}/p)$ under the boundary of $0 \to \mathbb{Z} \xrightarrow{p} \mathbb{Z} \to \mathbb{Z}/p \to 0$ gives periodicity isomorphisms $H^n(\Gamma, A) \xrightarrow{\sim} H^{n+2}(\Gamma, A)$ for $n > d$. The limit groups $H^{\mathrm{ev}}(\Gamma, A) := H^{2n}(\Gamma, A)$ and $H^{\mathrm{odd}}(\Gamma, A) := H^{2n+1}(\Gamma, A)$ (for $n \gg 0$) are easily made explicit. I only mention the following particular cases:

1) If Δ acts trivially on A then $\mathrm{Hom}_\Delta(\mathfrak{T}, A)$ can be identified as a G-module with $C(\mathfrak{T}/\Delta, A)$; and so $H_\Gamma^n(\mathfrak{T}, A) = C(\mathfrak{T}/\Delta, H^n(G, A))$, hence $H^\epsilon(\Gamma, A) = C(\mathfrak{T}/\Delta, H^\epsilon(G, A))$, $\epsilon \in \{\mathrm{ev, odd}\}$;

2) if $p = 2$ and A is a discrete $\mathbb{F}_2\Gamma$-module then

$$H^\infty(\Gamma, A) := H^{\mathrm{ev}}(\Gamma, A) = H^{\mathrm{odd}}(\Gamma, A) = \mathrm{coker}\Big(\mathrm{Hom}_\Delta(\mathfrak{T}, A) \to \mathrm{Hom}_\Gamma(\mathfrak{T}, A)\Big)$$

where the map on the right is the "norm" which sends $f \in \mathrm{Hom}_\Delta(\mathfrak{T}, A)$ to the Γ-map

$$\mathfrak{T} \longrightarrow A, \quad T = \langle t \rangle \longmapsto (1 + t)f(T).$$

(Of course these remarks generalize Corollary (9.9).)

(12.20.5) (Compare Remark (11.2.6.3).) Consider again the case where Δ is an open normal subgroup of Γ of index p, and put $G = \Gamma/\Delta$. Then $E := (\Delta\text{-sets})$ is a G-topos, and it was shown in Sect. 11.2 that the fixtopos F is equivalent to $\widetilde{\mathfrak{K}}(\Delta)$, the topos of Δ-sheaves on the space $\mathfrak{K}$ of splittings of $1 \to \Delta \to \Gamma \to G \to 1$. Assume that $d := \mathrm{cd}_p(\Delta)$ is finite. Then $\mathfrak{K} = \mathfrak{T}$ is the space of all finite p-subgroups of Γ, and the topos $F(G)$ of G-objects in F is equivalent to $\widetilde{\mathfrak{T}}(\Gamma)$ (8.8.4). Under this equivalence the functor $\nu(G)^*: E(G) = (\Gamma\text{-sets}) \to F(G) = \widetilde{\mathfrak{T}}(\Gamma)$ gets identified with the functor which sends a Γ-set A to the Γ-sheaf $\mathfrak{T} \times A \to \mathfrak{T}$ on $\mathfrak{T}$. Finally the map (11) of (12.13) is nothing but the map $H^n\big(E(G), A\big) \to H^n\big(F(G), \nu(G)^*A\big)$ induced by $\nu(G)^*$. As explained in Sect. 14, the fact that this map is an isomorphism for $n > d$ implies (and is, *cum grano salis*, equivalent to the fact) that the glued topos $E/G := \big(F, E(G); \rho\big)$ has $\mathrm{cd}_p(E/G) \leq d + 1$. This glued topos is to be regarded as the topos-theoretic version of the quotient space of a topological space with G-action, see Sect. 14.

13. Group actions on spaces: Topological versus topos-theoretic constructions

Let a (discrete) group G act on a topological space T. On the one hand one can form the direct resp. the inverse limit of this action in the category of spaces, which is the quotient space $T/G =: Y$ resp. the space of fixpoints $T^G =: Z$. On the other, one can study the G-topos $\widetilde{T}$ associated with the action (10.10) and form these limits in the sense of G-toposes. Below these two procedures are compared. It is shown (13.2) that if G is finite and T is Hausdorff (say), both yield the same result; i.e. the fixtopos $\underleftarrow{\operatorname{Limtop}}(\widetilde{T}/G)$ is equivalent to the topos of sheaves on the space of fixpoints $T^G = Z$. The second result (Proposition (13.4.1)) concerns the case when G is cyclic of prime order. In analogy with the construction of the b-topology of a scheme, one can glue the toposes $\widetilde{T}(G)$ and $\widetilde{Z}$, to arrive at a new topos denoted $\widetilde{T}/G$. It is shown that, although this $\widetilde{T}/G$ is generally not equivalent to $(T/G)^{\sim} = \widetilde{Y}$, both toposes have the same cohomology. Thus $\widetilde{T}/G$ can be seen as a topos-theoretic substitute for the formation of the quotient space T/G. See Sect. 14 for a complementary discussion.

(13.1) For the notion of G-toposes, and in particular for the inverse limit of a G-topos, I refer to Sect. 10.

Recall that an action of a group G on a topological space T is said to be *discontinuous* if every $x \in T$ has a neighborhood U such that $g \in G$ and $g \cdot x \neq x$ imply $g \cdot U \cap U = \emptyset$ [Gr1, p. 203]. If G is finite then this condition is automatically satisfied under either of the following two conditions: 1) T is Hausdorff; 2) T is a normal spectral space [CaC], e.g. the real spectrum of a separated quasi-compact scheme.

(13.2) **Proposition.** — *Let a finite group G act on a topological space T, and assume that G acts discontinuously (e.g. that T is Hausdorff). Let $Z := T^G$ be the subspace of fixpoints and let $\nu \colon Z \subset T$ be the inclusion. Then the topos morphism ν from $\widetilde{Z}$ to $\widetilde{T}$ identifies $\widetilde{Z}$ with the fixtopos of the G-topos $\widetilde{T}$:*

$$\widetilde{Z} = \underleftarrow{\operatorname{Limtop}}(\widetilde{T}/G).$$

Proof. I follow the construction of the fixtopos as given in Sect. 10. So let $\pi^*, \pi_* \colon \widetilde{T} \leftrightarrows \widetilde{T}(G)$ be the canonical functors. In the following, consider sheaves on T always as espaces étalés over T. Let τ be the coarsest topology on $\widetilde{T}(G)$ which makes π_* continuous. If $A \to T$ is a G-sheaf on T, denote by A^G the subspace of G-fixpoints in A. The restriction of $A^G \to T$ to Z is an espace étalé over Z which corresponds to the sheaf $\Gamma_G(A|_Z)$ on Z.

I claim that a family $\{A_\alpha \to A\}$ in $\widetilde{T}(G)$ is a covering for τ if and only if the family $\{A_\alpha^G \to A^G\}$ is surjective. From this description it is easy to see that the category of sheaves on the site $(\widetilde{T}(G), \tau)$ is identified with $\widetilde{Z}$. So it suffices to prove the claim. One checks easily that the notion of coverings just described defines a topology on $\widetilde{T}(G)$ which makes π_* continuous. So it remains to prove that any family $\mathfrak{U} = \{f_\alpha : A_\alpha \to A\}$ in $\widetilde{T}(G)$ for which $\{A_\alpha^G \to A^G\}$ is surjective is a τ-covering. The strategy is similar to Sect. 11. Let $\mathfrak{U}$ be given, let $p : A \to T$ be the espace étalé of A. It suffices to find a τ-covering $\{g_\beta : B_\beta \to A\}$ such that every g_β factors through some f_α. For this consider open subspaces U of the underlying espace étalé of A (i.e. subsheaves U of $\pi^* A$) and form the fibre products

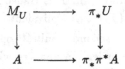

in $\widetilde{T}(G)$. Then M_U is the sub-G-sheaf of A whose espace étalé is $\bigcap_{g \in G} g \cdot U$. Fix now a point $a \in A$ and distinguish the following three cases. Firstly, if $p(a) \notin T^G$ then a has a neighborhood U in A such that $M_U = \emptyset$. This is true since G acts discontinuously on T. Secondly, if $a \in A^G$ then a has arbitrarily small G-invariant open neighborhoods U in A for which $p|_U : U \to p(U)$ is an isomorphism of G-spaces. In particular, in both these cases one finds a neighborhood U of a such that $M_U \to A$ factors through $f_\alpha : A_\alpha \to A$ for some α.

The remaining case is when $p(a) \in T^G$ but $a \notin A^G$. Here one sees that a has a neighborhood U in A such that $M_U \cap p^{-1}(T^G) \subset A^G$. (It suffices to take U so small that $p|_U$ is injective.) Thus the third case is eliminated when passing to the G-sheaf M_U.

If $\{U_\lambda\}$ is any open covering of A then the family $\{M_{U_\lambda} \to A\}$ is a τ-covering. Use open sets U_λ as described in the above three cases, and refine $\{M_{U_\lambda} \to A\}$ again in the third case, hereby using only the first two cases. The resulting τ-covering $\{g_\beta : B_\beta \to A\}$ has the property that every g_β factors through some f_α, as desired.   □

(13.3) Denote as before by $\nu : T^G = Z \subset T$ the inclusion of the fixpoints, and consider the functor $\rho := \Gamma_G \circ \nu(G)^* : \widetilde{T}(G) \to \widetilde{Z}$ which sends a G-sheaf A on T to the sheaf $\Gamma_G(A|_Z)$ of G-invariants of $A|_Z$. Identifying $\widetilde{Z}$ with the category of sheaves on the site $(\widetilde{T}(G), \tau)$ as in the proof above, ρ becomes the associated sheaf functor ϵ_τ. Furthermore one sees that the restriction functor $\nu^* : \widetilde{T} \to \widetilde{Z}$, $\nu^* A = A|_Z$, factors as $\nu^* = \rho \pi_*$. Note how this is completely analogous to the situation encountered for the G-topos $\widetilde{X}'_{et}$, for X a scheme over $\mathbb{Z}[\frac{1}{2}]$. Actually these relations are true for any G-topos, cf. (10.19.1).

(13.4) Let now G be a finite group of *prime* order which acts on a topological space T. Let $p: T \to Y := T/G$ be the topological quotient space. Let $\nu: Z = T^G \subset T$ be the inclusion of the fixpoints and consider the functor $\rho: \widetilde{T}(G) \to \widetilde{Z}$, $A \mapsto \Gamma_G(A|_Z)$ (13.3). Since ρ is left exact one can glue the topos $\widetilde{Z}$ to $\widetilde{T}(G)$ along ρ, i.e. ([SGA4 IV.9.5], cf. Sect. 2) form the topos $(\widetilde{Z}, \widetilde{T}(G), \rho)$ of triples $(C, A, \phi: C \to \rho A)$. This topos will be denoted by $\widetilde{T}/G$.

(13.4.1) Proposition. — *(G finite of prime order) There is a canonical topos morphism $\phi: \widetilde{T}/G \to (T/G)^{\sim} = \widetilde{Y}$. The inverse image functor ϕ^* is fully faithful; and if the G-action is discontinuous then ϕ_* on abelian sheaves is exact. However ϕ is an equivalence only if G acts freely on T, in which case $\widetilde{T}/G = \widetilde{T}(G)$ and ϕ is (p_G^*, p_*^G) (8.1.3).*

Proof. Let $i: \widetilde{Z} \to \widetilde{T}/G$ and $j: \widetilde{T}(G) \to \widetilde{T}/G$ be the canonical topos embeddings. To give a topos morphism ϕ from $\widetilde{T}/G$ to $\widetilde{Y}$ is in a functorial way the same as to give topos morphisms $\varphi: \widetilde{Z} \to \widetilde{Y}$ and $\psi: \widetilde{T}(G) \to \widetilde{Y}$ together with a morphism of functors $\varphi^* \to \rho \psi^*$; see [SGA4 IV.9.5.11]. Take φ to be the topos morphism induced by the inclusion $p \circ \nu: Z \hookrightarrow Y$, and ψ to be $r_Y \circ p(G) = (p_G^*, p_*^G)$, cf. (8.1.3). Note that $\varphi^* = (p \circ \nu)^*$ and $\rho \circ \psi^* = \rho \circ p_G^*$ are canonically isomorphic. Thus a topos morphism $\phi = (\phi^*, \phi_*): \widetilde{T}/G \to \widetilde{Y}$ has been defined. By construction ϕ^* is given by $\phi^* B = \left(B|_Z, p_G^* B, B|_Z \xrightarrow{\sim} \rho p_G^* B \right)$ $(B \in \widetilde{Y})$, and thus

$$\phi^* = j_* \circ p_G^* \tag{1}$$

holds (up to isomorphism). To describe ϕ_* let $F = (C, A, C \xrightarrow{f} \rho A)$ be an object of $\widetilde{T}/G$. Then F is the fibre product of the diagram

$$i_* C \xrightarrow{i_*(f)} i_* \rho A = i_* i^* j_* A \xleftarrow{\text{adj}} j_* A$$

in $\widetilde{T}/G$. So ϕ_* sends F to the fibre product of

$$(p\nu)_* C \xrightarrow{(p\nu)_* f} (p\nu)_* \rho A \xleftarrow{\text{can}} p_*^G A \tag{2}$$

in $\widetilde{Y}$. (The arrow "can" is again an adjunction map since $\rho = (p\nu)^* p_*^G$.)

The functor p_G^* is fully faithful (8.1.4), and so is j_*. Hence ϕ^* is fully faithful by (1). Moreover ϕ^* is an equivalence if and only if both j_* and p_G^* are equivalences; and j_* is one iff $Z = \emptyset$. It remains to prove that $\phi_*: \text{Ab}(\widetilde{T}/G) \to \text{Ab}(Y)$ is right exact. If $F = (C, A, C \to \rho A)$ is an abelian group object in $\widetilde{T}/G$ one sees from (2) that $(\phi_* F)|_{Y-Z} = (p_*^G A)|_{Y-Z}$ and $(\phi_* F)|_Z = C$. Assume that G acts discontinuously on T. Then $\text{Ab}_G(T) \to \text{Ab}(Y-Z)$, $A \mapsto (p_*^G A)|_{Y-Z}$ is an exact functor — this uses that G acts freely on $T-Z$, which is true since G has prime order! —, and one concludes that ϕ_* is exact. □

(13.4.2) Corollary. — *If G acts discontinuously then the glued topos $\widetilde{T}/G = (\widetilde{Z}, \widetilde{T}(G), \rho)$ has the same cohomology as the quotient space $Y = T/G$, in the sense that ϕ^* induces a canonical isomorphism*

$$H^*(Y, B) \xrightarrow{\sim} H^*(\widetilde{T}/G, \phi^* B)$$

for every abelian sheaf B on Y.

Proof. $B \xrightarrow{\sim} \phi_* \phi^* B$ plus exactness of ϕ_*. $\qquad\qquad\qquad\qquad\qquad\qquad\qquad$ □

(13.5) Remark. In the situation of (13.4) write u for the open inclusion $Y - Z \subset Y$. Let A be an abelian G-sheaf on T. Then by (2), $\phi_*(j_! A) \in \mathrm{Ab}(Y)$ is the kernel of the sheaf map $p_*^G A \to (p_*^G A)|_Z$ on Y, so $\phi_*(j_! A) = u_! u^* (p_*^G A)$, the sheaf $p_*^G A$ restricted to $Y - Z$ and extended by zero outside. Hence it follows that under ϕ_* the exact sequence

$$0 \longrightarrow j_! A \longrightarrow j_* A \longrightarrow i_* \rho A \longrightarrow 0 \tag{3}$$

in $\mathrm{Ab}(\widetilde{T}/G)$ is transformed into the exact sequence

$$0 \longrightarrow (p_*^G A)|_{Y-Z} \longrightarrow p_*^G A \longrightarrow (p_*^G A)|_Z \longrightarrow 0 \tag{4}$$

on Y. If G acts discontinuously it follows that the long exact cohomology sequences for (3) (on $\widetilde{T}/G$) and for (4) (on Y) are canonically isomorphic. If for example T is a CW complex on which G acts cellularly and M is an abelian group, this means that the long exact sequence

$$\cdots \longrightarrow H^n(\widetilde{T}/G, j_! M) \longrightarrow H^n(\widetilde{T}/G, M) \longrightarrow H^n(Z, M) \longrightarrow \cdots$$

arising from the glued topos $\widetilde{T}/G$ is canonically isomorphic to the long exact sequence in singular cohomology for the pair (Y, Z). (See the following remark.)

(13.6) Remark. This is also a suitable place to recall some well known facts about the comparison between singular and sheaf cohomology. Since they are used only in some informal discussions (cf. Sect. 14) I have in no way attempted to be complete here. This remark is included just because I was unable to find a convenient direct reference for these facts.

A suitable class of spaces for which these cohomologies agree is formed by the CW complexes. Let T be one, and let M be an abelian group. By sheafifying the presheaves $U \mapsto C_{\mathrm{sing}}^q(U, M)$ of singular cochains ($q \geq 0$) one gets a complex $S^\bullet(M)$ of sheaves on T, augmented over the constant sheaf $\underline{M}$.

Every CW complex is paracompact [LW II.4.2]. Hence singular cohomology $H_{\mathrm{sing}}^*(T, M)$ is also cohomology of the complex $H^0(T, S^\bullet(M))$ [Bd1, p. 19]. The sheaves $S^q(M)$ on T are soft [Go, p. 161], hence acyclic. Moreover, since every CW complex is locally contractible [LW II.6.6] the complex $S^{\cdot}(M)$ is a resolution of $\underline{M}$. Thus one has a canonical isomorphism $H^*(T, \underline{M}) \cong H^*_{\mathrm{sing}}(T, M)$. Of course, if Y is a subcomplex of T and $u : T - Y \subset T$ is the inclusion of its open complement, one has also a canonical isomorphism $H^*_{\mathrm{sing}}(T, Y; M) \cong H^*(T, u_! \underline{M})$ between relative singular cohomology and the cohomology of the constant sheaf $\underline{M}$ on $T - Y$, extended by zero on Y.

Now let G be a discrete group which acts cellularly on the CW complex T. There is a contractible CW complex EG with a free and cellular G-action, for example the geometric realization of the simplicial set E of (8.2.2). The diagonal action of G on $EG \times T$ is free and cellular. One writes $T_G := EG \times_G T :=$ $(EG \times T)/G$ for the quotient and defines equivariant singular cohomology on T by

$$H^*_{G, \mathrm{sing}}(T, M) := H^*_{\mathrm{sing}}(T_G, M), \tag{5}$$

M any abelian group. Compare [Bd2 VII.1], [tD III.1]. The projection from T_G to $EG/G =: BG$ is a fibration with fibre T. The Leray-Serre spectral sequence for this fibration has E_2-terms $E_2^{pq} = H^p_{\mathrm{sing}}(BG, H^q_{\mathrm{sing}}(T, M))$, with twisted coefficients $H^*_{\mathrm{sing}}(T, M)$ on BG. There is a canonical morphism between this spectral sequence and the spectral sequence

$$E_2^{pq} = H^p(G, H^q(T, \underline{M})) \implies H^{p+q}_G(T, \underline{M})$$

for equivariant *sheaf* cohomology. It is an isomorphism on E_2 terms, and hence gives an isomorphism on the abutments: $H^*_{G, \mathrm{sing}}(T, M) \cong H^*_G(T, \underline{M})$.

14. Quotient topos of a G-topos, for G of prime order

Let G be a cyclic group of *prime* order, and let $\mathfrak{E}$ be a G-topos with fibre E. In this section I consider the topos $\mathfrak{E}/G = E/G$ which is the glueing of the fixtopos $F = \underleftarrow{\mathrm{Lim}}\mathrm{top}(\mathfrak{E}/G)$ and the topos $E(G)$ of G-objects in E, along a canonical functor ρ. For the construction and some basic properties, most notably the long exact sequence, one proceeds by mimicking the case of the glued topos $X_{\mathfrak{h}}$ of a scheme X. The idea is that assigning the topos E/G to the G-topos $\mathfrak{E}$ should be the topos-theoretic equivalent of assigning to a topological G-space T its quotient space T/G. The usual topos-theoretic quotient construction, i.e. $\underrightarrow{\mathrm{Lim}}\mathrm{top}(\mathfrak{E}/G) = E(G)$, corresponds to T/G only if the G-action is free: Indeed, the topos $\widetilde{T}(G)$ of G-sheaves on T corresponds to the free-made quotient T_G (10.10), which is very different from T/G if the G-action has fixpoints. "Very different" means, for example, that cohomology with constant coefficients is different. On the other hand, the glueing construction proposed here yields the correct result in this case, as was shown in Sect. 13 (say T is Hausdorff): The glued topos $\widetilde{T}/G$ contains $(T/G)^{\sim}$ as a full subcategory, in such a way that the cohomology of every abelian sheaf on T/G is the same in both toposes. This fact justifies regarding E/G as a generalization of the topological quotient space. More evidence is provided by the "fundamental long exact sequence", which exists under mild hypotheses on the topos E (Proposition (14.6)) and which generalizes a basic sequence from equivariant singular cohomology, see below. The latter holds for constant coefficients and is constructed by means of the classifying space BG; whereas the glueing construction studied here applies to any sheaf of coefficients.

At the end of the section there is a discussion of the relations between high-dimensional G-equivariant cohomology on E and on F, and of the cohomological dimension of the quotient topos E/G. These two questions are connected through the long exact sequence.

The construction of the quotient topos E/G given below allows to build up the formalism of Smith theory, parallel to the classical situation. However, this topic will not be pursued here.

It should be interesting to define and study, in a similar vein, quotient toposes for more general finite groups G. This would involve more complicated glueing procedures, namely along diagrams of subgroups of G.

For the entire section let G be a finite group of *prime* order.

(14.1) Let $\mathfrak{E}$ be a G-topos with fibre E (Sect. 10). The fixtopos $\varprojlim\text{top}(\mathfrak{E}/G)$ will always be denoted by F. Let $\pi: E \to E(G)$ and $\nu: F \to E$ denote the canonical topos morphisms. Recall that F can be identified with the topos of sheaves on $(E(G), \tau)$ where τ is the coarsest topology which makes π_* continuous (10.16). Write $\rho := \epsilon_\tau$ for the functor "associated τ-sheaf" from $E(G)$ to F. I refer to (10.19) for a systematic display of the various functors between E, F and $E(G)$. In particular recall (10.19.1) that ρ is isomorphic to the composite functor $r_{F*}\nu(G)^* = \Gamma_G\nu(G)^*$.

(14.2) **Definition.** Since ρ is a left exact functor one can form the glued topos $(F, E(G), \rho)$ [SGA4 IV.9.5], cf. also Sect. 2. This topos will be called the *quotient topos* of the G-topos $\mathfrak{E}$ and denoted by $\mathfrak{E}/G$ or E/G. I use $i: F \to \mathfrak{E}/G$ and $j: E(G) \to \mathfrak{E}/G$ for the canonical topos immersions. Note that the definition may be written in the curious form

$$\mathfrak{E}/G = \left(\varprojlim\text{top}(\mathfrak{E}/G), \varinjlim\text{top}(\mathfrak{E}/G), \rho\right)!$$

(14.3) Let $\mathfrak{E}' \to \mathfrak{E}$ be a morphism of G-toposes. Let $m: E' \to E$ be the topos morphism in the fibre and $f: F' \to F$ the induced topos morphism between the fixtoposes. Then m induces a natural topos morphism $m/G: \mathfrak{E}'/G \to \mathfrak{E}/G$ between the quotient toposes such that

$$
\begin{array}{ccccc}
E'(G) & \xrightarrow{\ j'\ } & \mathfrak{E}'/G & \xleftarrow{\ i'\ } & F' \\
{\scriptstyle m(G)}\big\downarrow & & {\scriptstyle m/G}\big\downarrow & & \big\downarrow{\scriptstyle f} \\
E(G) & \xrightarrow{\ j\ } & \mathfrak{E}/G & \xleftarrow{\ i\ } & F
\end{array}
$$

commutes. Indeed, to give m/G means to give a morphism of functors $f^* i^* \longrightarrow \rho' \circ m(G)^* j^*$ [SGA4 IV.9.5.11]. Now

$$f^*\rho = f^* r_{F*}\nu(G)^* \underset{(*)}{=} r_{F'*}f(G)^*\nu(G)^* = r_{F'*}\nu'(G)^* m(G)^* = \rho' m(G)^*,$$

where $(*)$ follows from left exactness of f^*. So the canonical morphism of functors $i^* \to \rho j^*$ induces a morphism of functors $f^* i^* \longrightarrow f^* \rho j^* = \rho' m(G)^* j^*$, as desired.

(14.4) For the derivation of the long exact sequence I need the additional hypothesis that the topos F has enough points, cf. (14.5). I do not know whether this condition can be dropped. If it is satisfied one can proceed as in Sect. 6.

Actually one can prove that under suitable conditions on E, the topos F always has sufficiently many points. (This works for any finite group G, not necessarily of prime order.) For example it suffices that E is a quasi-separated topos in the sense of [SGA4 VI.2.3]. The proof is quite similar to the proof of Theorem (11.1.1). I do not include it here since this result is not used outside this section.

(14.5) Lemma. — *Assume that F has sufficiently many points. For every $A \in$ $\mathrm{Ab}(E(G))$ one has $\mathrm{H}_G^0(j_! \pi(G)_* A) = j_! A$ and $\mathrm{H}_G^n(j_! \pi(G)_* A) = 0$ for $n \geq 1$, as elements of $\mathrm{Ab}(E/G)$. Hence there are natural isomorphisms*

$$H_G^n \Big(E/G, \, j_! \pi(G)_* A \Big) \cong H^n(E/G, \, j_! A)$$

for $n \geq 0$.

Proof. Compare with the proof of (6.5). Again it is clear that $\mathrm{H}_G^0(j_! \pi(G)_* A) = j_! A$, and it suffices to show $\mathrm{H}_G^n(j_! \pi(G)_* A) = 0$ for $n \geq 1$. This in turn is equivalent to the vanishing of both $i^* \mathrm{H}_G^n(j_! \pi(G)_* A)$ and $j^* \mathrm{H}_G^n(j_! \pi(G)_* A)$. For the first object this can be checked stalkwise, by the hypothesis on F. For every fibre functor p^* of F one has by (6.2):

$$p^* \mathrm{H}_G^n(j_! \pi(G)_* A) = H^n \Big(G, \, p^* i^* j_! \pi(G)_* A \Big).$$

Since $i^* j_! = 0$, this is zero. On the other hand $j^* \circ \mathrm{R}\Gamma_G = \mathrm{R}\Gamma_G \circ j^*$ since j^* is exact and preserves injectives. So $j^* j_! = \mathrm{id}$ implies $j^* \circ \mathrm{R}\Gamma_G \circ j_! \pi(G)_* = \mathrm{R}\Gamma_G \circ \pi(G)_* = \mathrm{R}(\Gamma_G \circ \pi(G)_*) = \mathrm{id}$. The lemma is proved. $\square$

(14.6) Proposition. — *Assume that F has enough points, for example that the topos E is quasi-separated. Then for every $A \in \mathrm{Ab}(E(G))$ the short exact sequence in $\mathrm{Ab}_G(E/G)$*

$$0 \longrightarrow j_! \pi(G)_* A \longrightarrow j_* \pi(G)_* A \longrightarrow i_* \rho \pi(G)_* A \longrightarrow 0 \qquad (1)$$

gives rise to the long exact sequence of cohomology groups

$$\cdots \longrightarrow H^n(E/G, \, j_! A) \longrightarrow H_G^n(E, A) \longrightarrow H_G^n(F, \, \nu(G)^* A) \longrightarrow \cdots \qquad (2)$$

Proof. From $\rho \pi_* = \nu^*$ it follows that $\rho \pi(G)_* = \nu(G)^*$. The additive functor $j_* \pi(G)_*$ is exact since both $j^* j_* \pi(G)_* = \pi(G)_*$ and $i^* j_* \pi(G)_* = \nu(G)^*$ are exact. So the isomorphism of functors $\mathrm{H}_G^0(E/G, -) \circ j_* \pi(G)_* \cong \mathrm{H}_G^0(E, -)$ shows that $H_G^*(E/G, \, j_* \pi(G)_* A) \cong H_G^*(E, A)$. In view of Lemma (14.5) this proves the proposition. $\square$

(14.6.1) It is easy to see (compare (6.6)) that the sequence (2) is functorial with respect to the G-topos $\mathfrak{E}$. So if $m: E' \to E$ is a G-equivariant topos morphism and if also $F' = \underleftarrow{\mathrm{Limtop}}(E'/G)$ has enough points, then for $A \in \mathrm{Ab}(E(G))$ there is a natural morphism of long exact sequences from (2) to

$$\cdots \longrightarrow H^n(E'/G, \, j_! B) \longrightarrow H_G^n(E', B) \longrightarrow H_G^n(F', \, \nu(G)^* B) \longrightarrow \cdots,$$

with $B := m(G)^* A$.

(14.7) Remark. Using that F has enough points one gets also the analogue of Proposition (6.9) (and its consequences): For every injective object I of $\mathrm{Ab}(E(G))$ one has $\mathbf{H}_G^n(\nu(G)^*I) = 0$ for $n \geq 1$. Indeed, let p^* be a fibre functor of F. By (6.2),

$$p^*\mathbf{H}_G^n(\nu(G)^*I) \;=\; H^n\big(G, \, p^*\nu(G)^*I\big). \tag{3}$$

Let $q^* := p^*i^*$, a fibre functor of E/G. Since $\nu(G)^*I = \rho\pi(G)_*I = i^*j_*\pi(G)_*I$ and $J := j_*\pi(G)_*I$ is injective in $\mathrm{Ab}_G(E/G)$, it follows that the G-module $p^*\nu(G)^*I = q^*J$ is injective, by the argument in the proof of (6.2). So (3) vanishes for $n \geq 1$.

This fact can be used to obtain analogues of the results in the second part of Sect. 6. In particular, there are natural isomorphisms

$$H_G^n(F, \nu(G)^*A) \;\cong\; H_F^{n+1}(E/G, j_!A)$$

($n \in \mathbb{Z}$) for $A \in \mathrm{Ab}(E(G))$, and the long exact sequence (2) can be interpreted as the sequence for cohomology of E/G with support in the closed subtopos F.

Moreover, from $R^n j_*A = i_*\mathbf{H}_G^n(\nu(G)^*A)$ ($n \geq 1$) one sees that $R^n j_*A = 0$ for $n \geq 1$ holds e.g. for all ℓ-primary torsion objects $A \in \mathrm{Ab}(E(G))$, with ℓ any prime different from $|G|$. Hence $H^q(E/G, j_*A) = H^q(E(G), A)$ for such A ($q \geq 0$), and this group is zero for $q > \mathrm{cd}_\ell(E)$.

(14.8) Remark. For every $A \in \mathrm{Ab}(E(G))$ one has the canonical "comparison" homomorphisms in cohomology, induced by $\nu(G)^*$,

$$H_G^n(E, A) \;\longrightarrow\; H_G^n\big(F, \nu(G)^*A\big), \quad n \geq 0. \tag{4}$$

Assume again that F has enough points, e.g. that E is quasi-separated. The amount by which the maps (4) fail to be isomorphisms is measured by the cohomology groups $H^n(E/G, j_!A)$, which can be thought of as "relative cohomology groups $H^n(E/G, F; A)$".

First have a look at the classical situation. Let T be a Hausdorff space on which G acts, let $Z := T^G$ be the space of fixpoints and $p\colon T \to Y := T/G$ the topological quotient. Consider $\widetilde{T}$ as a G-topos. By (13.2) the topos F is canonically equivalent to $\widetilde{Z}$, and the maps (4) are simply the restriction maps in cohomology, $H_G^n(T, A) \to H_G^n(Z, A|_Z)$. In (13.4.1) a topos morphism ϕ from $\widetilde{T}/G$ to $(T/G)^\sim = \widetilde{Y}$ was constructed for which ϕ^* is fully faithful and ϕ_* (on abelian sheaves) is exact. Using also (13.5), in particular the isomorphisms $H^n(\widetilde{T}/G, j_!A) \cong H^n(Y, (p_*^G A)|_{Y-Z})$ for A an abelian G-sheaf on T, one sees that the cohomological (ℓ-) dimensions of $\widetilde{T}/G$ and Y agree.

Assume that T is a CW complex of dimension d, and that G acts cellularly. Then also Y is a d-dimensional CW complex. Any CW complex of dimension d has covering dimension d, at least if it has at most countably many cells [Pe], and so its sheaf cohomological dimension is $\leq d$ [Go II.5.12]. (Actually it is easy to see that the latter dimension is equal to d.) So if A is any abelian G-sheaf on T (and T is countable), then the maps (4) are isomorphisms for $n > d$, by (2) and the remarks just made.

The long exact sequence (2) generalizes an important exact sequence from equivariant (bundle) cohomology. For simplicity assume again that T is a CW complex with cellular G-action. Recall from (13.6) the definition of equivariant singular cohomology. Using the long exact sequence in singular cohomology for the closed pair (T_G, Z_G), plus the fact that G acts freely on $T-Z$, together with tautness properties of singular cohomology, one derives a long exact sequence (M any abelian group)

$$\cdots \longrightarrow H^n_{\mathrm{sing}}(T/G, Z; M) \longrightarrow H^n_{G,\mathrm{sing}}(T, M) \longrightarrow H^n_{G,\mathrm{sing}}(Z, M) \longrightarrow \cdots, \quad (5)$$

see [Bd2, p. 371]. The middle and the right groups in (5) are isomorphic to the corresponding G-equivariant sheaf cohomology groups (13.6). The group on the left is isomorphic to $H^n(\widetilde{T}/G, j_! \underline{M})$, as remarked in (13.5). In summary, the sequence (5) is canonically isomorphic to the sequence

$$\cdots \longrightarrow H^n(\widetilde{T}/G, j_! \underline{M}) \longrightarrow H^n_G(T, \underline{M}) \longrightarrow H^n_G(Z, \underline{M}) \longrightarrow \cdots$$

which is the long exact sequence (2) for the G-topos $E = \widetilde{T}$ and for $A = \underline{M}$, the constant G-sheaf on T.

Recall (cf. the Introduction) that the Cox exact sequence for an algebraic variety X over $\mathbb{R}$ was obtained in [Co] by an application of (5), as a corollary to the main result of this paper. This main result of [Co] identifies, in a weak sense, the étale homotopy type of X with the homotopy type of the space $X(\mathbb{C})_G = EG \times_G X(\mathbb{C})$.

(14.9) Remark. Guided by this topological situation one wonders whether similar relations between cohomological dimensions might hold for G-toposes of more general nature. (G is always a group of prime order.) Specifically, it is natural to ask if one or several of the following properties hold, for $\mathfrak{E}$ a G-topos with fibre E and fixtopos F:

(P1): coh.dim.$(F) \leq$ coh.dim.(E). Or even stronger:

(P1'): The direct image functor $\nu_*: \mathrm{Ab}(F) \to \mathrm{Ab}(E)$ is exact.

(P2): The maps (4) are isomorphisms for $n \geq$ coh.dim.(E). Or, closely related:

(P2'): coh.dim.$(E/G) \leq 1 +$ coh.dim.(E) (or even $=$ coh.dim.(E)).

Here a sensible notion of cohomological dimension should be used, and accordingly one should restrict (P2) to suitable "sheaves". For example, on the étale site of a scheme one would usually consider only torsion sheaves. I have written "coh.dim." for such a notion. Assume always that F has enough points, so that one disposes over the long exact sequences (2).

One has coh.dim.$(E) \leq$ coh.dim.(E/G) since $j_*\pi_*$: Ab$(E) \to$ Ab(E/G) is an exact functor, and $(P2')$ would mean that equality holds, resp. fails by at most 1. (Such remarks make some implicit assumptions on the definition of the coh.dim.'s, like here the one that the class of relevant sheaves is preserved under $j_*\pi_*$; such comments will be suppressed in the following informal discussion.) Assume that $(P1)$ is true. Let N be the smallest integer such that (4) is always an isomorphism for $n > N$. Then one sees easily that coh.dim.(E/G) is N or $N + 1$. In so far, $(P2')$ is almost equivalent to the conjunction of $(P1)$ and $(P2)$.

If the objects considered for coh.dim. are prime-to-$|G|$ torsion (e.g. if coh.dim. $=$ cd$_\ell$, $\ell \neq |G|$ prime), then $(P1)$ actually implies $(P2')$ (in its weak form), by what has been said in Remark (14.7). As in (7.20.1) one can also deduce $(P2)$; and the strong form of $(P2')$ holds if coh.dim.$(F) <$ coh.dim.(E). All this is analogous to the situation considered in Sect. 7 for *odd* torsion sheaves.

Let us review the two main examples studied in this treatise. Let X be a scheme over $\mathbb{Z}[\frac{1}{2}]$ and $\mathfrak{E}$ the G-topos $\widetilde{X}'_{et}$, so $G = \mathbb{Z}/2$ here. Then $E(G) = \widetilde{X}_{et}$, and F was identified with the real étale topos $\widetilde{X}_{ret}$ in (11.1.1). So E/G is the topos $\widetilde{X}_b$ introduced in Sect. 2. The direct image functor ν_* was studied in Sect. 5 and found to be exact (on abelian sheaves). The proof of this fact was non-trivial, it depended on a study of the real spectrum of Weil restrictions of strictly henselian rings. So properties $(P1)$ and $(P1')$ are satisfied. By the main results of Part One, also $(P2)$ holds (with coh.dim. $=$ cd$_\ell$, ℓ any prime), provided that X is quasi-compact and quasi-separated (7.20); and dito for $(P2')$ in its weaker form, cf. (7.4), (7.18). The stronger form of $(P2')$ need not hold.

On the other hand consider an extension

$$1 \longrightarrow \Delta \longrightarrow \Gamma \longrightarrow G \longrightarrow 1 \tag{6}$$

of discrete or profinite groups (with $|G| =$ prime), and the G-topos $\mathfrak{E} = (\Delta$-sets). Here $E(G) = (\Gamma$-sets), and F was identified in Sect. 11.2 with the topos of Δ-sheaves on $\mathfrak{K}$, the space of splittings of (6). Certainly ν_* is an exact functor here (11.2.4). And also $(P2)$ holds for cohomological ℓ-dimension, ℓ any prime (only $\ell = |G|$ is relevant): This is a particular case of Brown's theorem in the discrete case, and of Theorem (12.13) in the profinite case; see (12.20.5). In neither case is the proof trivial or follows from "general principles". Not even in the easier case where F is the empty topos, i.e. where (6) does not split: Here $(P2)$ is precisely Serre's famous theorem ([Se2], [Se3])! Returning to the general case, $(P2')$ in its stronger form may again fail, but the weaker form holds in any case, by $(P1)$ and $(P2)$.

The above remarks seem to suggest the following. For a G-topos $\mathfrak{E}$ it is reasonable to ask whether properties $(P1)$–$(P2')$ above are true, maybe with slight modifications. In the three cases reviewed here — topological spaces with G-action, the G-site $\widetilde{X}'_{et}$ (with $|G| = 2$), and discrete or profinite group extensions — this is more or less so, as long as suitable hypotheses are added. It would be very interesting to understand whether there are common general reasons for this phenomenon. If such reasons exist, i.e. proofs which only use topos-theoretic arguments, they cannot be very obvious, since the proofs for the above facts use techniques which are quite specific to the respective theories.

Part Three

15. Comparison theorems

It is well known that étale cohomology of a separated complex algebraic variety Y (with finite constant coefficients, say) is just classical (Betti) cohomology of the space $Y(\mathbb{C})$. More generally, such a comparison isomorphism holds for torsion sheaves on the étale site. Actually the same isomorphism exists over any algebraically closed field of characteristic 0, after a subfield R of index 2 has been specified (so that semi-algebraic topology over R can replace classical topology). This has been shown by R. Huber in his 1984 Regensburg thesis [Hu1].

In this section, analogous comparison theorems are obtained for algebraic varieties X over a *real* closed field R. It is shown that étale cohomology of X is G-equivariant cohomology on the semi-algebraic space $X(C)$, and that cohomology of the site X_b is cohomology on the semi-algebraic quotient space $X(C)/G$ (with suitable torsion coefficients in each case). These results are essentially corollaries to the "complex" Comparison Theorem and to results from Sect. 13. In particular a very easy deduction of the Cox sequence is given which does not need étale homotopy theory.

By making use of limit theorems it is possible to extend these comparison isomorphisms to quite general schemes X over a real closed field R, if one replaces $X(C)$ by the complex spectrum of X' over R. This is sketched in the second part of this section.

(15.1) For the entire section let R be a real closed field and write $C = R(\sqrt{-1})$ for its algebraic closure. G denotes always the Galois group of C over R.

If M is a semi-algebraic space over R [DK1] then the category of sheaves on the site M will be denoted by $\mathrm{Sh}(M)$. The notation $\tilde{M}$ is traditionally used for the topological space "real spectrum of M". It is a well known fact that the topos $\mathrm{Sh}(M)$ is equivalent to the category of sheaves on the topological space $\tilde{M}$ [BCR]. Sheaf cohomology on the site M will be denoted by $H^*_{\mathrm{sa}}(M, -)$.

(15.2) Fix R and C as above (15.1), and let Y be a separated scheme of finite type over $\mathrm{spec}\, C$. The set $Y(C)$ of C-rational points of Y has a canonical structure of a semi-algebraic space over R. If Y/C is quasi-projective, this structure may be obtained by forming the Weil restriction $Z := \mathrm{Res}_{\mathrm{spec}\,C/\,\mathrm{spec}\,R}(Y)$, a quasi-projective scheme over R, and using the canonical bijection $Y(C) \approx Z(R)$. For general Y one obtains the semi-algebraic structure on $Y(C)$ by glueing together affine open pieces.

Let Y_{sa} denote the following site: The underlying category is the category of semi-algebraic spaces M over $Y(C)$ whose structure map $M \to Y(C)$ is locally (with respect to M) a semi-algebraic homeomorphism; the topology is generated by the pretopology of surjective families. The natural site morphism from Y_{sa} to $Y(C)$ (which is given by viewing open semi-algebraic subsets of $Y(C)$ as objects of Y_{sa}) induces an equivalence between the associated toposes, as is immediate from the Comparison Lemma. On the other hand, if $V \to Y$ is an étale Y-scheme of finite type then $V(C) \to Y(C)$ is an object of Y_{sa} (1.8). This defines a site morphism from Y_{sa} to the restricted étale site of Y (which consists of the étale Y-schemes of finite type). The topos of sheaves on the latter is $\widetilde{Y}_{et}$ [SGA4 VII.3.2], so that one has obtained topos morphisms

$$\mathrm{Sh}(Y(C)) \xleftarrow{\ f\ } \widetilde{Y}_{\mathrm{sa}} \xrightarrow{\ g\ } \widetilde{Y}_{et} \tag{1}$$

of which the first one is an equivalence. Write $\Phi = (\Phi^*, \Phi_*)$ for the topos morphism $(f_* g^*, g_* f^*)$ from $\mathrm{Sh}(Y(C))$ to $\widetilde{Y}_{et}$. In [Hu1] the following is proved:

(15.2.1) Theorem. — *For every abelian torsion sheaf A on Y_{et} the maps*

$$H^n(Y_{et}, A) \longrightarrow H^n_{\mathrm{sa}}(Y(C), \Phi^* A)$$

induced by Φ^* are isomorphisms, $n \geq 0$. $\qquad\square$

(15.3) This result will now be extended to a comparison of étale resp. b-cohomology of algebraic R-varieties with semi-algebraic cohomology. Let X be a separated scheme of finite type over $\mathrm{spec}\,R$, put $X' = X \times_{\mathrm{spec}\,R} \mathrm{spec}\,C$ and let $\pi \colon X' \to X$ be the projection. The conjugation action of G on X' over X induces actions on the sites $X'(C) = X(C)$ and X'_{sa}. In this way $\mathrm{Sh}(X(C))$, $\widetilde{X}'_{\mathrm{sa}}$ and $\widetilde{X}'_{et}$ become (the fibres of) G-toposes, and it is clear that the diagram (1) (with X' instead of Y) consists of G-equivariant topos morphisms. In particular also Φ is G-equivariant. Hence (10.6.2) Φ induces a topos morphism $\Phi(G) =: \Psi$ from $\mathrm{Sh}_G(X(C))$ to $\widetilde{X}'_{et}(G) = \widetilde{X}_{et}$, where $\mathrm{Sh}_G(X(C))$ denotes the category of G-sheaves on the semi-algebraic space $X(C)$. Writing $H^*_{G,\,\mathrm{sa}}(X(C), -)$ for G-equivariant semi-algebraic cohomology on $X(C)$ one has

(15.3.1) Corollary. — *For every abelian torsion sheaf A on X_{et} the maps*

$$H^n(X_{et}, A) \longrightarrow H^n_{G,\,\mathrm{sa}}(X(C), \Psi^* A)$$

induced by Ψ^* are isomorphisms, $n \geq 0$.

The proof is immediate: The equivariant topos morphism Φ induces a morphism between the "Hochschild-Serre" spectral sequences for equivariant cohomology (10.9). The E_2^{pq}-component of this morphism is $H^p\big(G, H^q(X'_{et}, \pi^*A)\big) \longrightarrow H^p\big(G, H^q_{sa}(X(C), \Phi^*\pi^*A)\big)$, which is an isomorphism by Theorem (15.2.1). Hence the maps $H^n(X_{et}, A) = H^n_G(X'_{et}, \pi^*A) \longrightarrow H^n_{G,sa}(X(C), \Psi^*A)$ between the abutments are isomorphisms as well. $\qquad\square$

(15.3.2) If $R = \mathbb{R}$, the "real" real numbers, then the semi-algebraic cohomology groups $H^n_{sa}(X(\mathbb{R}), M)$ (with constant coefficients M, say) coincide with the singular (Betti) cohomology groups $H^n_{sing}(X(\mathbb{R}), M)$. The proof of (15.3.1) shows therefore that the Cox long exact sequence (for a variety $X/\mathbb{R}$, see the Introduction) is actually a corollary to the more classical Comparison Theorem for étale cohomology of complex varieties. In particular, it gives an easy deduction of this sequence which does not use étale homotopy theory.

(15.4.1) We continue to consider a separated R-variety X. The G-equivariant topos morphism $\Phi\colon \mathrm{Sh}\big(X(C)\big) \to \widetilde{X}'_{et}$ induces also topos morphisms between the fixtoposes $\underleftarrow{\mathrm{Lim}}\mathrm{top}(-/G)$ and between the quotient toposes $-/G$: (10.15.1), (14.3). Now the fixtopos of $\mathrm{Sh}\big(X(C)\big)$ is $\mathrm{Sh}\big(X(R)\big) \sim \widetilde{X}_r$, proof as for Proposition (13.2) (or by direct citation of this result, which is possible since the G-action on the real spectrum of $X(C)$ is discontinuous and has X_r as its space of fixpoints). And $\underleftarrow{\mathrm{Lim}}\mathrm{top}\big(\widetilde{X}'_{et}/G\big) = \widetilde{X}_{ret} \sim X_r$ was shown in (11.1.1). The topos morphism from $\mathrm{Sh}\big(X(R)\big)$ to $\widetilde{X}_{ret}$ induced by Φ is an equivalence.

Write E for the glued topos $E := \mathrm{Sh}\big(X(C)\big)/G := \big(\mathrm{Sh}\big(X(R)\big), \mathrm{Sh}_G\big(X(C)\big), \rho_1\big)$ (14.2) where ρ_1 is the functor $\mathrm{Sh}_G\big(X(C)\big) \to \mathrm{Sh}\big(X(R)\big)$ which sends a G-sheaf B on $X(C)$ to the sheaf $\Gamma_G\big(B|_{X(R)}\big)$ of its G-invariants on $X(R)$. By functoriality of the quotient construction (14.3), Φ induces a topos morphism $\Phi/G\colon E \to \widetilde{X}_b$. On the other hand there is a canonical topos morphism ϕ from E to $\mathrm{Sh}(Y)$, where $Y := X(C)/G$ is the semi-algebraic quotient space; and ϕ^* is fully faithful and ϕ_* is exact on abelian sheaves. This is proved in completely the same way as Proposition (13.4.1). (Again one may directly cite this proposition here if one prefers, since it is easily seen that the real spectrum functor commutes with the formation of the quotient mod G.) The topos morphisms considered so far can be displayed as follows:

$$
\begin{array}{ccccccccc}
\mathrm{Sh}(X(R)) & \longrightarrow & \mathrm{Sh}(X(C)) & \overset{\pi}{\longrightarrow} & \mathrm{Sh}_G(X(C)) & \overset{j}{\longrightarrow} & E & \overset{\phi}{\longrightarrow} & \mathrm{Sh}(Y) \\
{\scriptstyle\sim}\downarrow & & {\scriptstyle\Phi}\downarrow & & {\scriptstyle\Psi}\downarrow & & {\scriptstyle\Phi/G}\downarrow & & \\
\widetilde{X}_{ret} & \overset{\nu}{\longrightarrow} & \widetilde{X}'_{et} & \overset{\pi}{\longrightarrow} & \widetilde{X}_{et} & \overset{j}{\longrightarrow} & \widetilde{X}_b. &
\end{array}
$$

(15.4.2) The topos morphisms Φ/G and ϕ allow now to compare sheaf cohomology on X_b and on the semi-algebraic space $Y = X(C)/G$. For this let A be an abelian sheaf on X_{et}. By functoriality of the long exact sequence (14.6.1), Φ induces a morphism

$$\cdots\ H^n(X_b, j_! A) \longrightarrow H^n(X_{et}, A) \longrightarrow H^n_G(X_{ret}, \nu(G)^* A) \quad \cdots$$
$$\downarrow \qquad\qquad\qquad \downarrow \qquad\qquad\qquad \downarrow$$
$$\cdots\ H^n(E, j_! \Psi^* A) \longrightarrow H^n_{G,\,sa}(X(C), \Psi^* A) \longrightarrow H^n_{G,\,sa}\Big(X(R), (\Psi^* A)|_{X(R)}\Big)\ \cdots$$
$$(2)$$

between the long exact sequences. The right vertical arrows are isomorphisms, $n \geq 0$. So if A is torsion then Corollary (15.3.1) implies that (2) is an isomorphism of long exact sequences. Hence for A torsion also the morphism of long exact sequences

$$\cdots\ H^n(X_b, j_! A) \longrightarrow H^n(X_b, j_* A) \longrightarrow H^n(X_{ret}, \rho A) \quad \cdots$$
$$\downarrow \qquad\qquad\qquad \downarrow \qquad\qquad\qquad \downarrow$$
$$\cdots\ H^n(E, j_! \Psi^* A) \longrightarrow H^n(E, j_* \Psi^* A) \longrightarrow H^n_{sa}(X(R), \rho_1 \Psi^* A)\ \cdots$$

is an isomorphism, by another application of the Five Lemma.

On the other hand let B be an abelian (semi-algebraic) G-sheaf on $X(C)$. Let $p : X(C) \to Y = X(C)/G$ be the quotient map. Then (13.5) shows that also

$$\cdots\qquad H^n(E, j_! B) \qquad\longrightarrow H^n(E, j_* B) \longrightarrow H^n_{sa}(X(R), \rho_1 B) \quad \cdots$$
$$\downarrow \qquad\qquad\qquad \downarrow \qquad\qquad\qquad \|$$
$$\cdots\ H^n_{sa}\Big(Y, (p^G_* B)|_{Y-X(R)}\Big) \longrightarrow H^n_{sa}(Y, p^G_* B) \longrightarrow H^n_{sa}\Big(X(R), (p^G_* B)|_{X(R)}\Big) \quad \cdots$$

is an isomorphism of long exact sequences. Putting together these isomorphisms one gets:

(15.5) Theorem. — *Let X be a separated scheme of finite type over R. Let A be an abelian torsion sheaf on X_{et} and let $F := p^G_* \Psi^* A$, an abelian sheaf on the semi-algebraic space $Y = X(C)/G$. Then there are canonical isomorphisms of long exact sequences between*

$$\cdots\ H^n(X_b, j_! A) \longrightarrow H^n(X_{et}, A) \longrightarrow H^n_G(X_{ret}, \nu(G)^* A)\ \cdots$$

and

$$\cdots\ H^n_{sa}\Big(Y, F|_{Y-X(R)}\Big) \longrightarrow H^n_{G,\,sa}(X(C), \Psi^* A) \longrightarrow H^n_{G,\,sa}\Big(X(R), (\Psi^* A)|_{X(R)}\Big)\ \cdots,$$

and between

$$\cdots\ H^n(X_b, j_! A) \longrightarrow H^n(X_b, j_* A) \longrightarrow H^n(X_{ret}, \rho A)\ \cdots$$

and

$$\cdots H_{sa}^n\left(Y, F|_{Y-X(R)}\right) \longrightarrow H_{sa}^n(Y,F) \longrightarrow H_{sa}^n\left(X(R), F|_{X(R)}\right) \cdots. \qquad \Box$$

(15.5.1) Remark. In particular, if M is an abelian torsion group, there are canonical isomorphisms $H^*(X_b, j_! M) \cong H_{sa}^*(Y, X(R); M)$ and $H^*(X_b, M) \cong H_{sa}^*(Y, M)$. It is now completely clear (cf. (13.6)) that the fundamental long exact sequence

$$\cdots H^n(X_b, j_! A) \longrightarrow H^n(X_{et}, A) \longrightarrow H_G^n(X_{ret}, \nu(G)^* A) \cdots$$

from (6.6) reduces for $A = \underline{M}$ to the exact sequence

$$\cdots H_{sa}^n(X(C)/G, X(R); M) \longrightarrow H^n(X_{et}, M) \longrightarrow H_{G, sa}^n(X(R), M) \cdots,$$

which Cox [Co] stated for any scheme $X/\mathbb{R}$ of finite type and any finite abelian group M.

(15.6) Using theorems for the cohomology of inverse limits of schemes it is possible to obtain the above comparison theorems for a much larger class of schemes, namely for schemes of quite general nature which are defined over some real closed field. This is what will be sketched in the remainder of this section.

Write $\mathrm{Res}_{C/R} := \mathrm{Res}_{\mathrm{spec}\,C/\,\mathrm{spec}\,R}$ in the following (cf. Sect. 5), where R and C are as in (15.1). If Y is a scheme over $\mathrm{spec}\,C$, denote the complex spectrum of Y over $\mathrm{spec}\,R$ by $Y_{cx} := Y_{cx/\,\mathrm{spec}\,R}$. Recall (5.6.2) that, at least locally, Y_{cx} is the real spectrum of the Weil restriction $\mathrm{Res}_{C/R}(Y)$. From Remark (5.15) one has a topos morphism

$$\Phi = (\Phi^*, \Phi_*): \ (Y_{cx})^\sim \longrightarrow \tilde{Y}_{et}. \tag{3}$$

If Y is of finite type over $\mathrm{spec}\,C$ this is nothing but the Φ of (15.2) in an abstract disguise: The real spectrum of $\mathrm{Res}_{C/R} Y$ is the real spectrum of the semi-algebraic space $Y(C)$, and the construction of Φ in (5.15) is just the "abstract" analogue of the construction in (15.2).

This more abstract version applies to more general C-schemes. In this way one gets

(15.6.1) Proposition. — *Let Y be a separated scheme which is finitely presented over some C-algebra, and consider Y as a scheme over $\mathrm{spec}\,C$. For every constructible torsion sheaf $A \in \mathrm{Ab}(Y_{et})$ the maps*

$$H^n(Y_{et}, A) \longrightarrow H^n(Y_{cx}, \Phi^* A) \tag{4}$$

induced by the topos morphism (3) are isomorphisms ($n \geq 0$).

Proof. One can write $Y = \varprojlim Y_\lambda$ as a filtering inverse limit with affine transition maps, where $\{Y_\lambda\}_{\lambda \in I}$ is a family of separated schemes of finite type over spec C (cf. the proof of (7.15)). One can assume that there is a final index λ_0 in I and a (constructible) étale torsion sheaf A_0 on Y_{λ_0} such that A is the pullback of A_0 to Y. Let A_λ be the pullback of A_0 to Y_λ and fix $n \geq 0$. For each index λ the map

$$\Phi^*\colon H^n\Big((Y_\lambda)_{et}, A_\lambda\Big) \longrightarrow H^n\Big((Y_\lambda)_{cx}, \Phi^* A_\lambda\Big) \tag{5}$$

is an isomorphism by (15.2.1). The maps (5) form a direct system indexed by I. The limit of this system is the map (4), by limit theorems for cohomology (Sect. 3) and since $\mathrm{Res}_{C/R}$ commutes with inverse limits (and preserves affineness of transition maps). $\qquad\square$

(15.6.2) Let now X be a separated scheme which is finitely presented over some R-algebra. Write $X' := X \otimes_R C$ and $U := (X')_{cx}$. Then G acts on X' (over X) and on U, and the topos morphism $\Phi\colon \widetilde{U} \to \widetilde{X}'_{et}$ is G-equivariant. Therefore it induces a topos morphism

$$\Psi := \Phi(G)\colon \ \widetilde{U}(G) \longrightarrow \widetilde{X}'_{et}(G) = \widetilde{X}_{et}. \tag{6}$$

Note that the space U^G of fixpoints is canonically identified with X_r (5.7.5).

(15.6.3) **Corollary.** — *Let X be an R-scheme as above and put $U = (X')_{cx}$. For any constructible torsion sheaf $A \in \mathrm{Ab}(X_{et})$ the maps*

$$H^n(X_{et}, A) \longrightarrow H^n_G(U, \Psi^* A)$$

induced by (6) are isomorphisms ($n \geq 0$).

Proof. Either as for (15.3.1), or by limit arguments from this corollary. $\qquad\square$

Finally one can generalize the comparison theorem for the b-topology:

(15.7) **Proposition.** — *Let X be an R-scheme as before and put $U := (X')_{cx}$. Let $p\colon U \to U/G =: W$ be the topological quotient space of the G-action on U. Let A be a constructible torsion sheaf on X_{et} and write $F := p^G_* \Psi^* A$, a sheaf on W. Then the long exact sequences*

$$\cdots \longrightarrow H^n(X_b, j_! A) \longrightarrow H^n(X_b, j_* A) \longrightarrow H^n(X_r, \rho A) \longrightarrow \cdots$$

and

$$\cdots \longrightarrow H^n(W, F|_{W-X_r}) \longrightarrow H^n(W, F) \longrightarrow H^n(X_r, F|_{X_r}) \longrightarrow \cdots$$

are canonically isomorphic.

Note that the composition $X_r \to U \to W$ is a closed embedding.

Proof. This follows from Theorem (15.5) by limit arguments as in the proof of (15.6.1). Note that one has a canonical homeomorphism $\widetilde{M}/G \approx (M/G)^\sim$ for M a semi-algebraic G-space, and that the formation of topological quotient spaces mod G commutes with filtering inverse limits. $\qquad\square$

So the complex spectrum $U = (X')_{cx}$ of X' over R is a topological model on which one can calculate étale or b-cohomology of X with suitable coefficients: Étale cohomology on X corresponds to G-equivariant cohomology on U, while b-cohomology on X corresponds to cohomology on the quotient space U/G. In particular:

(15.7.1) Corollary. — *If M is a torsion abelian group then there are canonical isomorphisms*

$$H^*(X_{et}, M) \cong H^*_G(U, M),$$
$$H^*(X_b, M) \cong H^*(U/G, M),$$
$$H^*(X_b, j_! M) \cong H^*(U/G, X_r; M). \qquad \square$$

16. Base change theorems

This section contains the proper and smooth base change theorems for the topologies b and ret. In both situations one sees by formal arguments that the corresponding base change theorem for the topology b is equivalent to the conjunction of the respective base change theorems for the étale and the real étale topology. The étale cases being classical, one is left with the real spectrum. Here the proper base change theorem is also well known (proved by Delfs), whereas the smooth base change theorem hasn't yet been considered, as far as I know.

Applications of both base change theorems can be found in the next two sections.

(16.1) Remarks. Consider *any* cartesian cube of schemes

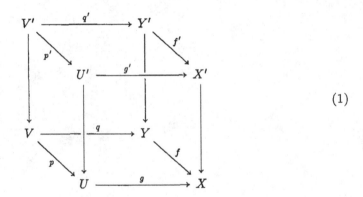

$$(1)$$

in which the bottom square is an arbitrary cartesian square, and the top square is the base extension of the bottom square by $\mathbb{Z} \to \mathbb{Z}[\sqrt{-1}]$.

(16.1.1) Let $t \in \{et, b, ret\}$ and consider the base change morphisms

$$\beta_t(F): \quad g_t^* R^n f_{t*} F \longrightarrow R^n p_{t*} q_t^* F \quad (n \geq 0) \tag{2}$$

and

$$\beta_t'(F'): \quad g_t'^* R^n f_{t*}' F' \longrightarrow R^n p_{t*}' q_t'^* F' \quad (n \geq 0), \tag{3}$$

for abelian sheaves F on Y_t and F' on Y_t'. From the commutation rules in (3.10) one sees that

$$\beta_b(i_* B) = i_* \beta_{ret}(B) \quad \text{and} \quad \beta_b(j_* \pi_* A') = j_* \pi_* \beta_{et}'(A'), \tag{4}$$

for $B \in \mathrm{Ab}(Y_{ret})$ and $A' \in \mathrm{Ab}(Y'_{et})$. (The second equality uses exactness of $j_* \pi_*$ (3.12d) plus finite étale base change by π, where for any scheme Z, π denotes invariantly the projection $Z \otimes_{\mathbb{Z}} \mathbb{Z}[\sqrt{-1}] \to Z$.) Moreover

$$j^* \beta_b(F) = \beta_{et}(j^* F) \tag{5}$$

for $F \in \mathrm{Ab}(Y_b)$ (3.12.2). This shows how (in the case of abelian sheaves) any base change theorem for the b-topology implies corresponding base change theorems for the topologies et and ret.

On the other hand, suppose that every sheaf $F \in \mathrm{Ab}(Y_b)$ injects into some $G \in \mathrm{Ab}(Y_b)$ for which the base change maps $\beta_b(G)$ are isomorphisms for all $n \geq 0$. Then one sees that $g_b^* f_{b*} F \xrightarrow{\sim} p_{b*} q_b^* F$ for every $F \in \mathrm{Ab}(Y_b)$, by looking at exact sequences $0 \to F \to G \to G'$ with such G and G'. Furthermore, from short exact sequences $0 \to F \to G \to F' \to 0$ it follows inductively (using the Five Lemma) that the $\beta_b(F)$ are isomorphisms for *every* $F \in \mathrm{Ab}(Y_b)$ and $n \geq 0$. Of course one can also apply this argument to appropriate subcategories of $\mathrm{Ab}(Y_b)$, such as categories of suitably torsion sheaves.

For an example where this is useful, suppose that suitable base change theorems have been proved for the étale and the real étale topology. Then one gets at once a corresponding base change theorem for the b-topology, by (4) and the observation that every $F \in \mathrm{Ab}(Y_b)$ injects into a direct sum $G = j_* \pi_* A' \oplus i_* B$ with $A' \in \mathrm{Ab}(Y'_{et})$ and $B \in \mathrm{Ab}(Y_{ret})$. (Take $A' = \pi^* j^* F$ and $B = i^* F$, for example.)

(16.1.2) For set-valued sheaves similar reductions are possible. Consider the base change morphism

$$\beta_t(F): \ g_t^* f_{t*} F \longrightarrow p_{t*} q_t^* F \tag{6}$$

for $F \in \widetilde{Y}_t$. Then

$$\beta_b(i_* B) = i_* \beta_{ret}(B) \quad \text{and} \quad \beta_b(j_* A) = j_* \beta_{et}(A)$$

for $B \in \widetilde{Y}_{ret}$ and $A \in \widetilde{Y}_{et}$, by (3.10). Since every sheaf on Y_b is the fibre product of a diagram

$$i_* B \longrightarrow i_* \rho A \longleftarrow j_* A,$$

and since the class of sheaves $F \in \widetilde{Y}_b$ for which $\beta_b(F)$ is an isomorphism is stable under fibre products, one sees how a base change theorem for the b-topology is equivalent to the conjunction of base change theorems for et and for ret (set-valued sheaves case).

(16.1.3) In summary, these arguments show that, under suitable natural conditions which I do not bother to make precise here, a base change theorem for the topology b (and certain classes of sheaves and of cartesian diagrams (1)) is *equivalent* to the conjunction of corresponding base change theorems for et and for ret.

(16.2) Theorem (Proper Base Change). — *Let* $t \in \{et, b, ret\}$, *and consider a cartesian square*

$$
\begin{array}{ccc}
Y_1 & \xrightarrow{\ g_1\ } & Y \\
{\scriptstyle f_1}\downarrow & & \downarrow{\scriptstyle f} \\
X_1 & \xrightarrow{\ g\ } & X
\end{array}
\tag{7}
$$

of schemes in which f *is a proper morphism.*

a) *For* $F \in \tilde{Y}_t$ *the base change map*

$$
g_t^* f_{t*} F \longrightarrow f_{1t*} g_{1t}^* F
\tag{8}
$$

is an isomorphism.

b) *Let* $F \in \mathrm{Ab}(Y_t)$. *If* $t = et$ *(resp. if* $t = b$*), assume that* F *(resp.* $j^* F$*) is a torsion sheaf. Then the base change maps*

$$
g_t^* R^q f_{t*} F \longrightarrow R^q f_{1t*} g_{1t}^* F
\tag{9}
$$

are isomorphisms, $q \geq 0$.

(16.2.1) For $t = et$ this is the object of exposés XII et XIII dans [SGA4]. Hence, by the technique of Remarks (16.1), it suffices to treat the case $t = ret$. Also this case is well known, and is due to Delfs [Df, II.7.8] (for abelian sheaves). There the proof is given in the broader context of abstract locally semialgebraic spaces (= real closed spaces in the sense of Schwartz), but essentially nothing new happens.

I give now a new proof for the real proper base change theorem. It can easily be transferred to the more general situation considered by Delfs, at least for usual cohomology. (Delfs is more generally working with cohomology with supports.)

(16.3.1) Notation. If S is a topological space and $S' \subset S$ is any subspace write $\mathrm{Gen}_S(S')$ or $\mathrm{Gen}\, S'$ for the subspace of S which is the intersection of all neighborhoods of S' in S. If $S' = \{s\}$ this notation simplifies to $\mathrm{Gen}_S(s)$ or $\mathrm{Gen}\, s$. The elements of $\mathrm{Gen}\, S'$ are called the *generalizations* of S', since $\mathrm{Gen}\, S' = \{x \in S: S' \cap \overline{\{x\}} \neq \emptyset\}$.

(16.3.2) Lemma. — *Let* $f: T \to S$ *be a spectral map between spectral spaces. Assume that the specializations of any* $t \in T$ *form a chain, i.e. that* $\overline{\{t\}}$ *is totally ordered by the specialization relation. If* f *is a closed map, then for any* $B \in \mathrm{Ab}(T)$ *and* $s \in S$ *the canonical maps*

$$
(R^n f_* B)_s \longrightarrow H^n\!\left(f^{-1}(s), B\big|_{f^{-1}(s)}\right), \quad n \geq 0,
$$

are isomorphisms. Also, for any $B \in \tilde{T}$ *and* $s \in S$ *the canonical map*

$$
(f_* B)_s \longrightarrow H^0\!\left(f^{-1}(s), B\big|_{f^{-1}(s)}\right)
$$

is bijective.

A typical example of a space T which satisfies the hypothesis of this lemma is the real spectrum of any ring, or more generally any pro-constructible subspace of such a real spectrum.

Proof. It is a fact which holds in general for spectral spaces and spectral maps, that

$$(R^n f_* B)_s \xrightarrow{\sim} H^n \left(f^{-1}(\text{Gen } s), B|_{f^{-1}(\text{Gen } s)} \right), \quad n \geq 0, \tag{10}$$

for $B \in \text{Ab}(T)$ and $s \in S$; for the easy proof see [Sch1 Prop. 3.3]. Since f is closed, $f^{-1}(\text{Gen } s) = \text{Gen } f^{-1}(s)$. Now the fibre $f^{-1}(s)$ is a closed subspace of $\text{Gen } f^{-1}(s)$ which contains every closed point of the latter space. Since on every normal spectral space (like, by hypothesis, $f^{-1}(s)$ or $\text{Gen } f^{-1}(s)$) the restriction of a sheaf to the subspace of closed points induces an isomorphism in cohomology [CaC, p. 231], it follows for any abelian sheaf F on $\text{Gen } f^{-1}(s)$ that the restriction maps

$$H^n \left(\text{Gen } f^{-1}(s), F \right) \longrightarrow H^n \left(f^{-1}(s), F|_{f^{-1}(s)} \right)$$

are isomorphisms ($n \geq 0$). Combined with (10) this gives the abelian sheaves case. For set-valued sheaves the arguments are the same. □

(16.3.3) Now let a cartesian square (7) of schemes be given and consider the topology $t = ret$. To prove that the base change maps (8) resp. (9) are isomorphisms one may argue stalkwise. Thus one can assume that $X_1 = \text{spec } S$ is the spectrum of a real closed field S. Let $X_1 \to x \to X$ be the (unique) factorization of g in which $x = \text{spec } R$ with R the algebraic closure of the image of $\mathcal{O}_X$ in S. Since f_r is a closed map the preceding lemma applies, and so one reduces to show the following: If $R \subset S$ is an extension of real closed fields and V is a proper scheme over $\text{spec } R$, then for every abelian sheaf (resp. every sheaf) F on V_r one has $H^*(V_r, F) \xrightarrow{\sim} H^*((V_S)_r, F_S)$ (resp. $H^0(V_r, F) \xrightarrow{\sim} H^0((V_S)_r, F_S)$), where F_S denotes the pullback of F to $(V_S)_r$.

Actually one doesn't need completeness any more here, so V could be any R-scheme of finite type. Suppose however that V is separated (from this one deduces readily the non-separated case). The proof is now easily completed: Since every (abelian) sheaf on V_r is a filtering direct limit of constructible (abelian) sheaves (Proposition (A.4)), it suffices to consider the case when F itself is constructible. First consider abelian sheaves. By Proposition (A.6) it suffices to treat the case $F = M_Z$, a constant sheaf on a closed constructible subset Z of V_r extended by zero on $V_r - Z$ (M is an abelian group). If K is the closed semi-algebraic subset of $V(R)$ which corresponds to Z, then K can be triangulated (in the semi-algebraic sense); and such a triangulation induces a triangulation of $K(S)$ (the base extension of K to S) which is combinatorially identical. Hence $H^*(K, M) \xrightarrow{\sim} H^*(K(S), M)$.

In the set-valued case one triangulates $V(R)$ in such a way that F is constant on every open simplex, and uses Proposition (17.15.1). (See (17.10) for more on semi-algebraic triangulation.)

The proof of Theorem (16.2) is thus complete.   □

(16.3.4) Let $f: Y \to X$ be a morphism of schemes which can be compactified, i.e. admits a factorization $Y \overset{h}{\hookrightarrow} \bar{X} \overset{\bar{f}}{\to} X$ with h an open immersion and $\bar{f}$ proper. As in the étale theory [SGA4 XVII] one can use the proper base change theorem to define (for the b-topology) higher direct images of f with proper support, for sheaves $F \in \mathrm{Ab}(Y_b)$ with j^*F torsion.

(16.4) The next topic is smooth base change. First a few phrases are introduced to facilitate the exposition. Only abelian sheaves are treated below. It is straightforward how to modify the following for the case of set-valued sheaves.

(16.4.1) Definitions. Let a topology $t \in \{et, b, ret\}$ be fixed. Also fix a set $\mathbb{L}$ of primes. In the following only schemes will be considered on which every $\ell \in \mathbb{L}$ is invertible.

a) Let X be (such) a scheme. An abelian sheaf F on X_t will be called *admissible* if it satisfies the following condition:
  - if $t = et$: F is an $\mathbb{L}$-primary torsion sheaf;
  - if $t = b$: j^*F is an $\mathbb{L}$-primary torsion sheaf;
  - if $t = ret$: no condition.

b) A morphism $g: Y \to X$ of schemes has *cohomological descent for admissible sheaves* (with respect to t) if for every admissible $F \in \mathrm{Ab}(X_t)$

$$F \overset{\sim}{\longrightarrow} g_{t*}g_t^*F \quad \text{and} \quad R^q g_{t*}(g_t^*F) = 0, \quad q \geq 1.$$

c) A morphism $g: Y \to X$ is *acyclic* for t if every base extension $g_1: Y_1 \to X_1$ of g by an étale morphism $X_1 \to X$ has cohomological descent for admissible sheaves. g is *locally acyclic* for t if for every t-point $\beta: y \to Y$ of Y the morphism $\tilde{g}: Y^\beta \to X^{g \circ \beta}$ induced by g is acyclic for t. Finally, g is *universally (locally) acyclic* for t if every base extension of g is (locally) acyclic for t.

Recall (Sect. 3) that by a t-point $\beta: y \to Y$ of Y one means a morphism of schemes where y is the spectrum of a field which is separably closed (if $t \in \{et, b\}$) or real closed (if $t \in \{b, ret\}$). Moreover Y^β is the strict resp. strict real localization of Y with respect to β (3.5).

(16.5) Remarks. Let $g: Y \to X$ be a morphism of schemes. Definitions c) above are modelled after [SGA4 XV.1].

(16.5.1) Using the results of Sect. 3 one sees easily as in [loc.cit., Prop. 1.6] that g is acyclic for t if and only if for every base extension $g_1: Y_1 = X_1 \times_X Y \to X_1$ of g

by an étale morphism $X_1 \to X$ and every admissible $F \in \mathrm{Ab}(X_{1t})$ the maps

$$H^n(X_{1t}, F) \longrightarrow H^n(Y_{1t}, g_{1t}^* F)$$

induced by g_{1t}^* are isomorphisms ($n \geq 0$).

(16.5.2) Also one can show as in [loc.cit., Prop. 1.10] that g is locally acyclic for t if and only if for each diagram

$$
\begin{array}{ccccc}
Y_2 & \xrightarrow{h_1} & Y_1 & \xrightarrow{f_1} & Y \\
{\scriptstyle g_2}\downarrow & & {\scriptstyle g_1}\downarrow & & \downarrow{\scriptstyle g} \\
X_2 & \xrightarrow{h} & X_1 & \xrightarrow{f} & X
\end{array}
$$

in which both squares are cartesian and f, h are étale, h finitely presented, the base change maps

$$g_{1t}^* R^q h_{t*} F \longrightarrow R^q h_{1t*} g_{2t}^* F$$

are isomorphisms for admissible $F \in \mathrm{Ab}(X_{2t})$ ($q \geq 0$). (The conditions on f and h may be weakened as in [loc.cit.].) I am only going to use the "only if" part of this assertion, which is the easier one.

(16.6) Lemma. — *Let $t \in \{et, b, ret\}$. Consider a cartesian square of schemes*

$$
\begin{array}{ccc}
Y_1 & \xrightarrow{g_1} & X_1 \\
{\scriptstyle f_1}\downarrow & & \downarrow{\scriptstyle f} \\
Y & \xrightarrow{g} & X
\end{array}
\tag{11}
$$

in which f is quasi-compact and quasi-separated and g is universally locally acyclic for t. Then the base change maps

$$g_t^* R^q f_{t*} F \longrightarrow R^q f_{1t*} g_{1t}^* F, \quad q \geq 0,$$

are isomorphisms for every admissible $F \in \mathrm{Ab}(X_{1t})$.

(16.6.1) The proof is the same as in the étale case [SGA4 XVI.1]. Therefore it will only be sketched. The assertion is local with respect to X, which one can therefore assume to be an affine scheme. If f is proper the assertion follows from the proper base change theorem, and if f is étale it is a particular case of Remark (16.5.2). By combining these two cases one gets the lemma if f is compactifiable (i.e. factorizes as $X_1 \xhookrightarrow{h} \bar{X} \xrightarrow{\bar{f}} X$ with h an open immersion and $\bar{f}$ proper), e.g. quasi-projective. Assume now that f is quasi-affine, so the X-scheme X_1 is an open subscheme of an affine X-scheme X_2. Writing X_2 as a filtering $\varprojlim$ of affine X-schemes of finite type one sees that X_1 is a filtering $\varprojlim$ of quasi-projective X-schemes $X_{1\lambda}$, with affine transition maps. For every abelian t-sheaf F on X_1 one has $R^q f_{t*} F = \varinjlim_\lambda R^q (f_\lambda)_{t*} (p_\lambda)_{t*} F$, where $X_1 \xrightarrow{p_\lambda} X_{1\lambda} \xrightarrow{f_\lambda} X$ is the canonical factorization of f. From this one deduces that the base change theorem holds for $f: X_1 \to X$. The final step in the general case is to cover X_1 by (finitely many) open affine subschemes U_i. Since base change holds for the restriction of f to every intersection of the U_i, one concludes that base change holds also for $f: X_1 \to X$. $\square$

(16.7) Lemma. — *A morphism $g\colon Y \to X$ of schemes is acyclic for the topology b if and only if g is acyclic for the real étale topology and $g'\colon Y' \to X'$ is acyclic for the étale topology. Here g' is the base extension of g by $\pi\colon X' = X \otimes_{\mathbb{Z}} \mathbb{Z}[\sqrt{-1}] \to X$.*

Proof. This equivalence is true even for the notion of cohomological descent for admissible sheaves. The verification is similar to the argument in Remarks (16.1). Using that

$$Rg_{b*}g_b^* i_* = i_* Rg_{ret*}g_{ret}^* \quad \text{and} \quad Rg_{b*}g_b^* j_* \pi_* = j_* \pi_* Rg'_{et*}g'^*_{et} \qquad (12)$$

it is clear how to conclude "from b to ret and et". The opposite direction follows from the fact that every admissible $F \in \mathrm{Ab}(X_b)$ injects into a direct sum $G = j_* \pi_* A' \oplus i_* B$, with $A' \in \mathrm{Ab}(X'_{et})$ and $B \in \mathrm{Ab}(X_{ret})$ admissible; one concludes as in (16.1.1).   □

(16.7.1) Corollary. — *g is locally acyclic for the topology b if and only if g is locally acyclic for both et and ret.*

Proof. If one uses the characterization of locally acyclic morphisms from (16.5.2), the corollary is an application of Remark (16.1). Here is a deduction from (16.7): Let g be locally acyclic for b and let $\eta\colon y \to Y$ be a b-point of Y; so $\tilde{g}\colon Y^\eta \to X^{g\circ\eta}$ is b-acyclic. If η is an étale point then $et = b$ on Y^η and on $X^{g\circ\eta}$ (a strictly henselian local ring has empty real spectrum), and so $\tilde{g}$ is et-acyclic. If η is a real point then $\tilde{g}\colon Y^\eta \to X^{g\circ\eta}$ is ret-acyclic by the lemma. Conversely let g be locally acyclic for et and ret, let $\eta\colon y \to Y$ be a b-point and $\tilde{g}\colon Y^\eta \to X^{g\circ\eta}$ the morphism induced by g. If η is an étale point then $\tilde{g}$ is acyclic for $b = et$. If η is a real point then $\tilde{g}$ is acyclic for ret, and $\tilde{g}'$ is acyclic for et since $\tilde{g}'$ corresponds to the étale point $\eta'\colon y' \xrightarrow{\pi} y \xrightarrow{\eta} Y$. By the lemma, $\tilde{g}$ is acyclic for b.   □

(16.7.2) Example (Homotopy invariance). — *For any scheme X and $n \geq 0$ the projection $\mathbb{A}_X^n \longrightarrow X$ is acyclic for all three topologies et, b and ret.*
For the étale topology this is proved in [SGA4 XV.2], so consider $t = ret$. It suffices to treat $g\colon \mathbb{A}_X^1 \to X$ with $X = \mathrm{spec}\, A$ an affine scheme. One has to show that the maps

$$H^q(\mathrm{sper}\, A,\, F) \longrightarrow H^q(\mathrm{sper}\, A[t],\, g_r^* F), \quad q \geq 0,$$

are isomorphisms for every abelian sheaf F on $\mathrm{sper}\, A$. Writing F as a filtering direct limit of constructible sheaves (A.4) one may assume that F itself is constructible. Then by Proposition (A.9) one can assume that A is finitely generated over $\mathbb{Z}$, or else, over a real closed field R. Finally Proposition (A.6) allows to assume $F = M_Z$, a constant sheaf on a closed constructible subset Z of $\mathrm{sper}\, A$. So one is reduced to show: If K is an affine semi-algebraic space over R and M is an abelian group then $H_{sa}^*(K, M) \xrightarrow{\sim} H_{sa}^*(K \times R, M)$. This is obvious.   □

(16.8) Theorem. — *Every smooth morphism* $g: Y \to X$ *of schemes is locally acyclic for the real étale topology.*

(16.9) By the main step in the proof of the étale smooth base change theorem, g is also locally acyclic for $t = et$ [SGA4 XV.2]. Hence g is locally acyclic for $t = b$ by (16.7.1). As one would expect, the proof of the theorem for the real étale topology is much easier than for the étale topology. Like in the étale case one shows that it suffices to prove that the "geometric fibres" of a strictly localized version $\tilde{g}: \tilde{Y} \to \tilde{X}$ of g are acyclic. (In the real case read "real spectrum fibres" instead of "geometric fibres".) While the proof of this acyclicity of "geometric fibres" requires considerable effort in the étale case [SGA4 XV, §§ 2 et 3] it is very easy in the real case. (The proof is a nice geometric argument, see (16.10.2)). On the other hand, also the way in which one reduces to this statement about geometric fibres is different in both cases. In the étale case one can reduce to noetherian schemes, and then use approximation of torsion sheaves by constructible sheaves, plus the fact that the latter embed into finite $\oplus$'s of direct images $\bar{x}_* M$, $\bar{x}$ geometric points. This last technique is not available in the real case. Instead one studies directly the topology of the real spectra here and uses (16.3.2) above (as a sort of Vietoris-Begle mapping theorem for real spectra).

(16.10) Definition. Let $f: T \to S$ be a map of topological spaces.
  a) f has *locally acyclic fibres* if for every $t \in T$ the restriction $f_t: \mathrm{Gen}_T(t) \to \mathrm{Gen}_S f(t)$ of f has acyclic fibres, i.e. if for every $s \in \mathrm{Gen}_S f(t)$ and every abelian group M one has $H^0\big(f_t^{-1}(s), M\big) = M$ and $H^n\big(f_t^{-1}(s), M\big) = 0$ for $n \geq 1$.
  b) f has *cohomological descent* if and only if $F \xrightarrow{\sim} f_* f^* F$ and $R^q f_*(f^* F) = 0$ ($q \geq 1$) for every abelian sheaf F on S.

Of course this definition of a map having locally acyclic fibres is tailored for (locally) spectral spaces, and is void e.g. for Hausdorff spaces.

(16.10.1) Proposition. — *Let* $g: Y \to X$ *be a morphism of schemes. Then g is locally acyclic for the real étale topology if and only if the map* $g_r: Y_r \to X_r$ *between the real spectra has locally acyclic fibres.*

(16.10.2) Assume the proposition for a moment. Then Theorem (16.8) is proved as follows. Since local acyclicity of $g: Y \to X$ is (Zariski) local with respect to Y one may assume that g factors as $Y \xrightarrow{h} \mathbb{A}_X^n \longrightarrow X$ with h étale [EGA IV, 17.11.4]. Either from the definitions (cf. (16.5.2)) or from Proposition (16.10.1) it is clear that h is locally acyclic for ret. Hence it suffices to treat the case where $Y = \mathbb{A}_X^1$ is the affine line over X. According to the proposition one has to check for any ring A that $p: \mathrm{sper}\, A[t] \to \mathrm{sper}\, A$ has locally acyclic fibres. So fix $\xi \in \mathrm{sper}\, A[t]$ and a generalization y of $p(\xi)$ in $\mathrm{sper}\, A$. Let $k(y)$ be the real closed field associated with

y, so that $p^{-1}(y)$ is canonically identified with the real spectrum of $k(y)[t]$. Under this identification, $p^{-1}(y) \cap \operatorname{Gen} \xi$ is a *connected* subspace of $\operatorname{sper} k(y)[t]$. It is clear that such a space has trivial (reduced) sheaf cohomology with constant coefficients.

The fact that $p^{-1}(y) \cap \operatorname{Gen} \xi$ is connected can be seen as follows. Nothing essential is changed if p is replaced by the projection $q: I_A \to \operatorname{sper} A$, where $I_A \subset \operatorname{sper} A[t]$ is the closed unit interval over $\operatorname{sper} A$. Let $y \succ x$ be a specialization in $\operatorname{sper} A$, and write $I_y = q^{-1}(y)$, $I_x = q^{-1}(x)$. For every $\eta \in I_y$ one has

$$\overline{\{\eta\}} \cap I_x = \{\xi', \xi\}$$

with $\xi' \succ \xi$, equality being possible. (The closure $\overline{\{\eta\}}$ is formed in I_A.) Define the map

$$\sigma: I_y \longrightarrow I_x, \quad \sigma(\eta) := \xi'.$$

The sets I_y, I_x are totally ordered in a natural way. The map σ is isotonic, i.e. satisfies $\eta' \leq \eta \Rightarrow \sigma(\eta') \leq \sigma(\eta)$. From this it follows that for $\xi \in I_x$ the subspace $I_y \cap \operatorname{Gen}(\xi)$ is order-convex in I_y, and so is connected since I_y is connected.

(16.10.3) To prove the proposition let $g: Y \to X$ be a morphism of schemes. Let $\beta: y \to Y$ be a real point of Y, write $\tilde{Y} := Y^\beta$ and $\tilde{X} := X^{g \circ \beta}$ for the corresponding strict real localizations and $\tilde{g}: \tilde{Y} \to \tilde{X}$ for the morphism induced by g. Consider an étale morphism $f: U \to \tilde{X}$ and the base extension h of $\tilde{g}$ by f:

$$
\begin{array}{ccccc}
V & \longrightarrow & \tilde{Y} & \longrightarrow & Y \\
\downarrow{\scriptstyle h} & & \downarrow{\scriptstyle \tilde{g}} & & \downarrow{\scriptstyle g} \\
U & \xrightarrow{\ f\ } & \tilde{X} & \longrightarrow & X.
\end{array}
\qquad (13)
$$

In Proposition (1.7) it was shown that (13) induces a homeomorphism $V_r \xrightarrow{\approx} U_r \times_{\tilde{X}_r} \tilde{Y}_r$ of the real spectra. Since f_r is a local homeomorphism (1.8) it follows that cohomological descent for $\tilde{g}_r$ will imply cohomological descent for h_r.

Let $y_0 \in Y_r$ be the point of the real spectrum determined by β, and put $x_0 := g_r(y_0) \in X_r$. The canonical morphisms $\tilde{Y} \to Y$ and $\tilde{X} \to X$ induce topological embeddings of the real spectra under which $\tilde{g}_r: \tilde{Y}_r \to \tilde{X}_r$ gets identified with the restriction $\operatorname{Gen}_{Y_r}(y_0) \to \operatorname{Gen}_{X_r}(x_0)$ of $g_r: Y_r \to X_r$. From the definition of locally acyclic morphisms one sees therefore that g is locally acyclic for ret if and only if the restriction $\operatorname{Gen}_{Y_r}(y_0) \to \operatorname{Gen}_{X_r}(x_0)$ of g_r has cohomological descent, for every choice of $y_0 \in Y_r$.

To simplify notations fix now $y_0 \in Y_r$ and write $p: T \to S$ for the restriction $\operatorname{Gen}_{Y_r}(y_0) \to \operatorname{Gen}_{X_r}(x_0)$ of g_r. It remains to show that p has cohomological descent if and only if the fibres of p have trivial cohomology with constant coefficients. For this note that p is a closed map. Indeed, it is equivalent that p be specializing, and this latter condition translates into a kind of convexity property for g_r which is well known to be true (and easy to prove, see e.g. [KS, Korollar 4 in § III.7]). From (16.3.2) it follows that

$$(R^q p_* p^* F)_s \xrightarrow{\sim} H^q(p^{-1}(s), F_s), \quad q \geq 0,$$

for any $F \in \mathrm{Ab}(S)$ and $s \in S$. From this one reads off immediately that p has co-homological descent if and only if all fibres of p have vanishing reduced cohomology with constant coefficients. □ □

(16.10.4) It is worthwhile to isolate the argument which was used in the last part of the proof. It says that if $f: T \to S$ is a spectral map between spectral spaces, if the specializations of every $t \in T$ form a chain and if f is closed, then

$$f \text{ has cohomological descent} \iff f \text{ has acyclic fibres.}$$

The smooth base change theorems for all three topologies under consideration may therefore be summarized as follows:

(16.11) **Theorem** (Smooth Base Change). — *Let* $t \in \{et, b, ret\}$. *Let*

$$
\begin{array}{ccc}
Y_1 & \xrightarrow{g_1} & X_1 \\
{\scriptstyle f_1}\downarrow & & \downarrow{\scriptstyle f} \\
Y & \xrightarrow{g} & X
\end{array}
$$

be a cartesian square of schemes in which f is quasi-compact and quasi-separated and g is smooth.
a) *For every $F \in \widetilde{X}_{1t}$ the base change map $g_t^* f_{t*} F \to f_{1t*} g_{1t}^* F$ is an isomorphism.*
b) *For every admissible abelian sheaf F on X_{1t} the base change maps $g_t^* R^q f_{t*} F \to R^q f_{1t*} g_{1t}^* F$ $(q \geq 0)$ are isomorphisms.*

Proof. b) follows from étale smooth base change, plus Theorem (16.8), (16.7.1) and (16.6). For a) one has to modify the preceding discussion in a straightforward way. □

An application of smooth base change will be given in (17.20).

(16.12) **Remark.** I want to point out a generalization of the smooth base change theorem to real closed spaces [Schw]. This remark is not used elsewhere.

Let $g: Y \to X$ be a morphism of real closed spaces. Say that g has universally locally acyclic fibres if every base extension of g has locally acyclic fibres (Definition (16.10)). Recall from [HS] the notion of universally convex morphisms (between real closed spaces): They form the abstract generalization of locally proper semi-algebraic maps, and hence should be thought of as the (abstract) relative version of the notion of a "locally compact" space. With these definitions one has:

(16.12.1) Theorem. — *Let a cartesian square of real closed spaces*

$$
\begin{array}{ccc}
Y' & \xrightarrow{\;g'\;} & X' \\
{\scriptstyle f'}\Big\downarrow & & \Big\downarrow{\scriptstyle f} \\
Y & \xrightarrow{\;g\;} & X
\end{array}
$$

be given. Assume that f is quasi-compact and quasi-separated, and that g has universally locally acyclic fibres and is universally convex. Then the base change maps $g^* R^n f_* F' \to R^n f'_* g'^* F'$ are isomorphisms, for $F' \in \mathrm{Ab}(X')$ and $n \geq 0$.

The proof proceeds similarly to the proof of (16.6). Instead of open embeddings f one has to study more generally constructible embeddings which are not in general locally closed. The hypothesis that g should be universally convex is needed to save the argument at the end of (16.10.3).

This theorem can be seen as a substitute and generalization of the real smooth base change theorem in a context where smoothness does not make sense. Note that the class of morphisms which are universally convex and have universally locally acyclic fibres is closed under composition.

17. Constructible sheaves and finiteness theorems

The object of this section is to prove finiteness theorems for the real étale and the b-topology, saying that the higher direct images of a constructible abelian sheaf under a proper morphism are again constructible. Also set-valued sheaves are considered. As an application one can prove a theorem on the smooth specialization of cohomology groups of proper schemes.

The section starts with some generalities on locally constant sheaves. Then the question is studied whether ρ and its derived functors preserve locally constant and/or constructible sheaves. Similarly to the preceding section, the proof of the finiteness theorem for the b-topology is reduced quite easily to the étale and the real étale topology. The proof of the real finiteness theorem is therefore the essential step here. One reduces to a semi-algebraic situation in which the theorem can be proved without any properness assumption. The main technical tool is semi-algebraic triangulation.

(17.1) Definition [SGA4 IX.2]. Let E be a topos. An object $x \in E$ is *locally constant* if there are an epimorphic family $\{e_i \to *\}_{i \in I}$ in E, and sets M_i and isomorphisms $\underline{M}_i \times e_i \xrightarrow{\sim} x \times e_i$ over e_i, for every $i \in I$. Here $*$ is the final object of E, and $\underline{M} = \coprod_M *$ is the *constant* object in E associated with the set M. If x is an abelian group object one requires (for x to be locally constant in the sense of abelian group objects) that the M_i are abelian groups and the isomorphisms preserve the group structures.

If $f \colon E' \to E$ is any topos morphism then the inverse image functor $f^* \colon E \to E'$ sends (locally) constant objects in E to (locally) constant objects in E'.

(17.2) Examples.

(17.2.1) If $E = \widetilde{C}$ is the topos of sheaves on a site C and $A \in \widetilde{C}$ is locally constant then there is a family $\{x_i\}_{i \in I}$ of objects of C such that $\{\epsilon x_i \to *\}$ is epi and A is constant over any of the x_i.

(17.2.2) Let T be a topological space and $A \in \widetilde{T}$. Then A is a locally constant object of $\widetilde{T}$ (in the sense just defined) if and only if A is a locally constant sheaf in the usual sense.

(17.2.3) Let T be a topological space and G a discrete group which acts *trivially* on T. Consider the topos $\widetilde{T}(G)$ of G-sheaves on T. Let $A \in \widetilde{T}(G)$, considered as an espace étalé $p \colon A \to T$ with G-action. Then A is locally constant in $\widetilde{T}(G)$ if and only if the sheaf A (stripped of the G-action) is locally constant on T. Thus the term "locally constant G-sheaf" is unambiguous. The necessity of the latter condition is obvious (consider the topos morphism π from $\widetilde{T}$ to $\widetilde{T}(G)$). For the sufficiency suppose that there are a set M and an isomorphism

$$A \xrightarrow{\sim} T \times M, \quad a \longmapsto \big(p(a),\, f(a)\big)$$

over T, not necessarily respecting the G-action. Then the map (of espaces étalés over T)

$$\varphi\colon G \times A \longrightarrow G \times T \times M, \quad \varphi(g,a) := \Big(g,\, p(a),\, f(g^{-1}a)\Big)$$

is G-equivariant, and is an isomorphism of G-sheaves on T.

(17.2.4) Keep the hypotheses of (17.2.3), and let A be a locally constant G-sheaf on T. One may ask whether it is true that for every $x \in T$ there are a neighborhood U of x and a G-set M such that $A|_U \cong U \times M$ over U, respecting the G-action. It is easy to see that, in this generality, the answer is negative. For example, suppose that T is an infinite boolean space. Let $x_0 \in T$ be a non-isolated point and let $\{K_i\}_{i \in I}$ be a fundamental system of clopen neighborhoods of x_0. Put $M := G \times I$, and let G act on the constant sheaf $\mathrm{pr}_1 \colon T \times M \to T$ by

$$g \cdot (x,h,i) := \begin{cases} (x,h,i) & \text{if } x \in K_i, \\ (x,gh,i) & \text{if } x \notin K_i. \end{cases}$$

The G-sheaf defined in this way is locally constant, but for every $x \neq x_0$ the fibres in x and in x_0 are not G-isomorphic. This motivates the following

(17.3) Definition. Let the discrete group G act trivially on the space T. A G-sheaf A on T is *strongly locally constant* if there are an open covering $\{U_i\}_{i \in I}$ of T and a family $\{M_i\}_{i \in I}$ of G-sets such that $A|_{U_i} \cong U_i \times M_i$ (as G-sheaves on U_i) for every $i \in I$. If A is an abelian G-sheaf one requires of course that the M_i are G-modules and the isomorphisms preserve the additive structure.

There is the following easy sufficient condition for this property:

(17.3.1) Lemma. — *Let G act trivially on the space T, and let A be a locally constant G-sheaf on T. Then A is strongly locally constant if either of the following conditions is satisfied:*

    1) *A has finite fibres and G is finitely generated;*

    2) *T is locally connected.*

The same conclusion holds if A is a sheaf of Λ-modules (Λ any ring) with finitely generated fibres, and G is finitely generated.

Proof. One can assume that the underlying sheaf of A is constant, i.e. that A corresponds to an espace étalé $\mathrm{pr}_1 \colon T \times M \to T$ with M a set. The map $G \times T \times M \to M$ arising from the G-action is continuous. This map is a family of G-actions on M parametrized by T, and can therefore be written as a map $T \to \mathrm{Hom}\big(G, \mathrm{Sym}(M)\big)$, $x \mapsto \sigma_x$, where $\mathrm{Sym}(M)$ is the symmetric group on M. For fixed $g \in G$ and $m \in M$ the map $T \to M$, $x \mapsto \big(\sigma_x g\big)(m)$ is locally constant. It is immediate that under any of the conditions of the lemma the map $x \mapsto \sigma_x$ is itself locally constant. $\quad\square$

A pleasant feature of strongly locally constant G-sheaves is that their sheaves of G-invariants are again locally constant. This is not true in general for locally constant G-sheaves, compare Example (17.2.4).

(17.3.2) Lemma. — *Let G act trivially on T. If A is a G-sheaf on T which is strongly locally constant then the sheaf $\Gamma_G A = \mathrm{H}^0_G(A)$ is locally constant. If A is an abelian G-sheaf on T which is strongly locally constant, and if G is finitely generated, then all sheaves $\mathrm{H}^n_G(A)$ are locally constant, $n \geq 0$.*

Proof. The assertion on $\mathrm{H}^0_G(A)$ is obvious. The case of $\mathrm{H}^n_G(A)$, $n \geq 1$, follows from the following more precise

(17.3.3) Lemma. — *Let the finitely generated group G act trivially on T. Let M be a G-module and $\underline{M}$ the associated "constant" abelian G-sheaf on T. Then $\mathrm{H}^n_G(\underline{M})$ is the constant abelian sheaf on T with stalks $H^n(G, M)$ $(n \geq 0)$.*

Proof. If I is an injective G-module then $\mathrm{H}^n_G(\underline{I}) = 0$ for $n \geq 1$ by Proposition (6.2). Therefore $\mathrm{H}^n_G(\underline{M})$ is the n-th cohomology object of

$$0 \longrightarrow \Gamma_G \underline{I}^0 \longrightarrow \Gamma_G \underline{I}^1 \longrightarrow \cdots$$

where $M \to I^\bullet$ is an injective resolution of G-modules. Now $\Gamma_G(\underline{I}^n)$ is the constant sheaf with stalks $H^0(G, I^n)$, and so $\mathrm{H}^n_G(\underline{M})$ is constant with stalks $H^n(G, M)$. $\quad\square$

From now on let G be the group of order two. Let X be a scheme. The functor $\rho \colon \widetilde{X}_{et} \to \widetilde{X}_{ret}$ clearly preserves constant sheaves (2.15.1). Does it also preserve locally constant sheaves?

(17.4) Proposition. — *Let X be any scheme.*
  a) *If A is a locally constant sheaf of sets on X_{et} with finite stalks, then $\nu(G)^* A$ is strongly locally constant as a G-sheaf on X_r, and ρA is a locally constant sheaf on X_r.*
  b) *If A is a locally constant abelian sheaf on X_{et} with finitely generated stalks, then $\nu(G)^* A$ is strongly locally constant as an abelian G-sheaf on X_r, and $R^n \rho A$ is a locally constant sheaf on X_r for every $n \geq 0$.*
If X is locally noetherian, or if X_r is locally connected, the hypotheses on the stalks of A in a) and b) may be dropped.

Proof. $\nu(G)^*A$ is locally constant by general reasons (17.1), so the first parts of a) and b) follow from (17.3.1). Since $\rho A = \Gamma_G \nu(G)^*A$ resp. $R^n \rho A = H^n_G(\nu(G)^*A)$ (6.9.1) the second parts follow from (17.3.2). In view of 2) of (17.3.1) it remains to show that for X locally noetherian and arbitrary locally constant $A \in \tilde{X}_{et}$ the G-sheaf $\nu(G)^*A$ is strongly locally constant.

Put $X' := X \otimes_{\mathbb{Z}} \mathbb{Z}[\sqrt{-1}]$, as usual, and let $\pi: X' \to X$ be the projection. One may assume that X is affine and that there is a surjective étale affine morphism $Y \to X$ such that $A|_Y$ is constant. There is an affine étale morphism $V \to X$ which is real surjective and for which $\mathrm{Hom}_X(V_{X'}, Y) \neq \emptyset$. For example one may take $V = \mathrm{Res}_{X'/X}(Y_{X'})$. (Here $V \to X$ is real surjective by (5.4).) In particular $A|_{V_{X'}}$ is constant. Since $V_r \to X_r$ is a surjective local homeomorphism (1.8) it suffices to show that the pullback of $\nu(G)^*A$ to V_r is strongly locally constant. But this G-sheaf on V_r is $\nu(G)^*(A|_V)$ (here $\nu(G): \tilde{V}_r(G) \to \tilde{V}_{et}$). In other words, one may assume $Y = X'$, i.e. that $A|_{X'}$ is constant.

Clearly one can assume that X is connected, since X is noetherian by assumption. Moreover $X_r \neq \emptyset$ without loss of generality. Thus also X' is connected.

Put $M := A(X')$. Then $A|_{X'}$ is constant with stalks M. The underlying sheaf of $\nu(G)^*A$ is $\nu^*\pi^*A = \nu^*(A|_{X'})$, so the map

$$\nu^*: M = A(X') \longrightarrow H^0\left(X_r, \nu^*(A|_{X'})\right) \tag{1}$$

gives a trivialization of this sheaf. For every $\xi \in X_r$ the G-action in the stalk at ξ is identified, under the trivialization (1), with the G-action on $M = A(X')$ given by the G-action on X' over X. Hence $\nu(G)^*A$ is the "constant" G-sheaf on X_r given by the G-set M. $\square$

The reader will have observed that the proof of the last assertion did not really use that X is locally noetherian. All that was needed was that every étale X-scheme is locally connected for the Zariski topology. Note that nevertheless the topology on the real spectrum X_r may be quite "wild" (think of $X =$ spectrum of a field), so this last assertion of the proposition is not a consequence of the general Lemma (17.3.1).

(17.4.1) **Remark.** Let X be a scheme. If M is a set then $\nu(G)^*\underline{M}_{et}$ is the constant sheaf $\underline{M}_r$ on X_r on which G acts trivially. Hence $\rho\underline{M}_{et} = \underline{M}_r$ (this was already observed in (2.15.1)). If M is an abelian group and $n \geq 0$ then $R^n\rho\underline{M}_{et}$ is the constant sheaf on X_r with stalks $H^n(G, M)$ (M is considered as a trivial G-module), as follows from Lemma (17.3.3). Note however that even if X is connected but $A \in \mathrm{Ab}(X_{et})$ is only locally constant (with finite stalks, say), the stalks of the locally constant sheaves $R^n\rho A$ ($n \geq 0$) need not be isomorphic on different connected components of X_r. (Easy examples are constructed with $X =$ spectrum of a field.)

(17.4.2) Example. Let A be a locally constant étale sheaf on a scheme X. In general it is *not* true that the sheaf ρA on X_r is locally constant (and *a fortiori*, the G-sheaf $\nu(G)^* A$ on X_r need not be strongly locally constant, although it is locally constant). Here is an example:

Let R be a ring such that $R/\mathfrak{p}$ is a real closed field for every prime ideal $\mathfrak{p}$ of R, and $\operatorname{spec} R$ is infinite. In particular, $\operatorname{spec} R$ is an infinite boolean space, and $\operatorname{sper} R = \operatorname{spec} R$. Such rings exist; for example one may take R to be the ring of all locally constant $\mathbb{R}$-valued functions on an infinite boolean space. If A is a sheaf on $(\operatorname{spec} R)_{et}$ then ρA is just the restriction of A to the Zariski site $(\operatorname{spec} R)_{zar} \approx \operatorname{sper} R$. Let $R' = R[\sqrt{-1}]$. Then $\operatorname{spec} R' \to \operatorname{spec} R$ is a homeomorphism. Let M be an infinite set, and suppose that G acts on the constant Zariski sheaf $\underline{M}$ on $\operatorname{spec} R'$, covering the trivial G-action on $\operatorname{spec} R'$. Call B the so-defined G-sheaf on the space $\operatorname{spec} R'$. One can find such an action for which $\Gamma_G B$ is not a locally constant sheaf, compare Example (17.2.4). By pullback to the étale site, B defines a G-sheaf on $(\operatorname{spec} R')_{et}$. This G-sheaf on $(\operatorname{spec} R')_{et}$ corresponds to a sheaf on $(\operatorname{spec} R)_{et}$ which will be called A (compare (10.12)). Clearly A is locally constant since its restriction to $(\operatorname{spec} R')_{et}$ is the constant sheaf $\underline{M}$. But under the identification $\operatorname{spec} R' \approx \operatorname{spec} R \approx \operatorname{sper} R$ the sheaf ρA is $\Gamma_G B$, the sheaf of G-invariants of B, and hence is not locally constant.

After these preliminaries recall now from [SGA4]:

(17.5) Definition [SGA4 IX.2.3]. Let X be a scheme. A set-valued sheaf (resp. an abelian sheaf) A on X_{et} is called *constructible* if every open affine subscheme U of X can be covered by (finitely many) constructible, locally closed reduced subschemes Y for which $A|_Y$ (a sheaf on Y_{et}) is locally constant with finite stalks (resp. with finitely generated stalks).

(17.5.1) Finiteness Theorem [SGA4 XIV]. — *Let $f\colon Y \to X$ be a finitely presented proper morphism of schemes. For any constructible sheaf $B \in \widetilde{Y}_{et}$ the direct image $f_* B$ on X_{et} is constructible. For any constructible abelian torsion sheaf B on Y_{et} all higher direct images $R^n f_* B$ $(n \geq 0)$ are constructible on X_{et}.*

In the following, analogues of this theorem will be proved for the topologies *ret* and *b* of a scheme. First consider the real étale topology. It is convenient here to treat $\widetilde{X}_{ret}$ as the category of sheaves on the real spectrum, i.e. on the topological space X_r. Since X_r is a locally spectral space one has a natural notion of constructible sheaves, see (A.3). For later use record the following consequence of Proposition (17.4):

(17.6) Proposition. — *Let X be any scheme. If A is a constructible sheaf of sets on X_{et} then ρA is a constructible sheaf on X_r. If A is a constructible abelian sheaf on X_{et} then $R^n \rho A$ is a constructible abelian sheaf on X_r for every $n \geq 0$.*

Proof. This follows from (17.4) and the fact that ρ and $R^n \rho$ commute with pullback of sheaves (3.10), (3.12.2).   □

The next goal is now to prove the following

(17.7) Theorem (Real Finiteness Theorem). — *Let $f: Y \to X$ be a finitely presented proper morphism of schemes. For any constructible abelian sheaf B on Y_r the higher direct images $R^n f_{r*} B$ ($n \geq 0$) are constructible on X_r. For any constructible sheaf of sets B on Y_r the sheaf $f_{r*} B$ on X_r is constructible.*

(17.7.1) Remark. In general, the real finiteness theorem does not hold without the properness hypothesis. For example, consider the case where X is affine and f is the inclusion of an open quasi-compact subscheme Y. For any non-zero abelian group M the support of $f_{r*} \underline{M}$ is the closure of Y_r in X_r. There are examples known where this closure is not a constructible subspace [Ga]. Hence $f_{r*} \underline{M}$ is not a constructible sheaf in these cases.

However, if A is an excellent ring then the closure of a constructible subset of sper A is again constructible [ABR Thm. 3.1]. It seems possible that, more generally, for any morphism $f: Y \to X$ of finite type between excellent schemes the direct image functor (resp. the higher direct image functors) of f_r send(s) constructible sheaves on Y_r to constructible sheaves on X_r. Below this will be proved to be true if the schemes are separated and of finite type over a real closed base field (Theorem (17.9)). The main tool for this proof is semi-algebraic triangulation.

(17.8) *Proof* of Theorem (17.7). The assertion being local on X one may assume that X is affine. There are a proper morphism $f_0: Y_0 \to X_0$ with X_0 affine of finite type over spec $\mathbb{Z}$ and a morphism $g: X \to X_0$ such that f is the base extension of f_0 by g [EGA IV, 8.8.2]. Write $X = \varprojlim X_\lambda$ as the filtering inverse limit of affine X_0-schemes of finite type, and correspondingly $Y = \varprojlim Y_\lambda$ with $Y_\lambda = X_\lambda \times_{X_0} Y_0$. Let B be a constructible (resp. constructible abelian) sheaf on Y_r. By Proposition (A.9) there are an index λ and a constructible (resp. constructible abelian) sheaf B_λ on $Y_{\lambda r}$ such that B is isomorphic to the pullback of B_λ to Y_r. By the proper base change theorem for the real étale topology (16.2) it suffices to prove the theorem for $Y_\lambda \to X_\lambda$ instead of f. In other words, one can assume that X is affine and of finite type over spec $\mathbb{Z}$. So Theorem (17.7) follows from the following theorem, in which no properness hypothesis is required:

(17.9) Theorem (Semi-algebraic Finiteness Theorem). — *Let R be a real closed field and let $f: L \to K$ be a semi-algebraic map between affine semi-algebraic spaces over R. For any constructible sheaf (resp. constructible abelian sheaf) F on L the direct image $f_* F$ (resp. all higher direct images $R^n f_* F$) are constructible on K.*

(17.9.1) Of course, by a sheaf on K I mean a sheaf on the semi-algebraic site of K, or equivalently, on the real spectrum $\tilde{K}$ of K. Such a sheaf is called constructible iff the corresponding sheaf on $\tilde{K}$ is constructible in the obvious sense (A.3). Recall that a semi-algebraic space is called *affine* if it is semi-algebraically homeomorphic to a semi-algebraic subspace of some R^n. For example, if X is any separated R-scheme of finite type then the semi-algebraic space $X(R)$ is affine in this sense.

(17.10) The proof of (17.9) relies on the fact that affine semi-algebraic spaces admit arbitrarily fine semi-algebraic triangulations. Therefore it seems in order to include here some reminders about these triangulations. Fix a real closed field R in the following; all semi-algebraic spaces are supposed to be over R.

The simplicial complexes in the semi-algebraic sense are different from the classical ones in that proper faces of simplices may be missing. In other words, an *abstract semi-algebraic complex* is a pair (V, Σ) where V is a set and Σ is a set of non-empty finite subsets of V such that V is the union of the $\sigma \in \Sigma$. The classical axiom about faces ("$\sigma \in \Sigma$, $\emptyset \neq \rho \subset \sigma \Rightarrow \rho \in \Sigma$") is absent.

This notion is the proper setting for semi-algebraic geometry. It allows to triangulate arbitrary semi-algebraic subsets of R^n, whether they are "compact" or not, with finitely many simplices. To be more precise, if the set V is *finite* one defines the *geometric realization* K of (V, Σ) in the straightforward way as a semi-algebraic subset of R^V. So this space K is the disjoint union of the open standard simplexes which correspond to the elements of Σ. These pieces are generally called the *open simplexes* of K, although most of them will not be open as subspaces of K. If X is any semi-algebraic space one calls *semi-algebraic triangulation* of X any pair consisting of an abstract semi-algebraic complex (V, Σ) and a semi-algebraic homeomorphism from the geometric realization of (V, Σ) to X. The important result of Delfs-Knebusch says: If X is any affine semi-algebraic space and $X_1, \ldots, X_N$ are finitely many semi-algebraic subspaces of X then there is a semi-algebraic triangulation of X such that each of the X_i is a union of open simplexes. See [DK2 §2].

Recall the notion of *star neighborhoods*: If X is a triangulated semi-algebraic space and K is a semi-algebraic subset of X then $\operatorname{star} K = \operatorname{star}_X K$ is defined to be the union of all open simplexes T with $\overline{T} \cap K \neq \emptyset$. Note that $\operatorname{star} K$ is an open neighborhood of K.

The following theorem on the semi-algebraic local triviality of semi-algebraic maps will be needed ("Hardt's theorem", see [BCR 9.3]):

(17.11) **Theorem.** — *Let R be a real closed field, let $L \subset R^m$ be a semi-algebraic subset and $f: L \to R^n$ a continuous semi-algebraic map. Then there is a covering of R^n by finitely many semi-algebraic subsets K_i with the following property: For every i there are a semi-algebraic space V_i and a semi-algebraic homeomorphism $f^{-1}(K_i) \xrightarrow{\approx} K_i \times V_i$ over K_i.* $\qquad\square$

(17.12) To prove Theorem (17.9) it obviously suffices to treat the two cases where f is a semi-algebraic embedding resp. where f is isomorphic to the projection $K \times I \to K$ with $I = [0,1]$. (Factorize f through its graph map.)

First the abelian sheaves case. By Propositions (A.5) and (A.6) it suffices to consider sheaves of the form $F = M_Z$ on L, with $Z \subset L$ a closed semi-algebraic subset and M an abelian group of finite type. (Recall that M_Z denotes the extension by zero of the constant sheaf M on Z.) Hence one reduces to the following two cases:

   Case 1: $f: L \subset K$ is a semi-algebraic embedding and $F = \underline{M}$ is a constant sheaf on L;

   Case 2: $f: L \longrightarrow K$ is proper of relative dimension ≤ 1, and $F = \underline{M}$ is constant on L.

In Case 1 one uses the following

(17.13.1) Lemma. — *Let $L \subset K$ be a pair of affine semi-algebraic spaces. Then there is a sequence of intermediate semi-algebraic sets*

$$L = K_0 \subset K_1 \subset \cdots \subset K_p = K$$

with the following property. For every $i = 1, \ldots, p$, writing $Z_i := K_i - K_{i-1}$, there are an open semi-algebraic neighborhood U_i of Z_i in K_i and a pointed semi-algebraic space (V_i, v_i) such that $(U_i, Z_i) \approx (V_i \times Z_i, v_i \times Z_i)$ (semi-algebraic homeomorphism of semi-algebraic pairs).

Proof. Triangulating the pair (K, L) in the semi-algebraic sense one sees that it is enough to assume that K is a triangulated semi-algebraic space, Z is an open simplex of K and $L = K - Z$. In this case it is easy to see that the star neighborhood star Z of Z in K satisfies $(\text{star}\, Z, Z) \approx (V \times Z, v \times Z)$ for a suitable pointed space (V, v). $\square$

(17.13.2) By the lemma one can assume for Case 1 that there are a pointed semi-algebraic space (V, v) and a semi-algebraic space Z such that f is the open embedding $(V - v) \times Z \subset V \times Z$. Let $h: Z \approx v \times Z \hookrightarrow V \times Z$ denote the inclusion of the closed complement. Then for every $n \geq 0$ the sheaf $h^* R^n f_* \underline{M}$ on Z is constant with stalks

$$H_v^n(V, M) := \varinjlim_{W \ni v} H_{\text{sa}}^n(W - v, M),$$

direct limit over the open neighborhoods W of v in V. Indeed, for every $z \in Z$ one has

$$\left(R^n f_* \underline{M}\right)_{(v,z)} = \varinjlim_{W \ni v, Y \ni z} H_{\text{sa}}^n((W - v) \times Y, M)$$

($\varinjlim$ over the open neighborhoods $W \times Y$ of (v, z) in $V \times Z$), which is $H_v^n(V, M)$ since there are arbitrarily small open neighborhoods Y of z in Z which are contractible. Thus the composition of the canonical map $H_v^n(V, M) \to H^0(Z, h^* R^n f_* \underline{M})$ with the stalk map $H^0(Z, h^* R^n f_* \underline{M}) \to (h^* R^n f_* \underline{M})_z$ is bijective for every $z \in Z$.

(17.14) In Case 2 one reduces, using Hardt's theorem (17.11) and proper base change (16.2), to the case where f is a projection $K \times J \to K$ with $J \subset I$ a closed semi-algebraic subset (and $F = \underline{M}$ constant). In this case clearly $f_* \underline{M}$ is constant and $R^n f_* \underline{M} = 0$ for $n \geq 1$.

(17.15) Now consider the case of set-valued constructible sheaves, which is slightly tricky. First the case is studied where f is a semi-algebraic embedding. For this I digress a little and give a description of constructible sheaves in terms of triangulations, which is of independent interest.

Let (V, Σ) be an abstract semi-algebraic complex, with V finite, and let X be its geometric realization. Let F be a sheaf (of sets) on X such that F is constant over every open simplex S in X. Let $W, Y \subset X$ be open simplexes with $Y \subset \overline{W}$, and let $y \in Y$. Since y has arbitrarily small open neighborhoods V in X with $W \cap V$ connected one can define a canonical map $F_y \to H^0(W, F|_W)$. The composition

$$r_{Y,W} \colon \quad H^0(Y, F|_Y) \longrightarrow F_y \longrightarrow H^0(W, F|_W) \tag{2}$$

is independent of the choice of y in Y since Y is connected. (Fixing $s \in H^0(Y, F|_Y)$ and varying y in Y the image of s under (2) remains obviously locally constant.) The construction is transitive in the sense that, if Z is another open simplex with $Z \subset \overline{Y}$ (hence also $Z \subset \overline{W}$) then $r_{Y,W} \circ r_{Z,Y} = r_{Z,W}$.

Call *sheaf on* Σ any functor $P \colon \Sigma \to \text{(sets)}$, where Σ is made into a category with the inclusions as arrows. It is easy to see that the above construction extends to the following characterization of constructible sheaves:

(17.15.1) **Proposition.** — *Let (V, Σ) be an abstract semi-algebraic complex with V finite, and X its geometric realization. The category of sheaves F on X which are constant over any open simplex is naturally equivalent to the category of sheaves on Σ. The global sections functor $F \mapsto H^0(X, F)$ for such sheaves on X corresponds to the inverse limit functor $P \mapsto \varprojlim P$ for sheaves on Σ.* □

Note that, for any constructible sheaf F on an affine semi-algebraic space X, there is some triangulation of X such that F becomes constant over any open simplex, and so the proposition applies.

(17.15.2) Without proof I mention that also the cohomology of constructible abelian sheaves can be expressed in this way: If A is an abelian sheaf on X which is constant over any open simplex then canonically $H^n_{\text{sa}}(X, A) \cong H^n(\Sigma, \tilde{A})$, $n \geq 0$. Here $\tilde{A}$ is the sheaf on Σ associated with A, and $H^n(\Sigma, -)$ is the n-th right derived functor of $H^0(\Sigma, -) = \varprojlim$.

Assume still that X is triangulated in the semi-algebraic sense.

(17.15.3) Lemma. — *Let F be a sheaf on X which is constant over any open simplex. Let U be an open semi-algebraic subset of X such that $T \cap U$ is connected for every open simplex T in X. Then the restriction map*

$$F(\operatorname{star} U) \longrightarrow F(U) \tag{3}$$

is bijective.

Proof. The map (3) is clearly injective. Let $s \in F(U)$. For every open simplex T with $T \cap U \neq \emptyset$ there is a unique global section $t_T \in H^0(T, F|_T)$ with $(t_T)|_{T \cap U} = s|_{T \cap U}$, since $T \cap U$ is connected. I claim that these t_T fit together to a global section of F over $\operatorname{star} U$. What has to be checked is that $r_{Z,T}(t_Z) = t_T$ for open simplexes T, Z with $Z \subset \overline{T}$ and $Z \cap U \neq \emptyset$. But this is clear since $r_{Z,T}$ can be represented using a point z in $Z \cap U$, as in (2).   □

(17.15.4) Proposition. — *Let X be a triangulated semi-algebraic space. Let $Y \subset X$ be a subcomplex, i.e. a union of open simplexes, and let F be a sheaf on Y which is constant over every open simplex of Y. Let $f: Y \subset X$ denote the inclusion. If T is any open simplex of X then $(f_* F)|_T$ is constant with stalks $F(Y \cap \operatorname{star}_X T)$.*

Proof. Since T is connected the proposition asserts that for every $x \in T$ both maps in

$$F(Y \cap \operatorname{star}_X T) = (f_* F)(\operatorname{star}_X T) \longrightarrow H^0(T, (f_* F)|_T) \longrightarrow (f_* F)_x \tag{4}$$

are bijective. For this it suffices that the composite map (4) is bijective. If W is a sufficiently small neighborhood of x in X then $\operatorname{star}_X W = \operatorname{star}_X T$. Moreover there are arbitrarily small such neighborhoods W such that $S \cap W$ is connected for every open simplex S. For such W one has by Lemma (17.15.3)

$$(f_* F)(W) = F(W \cap Y) = F(\operatorname{star}_Y (W \cap Y)).$$

But $\operatorname{star}_Y (W \cap Y) = Y \cap \operatorname{star}_X (W \cap Y) = Y \cap \operatorname{star}_X T$. Indeed, "$\subset$" in the second equality follows from $\operatorname{star}_X W = \operatorname{star}_X T$. If conversely $S \subset Y$ is an open simplex with $\overline{S} \cap T \neq \emptyset$ then $x \in \overline{S}$, and so $\emptyset \neq S \cap W = S \cap W \cap Y$. The proposition is proved.   □

(17.16) If the sheaf F in the last proposition is constructible, i.e. has finite stalks, then obviously the same is true for $f_* F$. Therefore (17.15.4) settles Theorem (17.9) for set-valued sheaves if f is a semi-algebraic embedding. It remains to consider the case where $f: K \times I \to K$ is the projection, $I = [0, 1]$.

Let F be a constructible sheaf on $K \times I$. Choose a covering $K \times I = M_1 \cup \cdots \cup M_s$ by semi-algebraic sets such that F is constant over each M_i. By the "saucissonnage theorem" [BCR §2.3] the following is true. There is a covering

$K = K_1 \cup \cdots \cup K_r$ by semi-algebraic sets, and for each i there are finitely many continuous semi-algebraic functions $0 = g_0^i < g_1^i < \cdots < g_{n_i}^i = 1$: $K_i \to I$ such that each $(K_i \times I) \cap M_j$ is a union of certain stripes

$$\{(x,t) \in K_i \times I: \ g_{\nu-1}^i(x) < t < g_\nu^i(x)\}$$

and certain graphs $\mathrm{graph}(g_\nu^i)$ (for suitable $\nu \in \{1, \ldots, n_i\}$). By proper base change one may replace K by one of the K_i. So one can assume that there are continuous semi-algebraic functions $0 = g_0 < g_1 < \cdots < g_n = 1$: $K \to I$ such that F is constant on every stripe $\{(x,t) \in K \times I: \ g_{\nu-1}(x) < t < g_\nu(x)\}$ and on every $\mathrm{graph}(g_\nu)$ ($\nu = 0, \ldots, n$). Assuming without loss of generality that K is connected one sees that $f_* F$ is constant on K. (Argue as for Proposition (17.15.1).)

The proofs of Theorems (17.9) and (17.7) are therefore complete.   □□

Finally a finiteness theorem for the topology b will be proved. The only reasonable way of defining constructible sheaves on X_b is the following:

(17.17) Definition. Let X be a scheme. A sheaf of sets (resp. an abelian sheaf) F on X_b will be called *constructible* if and only if both j^*F on X_{et} and i^*F on X_r are constructible (resp. constructible abelian) sheaves in the respective senses.

(17.17.1) Remark. Let F be a constructible sheaf on X_b; say X is affine. In general X does *not* have a covering by locally closed subschemes Y for which F is locally constant on Y_b. This failure is detected by the simplest example possible, namely $X = \mathrm{spec}\,R$ with R a real closed field: The topos $\widetilde{X}_b$ is the category of all triples (B, A, ϕ) where B is a set, A is a G-set and $\phi: B \to A^G$ is a map. The triple (B, A, ϕ) is locally constant iff it is constant, i.e. iff G acts trivially on A and ϕ is bijective. But (B, A, ϕ) constructible just means that the sets B and A are finite.

The class of sheaves on X_b which become locally constant on the pieces of a covering by locally closed subschemes would be too narrow to get any interesting results. For example consider a separated algebraic variety X over a real closed field R, and let $f: X \to \mathrm{spec}\,R$ be the structure map. If M is a finite abelian group then

$$j^* R^n f_{b*} M \ = \ H^n(X_{et}', M) \ = \ H_{sa}^n(X(C), M)$$

(which is a G-module) but

$$i^* R^n f_{b*} M \ = \ H^n(X_b, M) \ = \ H_{sa}^n(X(C)/G, M),$$

cf. (15.5.1). These two groups do rarely coincide, and so $R^n f_{b*} M$ is rarely a (locally) constant sheaf on $(\mathrm{spec}\,R)_b$.

The following properties of constructible sheaves on X_b are immediate consequences of corresponding properties for X_{et} and X_r:

(17.17.2) Proposition. — *Let X be a scheme.*

a) *The class of constructible sheaves on X_b is closed in $\widetilde{X}_b$ under finite direct and inverse limits.*

b) *Kernel and cokernel of a morphism between constructible sheaves in $\mathrm{Ab}(X_b)$ are again constructible. If $0 \to F' \to F \to F'' \to 0$ is exact in $\mathrm{Ab}(X_b)$ and F', F'' are constructible then so is F.*

c) *The functors $j_!$, j_* and i_* send constructible (abelian) sheaves to constructible (abelian) sheaves.*

d) *If $f: Y \to X$ is a morphism of schemes and F is a constructible (abelian) sheaf on X_b then $f_b^* F$ is a constructible (abelian) sheaf on Y_b.*

e) *If X is an inverse limit of a filtering system of quasi-compact and quasi-separated schemes X_λ with affine transition maps, then every constructible (abelian) sheaf on X_b is a pullback of a constructible (abelian) sheaf on $X_{\lambda b}$, for some index λ.* $\qquad\qquad\square$

(17.18) Theorem. — *Let $f: Y \to X$ be a proper and finitely presented morphism of schemes. For any constructible sheaf of sets F on Y_b the sheaf $f_{b*} F$ on X_b is constructible. For any constructible abelian sheaf F on Y_b for which $j^* F$ is torsion the abelian sheaves $\mathrm{R}^n f_{b*} F$ on X_b are constructible, $n \geq 0$.*

Proof. Actually the theorem is a corollary to the finiteness theorems for the étale and the real étale topologies. Constructibility of the direct image $f_{b*} F$ is seen as follows (both in the set-valued and in the abelian case): F is the fibre product of a diagram

$$i_* B \longrightarrow i_* \rho A \longleftarrow j_* A$$

with B constructible on Y_{ret} and A constructible on Y_{et}. Since f_{b*} preserves fibre products, since ρA is constructible and since $f_{b*} i_* = i_* f_{ret*}$ and $f_{b*} j_* = j_* f_{et*}$, it is clear that $f_{b*} F$ is again constructible (17.17.2a). Let now F be a constructible abelian sheaf on Y_b. If $F = i_* B$ with $B \in \mathrm{Ab}(Y_{ret})$ constructible then the $\mathrm{R}^n f_{b*} F = i_* \mathrm{R}^n f_{ret*} B$ are constructible by the real case, $n \geq 0$. From the usual short exact sequences together with (17.17.2b) it follows that it suffices to show for every constructible torsion sheaf $A \in \mathrm{Ab}(Y_{et})$ that the $\mathrm{R}^n f_{b*} j_! A$ are constructible, $n \geq 0$. Since $j^* \mathrm{R}^n f_{b*} j_! A = \mathrm{R}^n f_{et*} A$ (3.12.2) is constructible by the étale case it remains to show that $i^* \mathrm{R}^n f_{b*} j_! A$ is constructible on X_r. For this one can assume that 2 is invertible on X since the question is local on X. By the relative version of the fundamental long exact sequence (Corollary (6.8)) it suffices to show that the sheaves $i^* \mathrm{R}^n (j_* f_{et*}) A = \mathrm{R}^n (\rho f_{et*}) A$ and $\mathrm{R}^n f_{ret*}^G \nu(G)^* A$ on X_{ret} are constructible. In the first case this follows from the étale finiteness theorem and the spectral sequence

$$E_2^{pq} = \mathrm{R}^p \rho \, \mathrm{R}^q f_{et*} A \implies \mathrm{R}^{p+q} (\rho f_{et*}) A,$$

using Proposition (17.6). The second case follows in the same way from the spectral sequence

$$E_2^{pq} = R^p f_{ret*} R^q \rho A \implies R^{p+q} f_{ret*}^G \nu(G)^* A$$

of Corollary (6.9.2), by the real finiteness theorem. $\qquad \square$

As an application I give a theorem on smooth specialization of cohomology groups of proper schemes. First however a remark on locally constant sheaves for the b-topology:

(17.19) Remark. — *Let X be a scheme and F a sheaf on X_b. The following conditions are equivalent:*
(i) *F is a locally constant object in $\widetilde{X}_b$;*
(ii) *there is $A \in \widetilde{X}_{et}$ with $F \cong j_* A$ such that there is a surjective and real surjective family $\{U_i \to X\}$ of étale X-schemes with $A|_{U_i}$ constant for each i;*
(iii) *there is a locally constant sheaf $A \in \widetilde{X}_{et}$ with $F \cong j_* A$ such that G acts trivially on the sheaf $\nu(G)^* A$.*

Proof. The constant sheaves on X_b are the sheaves $j_* \underline{M}_{et}$ with M a set (2.15.1). Hence the equivalence of (i) and (ii) is clear. Also (ii) $\Rightarrow$ (iii) is clear. To prove (iii) $\Rightarrow$ (ii), observe that for every $\xi \in X_r$ there is an étale morphism $f: U \to X$ such that ξ lifts to U and $A|_U$ is constant. Hence $f_b^* F = j_*(A|_U)$ is constant on U_b. There is also a surjective family of étale morphisms $\{V_i \to X\}$ such that $A|_{V_i}$ is constant for every i. One can assume that none of the V_i has a real point. The family consisting of the $V_i \to X$ and the $U \to X$ as before is a covering for the topology b, and F is constant on each of its members. $\qquad \square$

(17.20) Corollary. — *Let $f: Y \to X$ be a proper and smooth morphism of schemes.*
a) *Let B be a constructible locally constant abelian sheaf on Y_r. Then the sheaves $R^n f_{r*} B$ are constructible and locally constant on X_r ($n \geq 0$).*
b) *Let F be a constructible locally constant abelian sheaf on Y_b for which $j^* F$ is torsion, prime to the residue characteristics of Y. Then the sheaves $i^* R^n f_{b*} F$ are constructible and locally constant on X_r ($n \geq 0$).*

Of course also the corresponding assertions for constructible locally constant sheaves of sets are true. Note that, although also $j^* R^n f_{b*} F = R^n f_{et*}(j^* F)$ is locally constant, one cannot expect $R^n f_{b*} F$ to be locally constant in b) (Remark (17.17.1)).

Proof. Only the proof of b) will be given, the proof of a) being similar. The proof proceeds similarly as in the étale case [SGA4 XVI.2]. By Theorem (17.18) the sheaves $i^* R^n f_{b*} F$ are constructible on X_r. Therefore it suffices to show that for every specialization $\eta \succ \xi$ in X_r the "cospecialization maps"

$$\left(i^* R^n f_{b*} F\right)_\xi \longrightarrow \left(i^* R^n f_{b*} F\right)_\eta \tag{5}$$

are isomorphisms. Choosing a real closed valuation ring B and a morphism $\operatorname{spec} B \to X$ which induces $\eta \succ \xi$ (see (1.5.2)) one can replace X by $\operatorname{spec} B$, by proper base change (16.2). So without loss $X = \operatorname{spec} B$ with B a real closed valuation ring.

Let $g : x_0 \to X$ be the inclusion of the generic point of X and let Y_0 be the generic fibre of f, so that one has a cartesian square

$$
\begin{array}{ccc}
Y_0 & \xrightarrow{\ h\ } & Y \\
{\scriptstyle f_0}\downarrow & & \downarrow{\scriptstyle f} \\
x_0 & \xrightarrow{\ g\ } & X.
\end{array}
$$

Since (5) is the map

$$ H^n(Y_b, F) \longrightarrow H^n(Y_{0b}, h_b^* F) $$

induced by h (3.11) the Leray spectral sequence for h_{b*} shows that it suffices to prove

$$ F \xrightarrow{\ \sim\ } h_{b*} h_b^* F \quad \text{and} \quad R^q h_{b*}(h_b^* F) = 0 \text{ for } q \geq 1. \tag{6} $$

The assertion (6) is local on Y with respect to the topology b. So one can replace $f : Y \to X$ by a composition $Y_1 \to Y \xrightarrow{f} X$ with $Y_1 \to Y$ étale for which $F\big|_{Y_1}$ is constant. In other words, one may assume that F is constant with finite stalks M, thereby giving up properness of f. Then

$$ R^q h_{b*}(h_b^* F) = R^q h_{b*} M = R^q h_{b*}(f_0)_b^* M \cong f_b^* R^q g_{b*} M, $$

the last isomorphism by smooth base change (16.11). Now g_{b*} is exact since x_0 is the spectrum of a real closed field. Since $g_{b*} M$ is the constant sheaf M on X_b the corollary is proved. $\qquad\square$

(17.21) Corollary. — *Let R be a real closed field and $f : Y \to X$ a proper and smooth morphism of schemes of finite type over $\operatorname{spec} R$. For $x \in X(R)$ let $Y_x = f^{-1}(x)$ be the fibre scheme over x. If M is any abelian group then the semi-algebraic cohomology groups $H^n_{\mathrm{sa}}(Y_x(R), M)$ and $H^n_{\mathrm{sa}}(Y_x(C)/G, M)$ depend only on the semi-algebraic connected component of x in $X(R)$.*

Indeed, these groups are the stalks of the sheaves $R^n f_{r*} M$ resp. $i^* R^n f_{b*} M$ in x, by proper base change and the comparison theorems of Sect. 15 (assume that M is a torsion group). By Corollary (17.20) these sheaves on X_r are locally constant, whence the assertion for M torsion. By the universal coefficient theorem this implies the corollary for arbitrary M. $\qquad\square$

18. Cohomology of affine varieties

Let $X/\mathbb{C}$ be a d-dimensional affine algebraic variety. Then $X(\mathbb{C})$ has the homotopy type of a d-dimensional (finite) CW complex. This fact is classical and can be proved using Morse theory. It has a well known counterpart in étale cohomology, which says that any d-dimensional affine variety over a separably closed field has cohomological dimension $\leq d$ (for torsion sheaves).

This section is devoted to the proof of the analogous fact for b-cohomology. That is, if X is a d-dimensional affine variety over a real closed base field then $H^n(X_b, F) = 0$ for $n > d$ and every sheaf F on X_b for which j^*F is torsion. Note that from the étale theorem one could immediately deduce that these groups vanish for $n > d + 1$, by the results of Sect. 7; so the only point here is to replace $d + 1$ by d! Unfortunately I could not find a direct proof for this fact. Instead the proof of the étale case will be mimicked. This means in particular that one studies more generally a relative affine situation $f: Y \to X$ between algebraic varieties over a field.

(18.1) Notation. Let X be a scheme, and fix $t \in \{et, b, ret\}$. For $x \in X$ write $d(x) := d_X(x) := \dim \overline{\{x\}}$ (Krull dimension), a non-negative integer or ∞. If F is an abelian sheaf on X for the topology t, one defines $d(F)$ as the supremum of all $d(x)$, where x is ranging over those points of X for which there is a t-point of X, centered at x, in which F has a non-zero stalk. If $F = 0$ then $d(F) = -\infty$.

(18.2) Theorem. — *Let k be a field and $f: Y \to X$ an affine k-morphism between schemes which are locally of finite type over k. Let F be an abelian sheaf on Y_b such that j^*F is torsion. Then $d(R^q f_{b*} F) \leq d(F) - q$ for $q \geq 0$.*

(18.2.1) This result is completely analogous to a similar theorem for étale cohomology [SGA4 XIV.3]. By this latter theorem it suffices in the proof of Theorem (18.2) to consider real stalks of $R^q f_{b*} F$, since $j^* R^q f_{b*} F = R^q f_{et*} j^* F$. Obviously it is enough to treat the two cases $F = i_* B$ with $B \in \mathrm{Ab}(Y_{ret})$, and $F = j_! A$ with $A \in \mathrm{Ab}(Y_{et})$ torsion. The first is easy:

(18.3) Lemma. — *Let k be a field and $f: Y \to X$ a k-morphism between schemes of finite type over k. For every abelian sheaf B on Y_r one has $d(R^q f_{r*} B) \leq d(B) - q$ for $q \geq 0$.*

Proof. Let $d := d(B)$. For every closed d-dimensional subscheme Z of Y let $B_Z \in \mathrm{Ab}(Y_r)$ be the sheaf $B|_{Z_r}$, extended by zero on $(Y - Z)_r$. These subschemes

Z form a directed family, and $B \cong \varinjlim_Z B_Z$. Hence it is enough to prove the lemma for B_Z instead of B. In other words, replacing Y by Z, it suffices to show $d(R^q f_{r*} B) \le \dim(Y) - q$ for $q \ge 0$. So fix $q \ge 0$ and let $\xi \in X_r$, put $x := \mathrm{supp}(\xi) \in X$. Then

$$(R^q f_{r*} B)_\xi \cong H^q\left(f_r^{-1}(\mathrm{Gen}\,\xi), B\big|_{f_r^{-1}(\mathrm{Gen}\,\xi)}\right). \tag{1}$$

Here $\mathrm{Gen}\,\xi := \{\eta \in X_r : \eta \succ \xi\}$. Formula (1) is easy to see since $f_r : Y_r \to X_r$ is a spectral map of spectral spaces; see [Sch1, Prop. 3.3] for the proof. Now $f_r^{-1}(\mathrm{Gen}\,\xi)$ is a pro-constructible subspace of the real spectrum of

$$Y \times_X \mathrm{spec}\,\mathcal{O}_{X,x} = f^{-1}(\mathrm{Gen}\,x),$$

where $\mathrm{Gen}\,x \approx \mathrm{spec}\,\mathcal{O}_{X,x}$ denotes the set of generalizations of x in X. For every y in $f^{-1}(\mathrm{Gen}\,x) \subset Y$ one has

$$d_Y(y) \underset{(*)}{\ge} d_X(f(y)) \ge d_X(x). \tag{2}$$

Here $(*)$ is true since X and Y are of finite type over a field. Clearly

$$d_Y(y) + \dim \mathcal{O}_{Y,y} \le \dim Y \tag{3}$$

(actually equality holds), and combining (2) and (3) one gets

$$\dim \mathcal{O}_{Y,y} \le \dim(Y) - d_X(x). \tag{4}$$

Hence

$$\dim f_r^{-1}(\mathrm{Gen}\,\xi) \le \dim f^{-1}(\mathrm{Gen}\,x) \le \dim(Y) - d_X(x). \tag{5}$$

Since the cohomological dimension of a spectral space does not exceed its Krull dimension ([Sch1], or [CaC] for normal spectral spaces which is sufficient here) it follows from (5) that (1) vanishes if $q > \dim(Y) - d_X(x)$; whence the lemma. $\square$

(18.4) Remarks.

(18.4.1) Lemma (18.3) holds under weaker assumptions. If one assumes only that $f : X \to Y$ is a quasi-compact and quasi-separated morphism of schemes, then the only non-trivial step in the proof is $(*)$ in (2). This is known to be true under weaker hypotheses. It suffices for example that f is an S-morphism between schemes locally of finite type over S, where S is a scheme which satisfies conditions $1°$, $2°$ and $3°$ of [EGA IV, 10.6.1], by (ii) of [loc.cit.].

(18.4.2) From a purely real point of view it might appear more natural to consider the following modified notion. For $\eta \in Y_r$ write $d_r(\eta) := \dim \overline{\{\eta\}}$ (closure in Y_r), and for $B \in \mathrm{Ab}(Y_r)$ put

$$d_r(B) := \sup \{d_r(\eta) : \eta \in Y_r, B_\eta \ne 0\}.$$

Thus clearly $d_r(B) \le d(B)$. However Lemma (18.3) fails if one uses d_r instead of d! The reason is that $d_r(f_r(\eta)) \le d_r(\eta)$ is not true in general. As a simple counterexample consider the inclusion $f : \mathbb{A}^1_R \subset \mathbb{P}^1_R$ over a real closed field R: If $B \ne 0$ is a sheaf on $\mathrm{sper}\,R[t]$ with non-zero stalks only at the points at infinity, then $d_r(B) = 0$, but $d_r(f_{r*}B) = 1$.

(18.5) To prove Theorem (18.2) it remains to treat the case $F = j_! A$, cf. (18.2.1). Here the proof follows very closely the proof in the étale case as given in [SGA4 XIV.4]. Unlike the base change theorems or the finiteness theorem (Sections 16 and 17), Theorem (18.2) does not seem to be a more or less formal consequence of this étale case plus the real case (18.3).

Fix an integer $d \geq 0$, and consider the following assertion (which is part of the theorem):

A(d): *For every affine k-morphism $f: Y \to X$ between schemes of finite type over a field k and every torsion abelian sheaf A on Y_{et} with $d(A) \leq d$, one has $d\big(i^* R^q f_{b*}(j_! A)\big) \leq d - q$ for $q \geq 0$.*

Observe that A(d), once known to be true, implies $d\big(R^q f_{b*}(j_! A)\big) \leq d - q$, since the étale analogue of Theorem (18.2) is known. Following [SGA4] I give next a local version of A(d) which turns out to be equivalent. Say that a strictly local resp. a strictly real local scheme is *of geometric origin* if it arises by strict (resp. strict real) localization from a scheme which is of finite type over a field.

B(d): *For every strictly real local scheme V of geometric origin, every affine morphism $g: W \to V$ of finite type and every torsion sheaf $A \in \text{Ab}(W_{et})$ with $\delta(A) \leq d$, one has $H^q(W_b, j_! A) = 0$ for $q > d$.*

Here $\delta(A)$ is defined as in [loc.cit.], i.e.

$$\delta(A) := \max\Big\{\delta(w): w \in W \text{ and } A_{\bar{w}} \neq 0\Big\}$$

where $\delta(w) := d_V(g(w)) + \text{trdeg}\big(\kappa(w)/\kappa(g(w))\big)$ for $w \in W$.

(18.6) Lemma. — A(d) *and* B(d) *are equivalent.*

The proof is the same as in the étale case (pp. 17–18 of [SGA4 XIV]). One only has to replace the use of Proposition 2.4 in [loc.cit.] by its analogue for strict real localizations (the proof is identical). Instead of [SGA4 X.3.3(i)] one uses

(18.6.1) Lemma. — *If X is a scheme of finite type over a field k and $\xi \in X_r$, there exist a real closed field R and a scheme Y of finite type over R such that X^ξ is isomorphic to the strict real localization of Y in an R-rational point.*

Proof. Recall that X^ξ denotes the strict real localization of X at ξ (3.7.3). One may assume that X is affine. Let $x = \text{supp}(\xi) \in X$, and let $Z = \overline{\{x\}}$ be the reduced closure. As in [loc.cit.] one finds by Noether normalization a k-morphism $f: X \to \mathbf{A}_k^n$ for some $n \geq 0$ such that $f|_Z: Z \to \mathbf{A}_k^n$ is finite and surjective. Let R be the real closure of the function field of $\mathbf{A}_k^n$ with respect to the ordering $f_r(\xi)$, and take $Y := X \times_{\mathbf{A}_k^n} \text{spec } R$ together with the canonical R-rational point in Y. $\square$

A particular case of B(d) is

> C(d): **For every strictly real local scheme** $V = \operatorname{spec}\Omega$ **of geometric origin with** $\dim V \leq d$, **and for every** $f \in \Omega$, **writing** $W := \operatorname{spec}\Omega_f = \operatorname{spec}\Omega[f^{-1}]$, **one has** $H^q(W_b, j_!A) = 0$ **for** $q > d$ **and** $A \in \operatorname{Ab}(W_{et})$ **torsion.**

(18.7) Lemma. — *If $d \geq 2$, then B(d') holds for all $d' \leq d$ if and only if C(d') holds for all $d' \leq d$.*

Proof. Assuming C(d') for $d' \leq d$ one has to prove B(d). (For $d \leq 1$, B(d) is proved independently in (18.9) below.) Let $V = \operatorname{spec}\Omega$ be as in B(d). First consider the case where $W = \mathbb{A}^1_V$ and $g : W \to V$ is the canonical morphism. One completes W by the projective line over V:

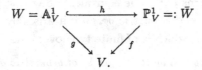

Let $V_\infty = \bar{W} - W$ be the section at infinity. Let $A \in \operatorname{Ab}(W_{et})$ be a torsion sheaf with $\delta(A) \leq d$. One can assume without any loss that A is constructible. In the Leray spectral sequence

$$E_2^{pq} = H^p\big(\bar{W}_b, R^q h_{b*}(j_!A)\big) \implies H^{p+q}(W_b, j_!A) \tag{6}$$

one has $E_2^{pq} = 0$ if both $p > 0$ and $q > 0$, since $R^q h_{b*}(j_!A)$ has support on $V_\infty \cong V$. Let v be the closed point of V and w the closed point of $V_\infty \subset \bar{W}$. If F is any abelian sheaf on $\bar{W}_b$ then $H^*(\bar{W}_b, F) \cong H^*\big((\mathbb{P}^1_v)_b, F|_{\mathbb{P}^1_v}\big)$, by proper base change (16.2). From Sect. 7 (or Sect. 15) it follows that the $\mathbb{P}^1$ over a real closed field has cohomological ℓ-dimension 2 for the b-topology (ℓ any prime). Hence $E_2^{pq} = 0$ for $p > 2$. I claim that moreover $E_2^{0q} = 0$ for $q > d$ (which in turn implies $H^i(W_b, j_!A) = 0$ for $i > d \geq 2$). Since A is constructible there is a closed subscheme Z of $\bar{W}$ with $\operatorname{supp}(A) \subset Z \cap W$ and $\dim Z \leq d$. For $q \geq 1$, writing $F := j_!A$,

$$E_2^{0q} = H^0(\bar{W}_b, R^q h_{b*}F) = (R^q h_{b*}F)_w.$$

If $Z \cap V_\infty = \emptyset$ this group is zero anyway, and otherwise it is $\big(R^q \tilde{h}_{b*}(F|_{Z \cap W})\big)_w$ where $\tilde{h}$ is the inclusion $Z \cap W \subset Z$. If $\tilde{Z} := Z^w$ (the strict real localization) and $\tilde{W} := (Z \cap W) \times_Z \tilde{Z}$ then

$$E_2^{0q} = \Big(R^q \tilde{h}_{b*}\big(F|_{Z \cap W}\big)\Big)_w = H^q(\tilde{W}_b, j_!\tilde{A}) \tag{7}$$

with $\tilde{A} = A|_{\tilde{W}}$. Now C($d$) applies to show that (7) vanishes.

To prove B(d) in general one can assume that $W = \mathbb{A}_V^n$. Proceeding inductively one factors g as

$$W = \mathbb{A}_V^n \xrightarrow{\quad h \quad} \mathbb{A}_V^{n-1} =: U$$

with g and f pointing to V.

Let $A \in \mathrm{Ab}(W_{et})$ be torsion with $\delta(A) \leq d$. From the case $n = 1$, applied to the stalks of $R^q h_{b*}(j_! A)$, one sees $\delta(R^q h_{b*}(j_! A)) \leq d - q$. Therefore, in the spectral sequence

$$E_2^{pq} = H^p\big(U_b, R^q h_{b*}(j_! A)\big) \implies H^{p+q}(W_b, j_! A)$$

the terms E_2^{pq} vanish for $p > d - q$, by B(d) applied to f. Hence $H^i(W_b, j_! A) = 0$ for $i > d$. $\qquad\square$

Next a very particular case of the theorem (in fact, of A(1)) is established:

(18.8) Lemma. — *Let R be a real closed field and X an affine curve over R. Then $\mathrm{cd}_\ell(X_b) = 1$ for any prime ℓ.*

Proof. It is known that $\mathrm{cd}_\ell(X'_{et}) = 1$ for all ℓ. Hence $\mathrm{cd}_\ell(X_b) \leq 2$ is clear, and it suffices to show $H^2(X_b, j_! A) = 0$ for any torsion sheaf $A \in \mathrm{Ab}(X_{et})$ (Sect. 7). One may assume that A is constructible. If $h: \tilde{X} \to X$ is the normalization of X then $j_! A \to h_{b*} h_b^*(j_! A)$ is an isomorphism outside a finite closed subscheme. Hence one can assume that X is smooth and connected. There are an affine étale X-scheme $f: U \to X$ and a sheaf epimorphism $f_{et!} \underline{M} \to A$, for some finite abelian group M [SGA4 IX.2.7]. One can assume $A = f_{et!} \underline{M}$. Hence $j_! A = j_! f_{et!} \underline{M} = f_{b!}(j_! \underline{M})$ (cf. (2.16) for $f_{b!}$). Let

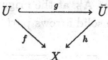

be a smooth completion of U, i.e. g is an open embedding, $\bar{U}/R$ is a smooth curve and the morphism h is finite. One has the sheaf homomorphisms on X_b

$$j_! A = f_{b!}(j_! \underline{M}) \longrightarrow f_{b*}(j_! \underline{M}) = h_{b*} g_{b*} g_b^*(j_! \underline{M}) \longleftarrow h_{b*}(j_! \underline{M}),$$

and both are isomorphisms outside finite closed subschemes of X. Hence it suffices to show that $H^2(X_b, h_{b*}(j_! \underline{M})) = H^2(\bar{U}_b, j_! \underline{M})$ vanishes. By the comparison theorem (cf. (15.5.1)) this group is the relative semi-algebraic cohomology group $H_{sa}^2(\bar{U}(C)/G, \bar{U}(R); M)$. Now $Y := \bar{U}(C)/G$ is a semi-algebraic 2-manifold with boundary $\bar{U}(R)$, and Y is not complete. Therefore $H^2(Y, \partial Y; M) = 0$, which proves the lemma. $\qquad\square$

(18.9) Proof of A(d) for $d \leq 1$. Let f and A be as in the statement of A(d). Let $\xi \colon x \to X$ be a real point of X. Let R be the real closure of k with respect to the ordering induced by ξ, and let $f_0 \colon Y_0 \to X_0$ be the base extension of f by $\operatorname{spec} R \to \operatorname{spec} k$. Then the stalks of $R^q f_{b*}(j_! A)$ and of $R^q (f_0)_{b*}(j_! A_0)$ in ξ resp. ξ_0 coincide ($A_0 := A|_{Y_0}$). Hence one can assume that the field k is real closed. To prove A(d) one can moreover assume that the sheaf A is constructible. Replacing Y by a closed subscheme of dimension $\leq d$ which contains $\operatorname{supp}(A)$, and X by a closed subscheme through which f factors as a dominant morphism, one gets $\dim X \leq \dim Y \leq d$ without loss of generality. If $d = 0$ then f_{b*} is exact. Let $d = 1$. Then there is a finite closed subscheme Z of X such that f is finite over $X - Z$, and hence $R^1 f_{b*}(j_! A)$ has support in Z. From Lemma (18.8) it follows that $R^q f_{b*}(j_! A) = 0$ for $q \geq 2$, which settles the case $d \leq 1$. $\qquad\square$

The proof of Theorem (18.2) is completed by

(18.10) Lemma. — *If* B(d') *holds for all* $d' < d$ *then* C(d) *holds.*

Proof. Let Ω be a strictly real local ring of geometric origin, $\dim \Omega \leq d$, let $f \in \Omega$ and $W := \operatorname{spec} \Omega_f$, and let $A \in \operatorname{Ab}(W_{et})$ be a torsion sheaf. One can assume that there are a real closed field R, an affine R-variety V_0 with $\dim V_0 \leq d$ and a point $x_0 \in V_0(R)$ such that $V := \operatorname{spec} \Omega$ is the strict real localization of V_0 at x_0 (Lemma (18.6.1)). Write $V = \varprojlim V_\lambda$ as a filtering inverse limit of affine V_0-schemes V_λ of dimension $\leq d$ ($\lambda \in I$). One may assume that there is $f_0 \in \Gamma(V_0, \mathcal{O}_{V_0})$ such that $f = f_0|_V$. Let f_λ be the pullback of f_0 to V_λ ($\lambda \in I$). Writing $W_\lambda := \{f_\lambda \neq 0\} \subset V_\lambda$ one has therefore $W = \varprojlim W_\lambda$. One can assume that A is constructible, and then, that there is a constructible sheaf A_0 on W_0 such that $A = A_0|_W$. Let $A_\lambda := A_0|_{W_\lambda}$ ($\lambda \in I$). Consider f_λ as a morphism $V_\lambda \to \mathbf{A}_R^1$. Let Y be the strict real localization of $\mathbf{A}_R^1$ at the origin, and let $\eta = \operatorname{spec} \kappa(Y)$ be the generic point of Y. Consider the solid arrows part of the following diagrams

$$
\begin{array}{ccccc}
W & \dashrightarrow & Z_\lambda & \xrightarrow{\ \varphi_\lambda\ } & \eta \\[2pt]
\big\uparrow & & \big\uparrow & & \big\uparrow \\[6pt]
V & \dashrightarrow & Y_\lambda & \longrightarrow & Y \\[2pt]
& \searrow & \big\downarrow & & \big\downarrow \\[6pt]
& & V_\lambda & \xrightarrow{\ f_\lambda\ } & \mathbf{A}_R^1
\end{array}
\qquad (8)
$$

in which the squares of the right column are supposed to be cartesian ($\lambda \in I$). The projection $V \to V_\lambda$ factors naturally through Y_λ as indicated (f is a non-unit in Ω, w.l.o.g.), and this induces also the upper dotted arrow in (8). The projection $W \to W_\lambda$ factors through $W \to Z_\lambda$. Since $H^*(W_b, j_! A) = \varinjlim_\lambda H^*(W_{\lambda b}, j_! A_\lambda)$, it suffices therefore to show that $H^q(Z_{\lambda b}, j_! B) = 0$ holds for all torsion étale sheaves

B on Z_λ and $q > d$ ($\lambda \in I$). Consider the Leray sequence

$$E_2^{pq} = H^p\Big(\eta_b, \, R^q\varphi_{\lambda b*}(j_! B)\Big) \implies H^{p+q}(Z_{\lambda b}, \, j_! B).$$

By Sect. 9 one has $E_2^{pq} = 0$ for $p > 1$, since $\kappa(\eta) = \kappa(Y)$ has transcendence degree one over R. Since $\dim Z_\lambda \leq d - 1$ and since $B(d-1)$, hence $A(d-1)$, holds by hypothesis, $R^q\varphi_{\lambda b*}(j_! B) = 0$ for $q \geq d$. This proves $H^i(Z_{\lambda b}, j_! B) = 0$ for $i > d$, as desired. $\qquad \square$

(18.11) Corollary. — *If X is a d-dimensional affine variety over a real closed field R then $\mathrm{cd}_\ell(X_b) \leq d$ for all primes ℓ. In particular, $H_{\mathrm{sa}}^q\Big(X(C)/G, M\Big)$ and $H_{\mathrm{sa}}^q\Big(X(C)/G, X(R); M\Big)$ vanish for $q > d$ and any abelian group M.*

Proof. The first assertion from Theorem (18.2), the second by the comparison results of Sect. 15. $\qquad \square$

19. Relations to the Zariski topology

The main theme of this book has been that three Grothendieck topologies were considered which are associated with any scheme X — the étale, the real étale and the b-topology —, and that the interplay between them was studied, mainly under cohomological aspects. The most basic Grothendieck topology, however, which comes along with X is the Zariski topology. Therefore it is in order to add some remarks on its relations to the other sites.

First it is shown that the direct image functor of the support map $X_r \to X$ is exact. This is an easy but important theorem which has already been used in former sections for several times. Here it allows to relate, for any étale sheaf A, the cohomology of the sheaves $R^n \rho A$ on X_r to the Zariski cohomology of the higher direct images $\mathcal{H}^n(A)$ of A on X_{zar}. These latter groups "approximate" the cohomology of A via a local-to-global (Leray) spectral sequence, but are often hard to calculate. On the other hand, the sheaves $R^n \rho A$ on the real spectrum and their cohomology are usually easily accessible. Using the main results of Sect. 7 one gets in this way very far reaching generalizations of the main result of Colliot-Thélène and Parimala [CTP]: (19.5), (19.6.1). Another application is given to the vanishing of differentials in the local-to-global sequence (19.8).

(19.1) If X is a scheme then X_{zar} denotes the Zariski site of X, i.e. the topological space X with the Zariski topology, considered as a site in the usual way. The category of sheaves (resp. of abelian sheaves) on this site is $\widetilde{X}_{zar}$ (resp. $\mathrm{Ab}(X_{zar})$).

The Zariski topology (on Et/X, say) is coarser than any of the topologies $t = et, b, ret$. Hence there is a canonical site morphism from X_t to X_{zar} (given by regarding open subschemes of X as étale X-schemes). In the sequel, these site morphisms will be denoted as indicated in the following commutative diagram:

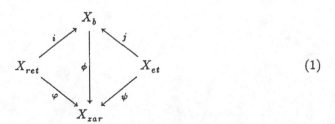

$$(1)$$

φ will also denote the support mapping $X_r \to X$ (0.4.1). This is justified since under the equivalence of $\widetilde{X}_{ret}$ and $\widetilde{X}_r$ the two usages of φ correspond to each other.

An easy but important fact is

(19.2) Theorem. — *On any scheme X the direct image functor φ_*: $\mathrm{Ab}(X_r) \to \mathrm{Ab}(X_{zar})$ is exact.*

Proof. The question is Zariski local, so one can assume that $X = \operatorname{spec} A$ is affine. Then φ: $\operatorname{sper} A \to \operatorname{spec} A$ is a spectral map of spectral spaces. Let F be an abelian sheaf on $\operatorname{sper} A$, and let $\mathfrak{p} \in \operatorname{spec} A$. Then canonically

$$(R^q \varphi_* F)_{\mathfrak{p}} \cong H^q \Big(\operatorname{sper} A_{\mathfrak{p}}, F\big|_{\operatorname{sper} A_{\mathfrak{p}}} \Big), \quad q \geq 0.$$

(See [Sch1, Prop. 3.3] for a proof.) So the theorem follows from

(19.2.1) Proposition. — *If A is a semilocal ring and F is any abelian sheaf on $\operatorname{sper} A$, then $H^n(\operatorname{sper} A, F) = 0$ for $n > 0$.*

I do not know whether this can be found explicitly in the literature, but certainly the result is well known to the experts. Nevertheless I include a proof, for the reader's convenience:

If A is any ring let $(\operatorname{sper} A)_{\max}$ denote the set of closed points of $\operatorname{sper} A$, with the subspace topology. This is a compact Hausdorff space. The restriction maps

$$H^n(\operatorname{sper} A, F) \longrightarrow H^n\Big((\operatorname{sper} A)_{\max}, F\big|_{(\operatorname{sper} A)_{\max}} \Big)$$

are bijective for all n and $F \in \mathrm{Ab}(\operatorname{sper} A)$; the (easy) proof is in [CaC]. Hence the proposition follows from

(19.2.2) Lemma. — *If A is a semilocal ring then the compact space $(\operatorname{sper} A)_{\max}$ is totally disconnected.*

Proof. Let ξ, η be two different closed points of $\operatorname{sper} A$. There is $a \in A$ such that $a(\xi) > 0$ and $a(\eta) < 0$. Since ξ is closed, the ordered residue field of $\operatorname{supp} \xi$ is archimedean over the image of A. By multiplying a with a suitable element $1 + b^2$ ($b \in A$) one can therefore get $a(\xi) > 1$ (and keep $a(\eta) < 0$). By the Chinese Remainder Theorem one finds $c \in A$ such that $u := a(1 + c^2) \not\equiv 1 \pmod{\mathfrak{m}}$ for every maximal ideal $\mathfrak{m}$ of A. Hence $u - 1$ is a unit in A, and so $\{u > 1\} \cup \{u < 1\}$ is a disjoint clopen decomposition of $\operatorname{sper} A$ which separates ξ and η. $\quad\square\square\square$

(19.3) If one wants information on the cohomology of an étale sheaf A on X, one sometimes uses the Leray spectral sequence of ψ_* to relate $H^*(X_{et}, A)$ to the Zariski cohomology of the higher direct image sheaves $R^q\psi_* A$. For the latter, the following notation is widely used in the literature:

(19.3.1) Definition. If A is an abelian étale sheaf on X, one writes $\mathcal{H}^n(A) := R^n\psi_* A$, $n \geq 0$.

Thus $\mathcal{H}^n(A)$ is the Zariski sheaf on X which is associated to the Zariski presheaf $U \mapsto H^n(U_{et}, A|_U)$ ($U \subset X$ open). The Leray (or "local-to-global") spectral sequence for A reads

$$E_2^{pq} = H^p(X_{zar}, \mathcal{H}^q(A)) \implies H^{p+q}(X_{et}, A). \tag{2}$$

For example, let X/k be a smooth algebraic variety over a perfect field k. Then the spectral sequence (2) for $A = \mu_n^{\otimes i}$ ($i \in \mathbb{Z}$, n prime to char k) coincides with the sequence studied by Bloch and Ogus in their famous paper [BO]. For example they showed that $E_2^{pq} = 0$ for $p > q$, a fact which has many important applications.

For a real closed ground field $k = R$ and the sheaf $A = \mu_2 = \mathbb{Z}/2$, the sequence (2) was studied by Colliot-Thélène and Parimala in [CTP]. The main result of this paper can be reformulated as saying that if X/R is smooth, the canonical map $\mathcal{H}^n(\mathbb{Z}/2) \to \varphi_* \mathbb{Z}/2$ of Zariski sheaves on X (see below (19.4) and the remark thereafter) is an isomorphism for $n > \dim X$.

I'm now going to state a far reaching generalization of this result. First observe:

(19.4) Proposition. — *Let X be any scheme and A any étale sheaf of abelian groups on X. There is a long exact sequence of Zariski sheaves on X:*

$$0 \longrightarrow R^1\phi_*(j_*A) \longrightarrow \mathcal{H}^1(A) \longrightarrow \varphi_* R^1 \rho A \longrightarrow \cdots$$
$$\cdots \longrightarrow R^n\phi_*(j_*A) \longrightarrow \mathcal{H}^n(A) \longrightarrow \varphi_* R^n \rho A \longrightarrow \cdots \tag{3}$$

Proof. Consider the (Leray) spectral sequence

$$E_2^{pq} = R^p\phi_*(R^q j_*A) \implies \mathcal{H}^{p+q}(A). \tag{4}$$

For $q \geq 1$ one has $R^q j_* A = i_* R^q \rho A$ (3.12b). Since $R^p\phi_* \circ i_* = R^p\varphi_*$ and φ_* is exact (19.2) it follows that $E_2^{pq} = 0$ if $p, q \geq 1$. $\qquad\square$

In particular one sees how natural sheaf maps $\mathcal{H}^n(A) \to \varphi_* R^n \rho A$ ($n \geq 0$) arise as edge homomorphisms of the spectral sequence (4). They are easy to describe in explicit terms: If $U \subset X$ is an open subscheme and $x \to U$ is a real point of U, the stalk of $R^n \rho A$ at x is $H^n(x_{et}, A|_x)$ (3.12c); the image of the local section of $\mathcal{H}^n(A)$ represented by $\alpha \in H^n(U_{et}, A)$ has in x the stalk $\alpha|_x$.

The sheaves $R^n \rho A$ and their cohomology are usually very easy to compute, and so (3) can be useful if one has control over the $R^n\phi_*$-sheaves. For example:

(19.5) Corollary. — *Assume that 2 is invertible on X, and that X has a Zariski basis of open subschemes U for which $\mathrm{cd}_2(U'_{et}) \leq d$. ($U' := U \otimes_{\mathbb{Z}} \mathbb{Z}[\sqrt{-1}]$, as usual.) Then*

$$\mathcal{H}^n(A) \xrightarrow{\sim} \varphi_* R^n \rho A \quad \text{and} \quad H^p(X_{zar}, \mathcal{H}^n(A)) \xrightarrow{\sim} H^p(X_r, R^n \rho A) \quad (p \geq 0),$$

for $n > d$ and every 2-primary torsion sheaf A on X_{et}. Moreover, if $d \geq 1$ then also $\mathcal{H}^d(A) \to \varphi_* R^d \rho A$ is surjective.

Proof. $R^n \phi_*(j_* A)$ vanishes for $n \geq d+1$ by Corollary (7.18) and Lemma (7.20.1). The isomorphism between the cohomology groups follows from the sheaf isomorphism since φ_* is exact. $\qquad\square$

(19.5.1) Corollary. — *For X as in (19.5), one has*

$$\mathcal{H}^n(\mathbb{Z}/2) \xrightarrow{\sim} \varphi_* \mathbb{Z}/2 \quad and \quad H^p(X_{zar}, \mathcal{H}^n(\mathbb{Z}/2)) \xrightarrow{\sim} H^p(X_r, \mathbb{Z}/2)$$

for every $p \geq 0$ and $n > d$. $\qquad\square$

For example, (19.5) and (19.5.1) apply to every d-dimensional scheme X locally of finite type over a real closed field R. The first isomorphism in Corollary (19.5.1), for X/R smooth, is the result of [CTP] alluded to above.

(19.5.2) Application of ϕ_* to the exact sequence $0 \to j_! A \to j_* A \to i_* \rho A \to 0$ yields a 5-term exact sequence

$$0 \longrightarrow \phi_*(j_! A) \longrightarrow \psi_* A \longrightarrow \varphi_* \rho A \longrightarrow R^1 \phi_*(j_! A) \longrightarrow R^1 \phi_*(j_* A) \longrightarrow 0 \qquad (5)$$

and isomorphisms $R^n \phi_*(j_! A) \cong R^n \phi_*(j_* A)$, $n \geq 2$. So one can splice together (3) and (5), and it follows that (3) remains exact if $j_* A$ is replaced by $j_! A$ throughout, and if one starts with R^0 instead of R^1.

(19.6) If $A \in \mathrm{Ab}(X_{et})$ is annihilated by 2 (and $\frac{1}{2} \in \mathcal{O}(X)$), the exact sequence $0 \to A \to \pi_* \pi^* A \to A \to 0$ gives rise to the long exact sequence of Zariski sheaves

$$\cdots \longrightarrow \mathcal{H}^n(A) \longrightarrow \mathcal{H}^n(\pi_* \pi^* A) \longrightarrow \mathcal{H}^n(A) \xrightarrow{(-1)} \mathcal{H}^{n+1}(A) \cdots \qquad (6)$$

in which the maps $\mathcal{H}^n(A) \to \mathcal{H}^{n+1}(A)$ are cup-product with (-1) on local sections (see (7.10.1)). The square

$$(7) \quad \begin{array}{ccc} \mathcal{H}^n(A) & \xrightarrow{(-1)} & \mathcal{H}^{n+1}(A) \\ \downarrow & & \downarrow \\ \varphi_* R^n \rho(A) & \longrightarrow & \varphi_* R^{n+1} \rho(A) \end{array}$$

commutes for every $n \geq 0$, where the vertical maps are those of (19.4) and the bottom map is the natural map defined in (3.12.1). Note that this bottom map is an isomorphism for $n \geq 1$.

(19.6.1) Proposition. — *Assume* $\frac{1}{2} \in \mathcal{O}(X)$, *and let* $A \in \mathrm{Ab}(X_{et})$ *with* $2A = 0$. *Then*

$$\lim_{n \to \infty} \mathcal{H}^n(A) \longrightarrow \lim_{n \to \infty} \varphi_* \mathrm{R}^n \rho(A) = \varphi_* \mathrm{R}^1 \rho(A) \tag{8}$$

is an isomorphism of Zariski sheaves, where the left hand limit is taken using $\cup(-1)$ *as transition maps. If* X *satisfies the condition of (19.5) then* $\mathcal{H}^n(\pi_* \pi^* A) = 0$ *and* $\mathcal{H}^n(A) \xrightarrow{\sim} \varphi_* \mathrm{R}^n \rho(A)$ *for* $n > d$.

Proof. The stalk of (8) at a point $x \in X$ is the map

$$\lim_{n \to \infty} H^n_{et}(\mathcal{O}_{X,x}, A) \longrightarrow \lim_{n \to \infty} H^0(\mathrm{sper}\, \mathcal{O}_{X,x}, \mathrm{R}^n \rho A). \tag{9}$$

But $H^0(\mathrm{sper}\, \mathcal{O}_{X,x}, \mathrm{R}^n \rho A) = H^n_G(\mathrm{sper}\, \mathcal{O}_{X,x}, \nu(G)^* A)$ for every $n \geq 0$, by the spectral sequence (14) of (6.9.2) and since $\mathrm{sper}\, \mathcal{O}_{X,x}$ has cohomological dimension 0 (19.2.1). Therefore (9) is an isomorphism by (7.19). Under the condition of (19.5), it is clear that $\mathcal{H}^n(\pi_* \pi^* A) = 0$ for $n > d$, which also (re)proves the last claim. $\square$

(19.7) Proposition. — *Assume that 2 is invertible on* X, *and let* $A \in \mathrm{Ab}(X_{et})$. *Then there is a natural morphism of spectral sequences from the local-to-global sequence* (2),

$$E_2^{pq} = H^p(X_{zar}, \mathcal{H}^q(A)) \implies H^{p+q}(X_{et}, A), \tag{10}$$

to the spectral sequence

$$\tilde{E}_2^{pq} = H^p(X_r, \mathrm{R}^q \rho A) \implies H^{p+q}_G(X_r, \nu(G)^* A) \tag{11}$$

of Corollary (6.9.2). On limit terms it is the canonical map h *(6.6.3), and on the* E_2*-level it consists of the maps in cohomology induced by the sheaf maps* $\mathcal{H}^q(A) \to \varphi_* \mathrm{R}^q \rho(A)$ *of (19.4).*

Proof. First consider the diagram of toposes and topos morphisms

$$
\begin{array}{ccc}
\tilde{X}_{ret} & \xrightarrow{\ i\ } & \tilde{X}_b \\
{\scriptstyle r_{ret}}\big\uparrow\big\downarrow{\scriptstyle q_{ret}} & & \big\uparrow{\scriptstyle j} \\
\tilde{X}_{ret}(G) & \xrightarrow{\ \nu(G)\ } & \tilde{X}_{et}
\end{array}
$$

in which r_{ret} and q_{ret} are the canonical topos morphisms (cf. (6.4.1)). Recall that $(r_{ret})_* = \Gamma_G$. For $F \in \tilde{X}_b$ there is the functorial map $i^* F \to \rho j^* F = \Gamma_G \nu(G)^* j^* F = (r_{ret})_* \nu(G)^* j^* F$ (cf. (6.4.2b)), which shows that there is a canonical morphism

$$j \circ \nu(G) \longrightarrow i \circ r_{ret} \tag{12}$$

between the two topos morphisms from $\tilde{X}_{ret}(G)$ to $\tilde{X}_b$. (If one composes both sides of (12) on the right with q_{ret} one gets the morphism $j\pi\nu \to i$ discussed in

Remark (5.5.2).) If one composes both sides of (12) with ϕ on the left, the resulting morphism

$$\psi \circ \nu(G) \longrightarrow \varphi \circ r_{ret} \tag{13}$$

is an *isomorphism* of topos morphisms from $\tilde{X}_{ret}(G)$ to $\tilde{X}_{zar}$. Indeed, if $H \in \tilde{X}_{zar}$ then the map $r_{ret}^*\varphi^*H \longrightarrow \nu(G)^*\psi^*H$ given by (13) is an isomorphism, both sides having in $\xi \in X_r$ the stalk $H_{\varphi(\xi)}$ with trivial G-action.

This being said, consider the adjunction map $A \to \nu(G)_*\nu(G)^*A$ on X_{et}. The induced maps in étale cohomology are the canonical maps $h\colon H^n(X_{et}, A) \to H^n_G(X_r, \nu(G)^*A)$, cf. Remark (6.7). The Leray sequence for the second sheaf (with respect to ψ_*) has E_2-terms

$$H^p\Big(X_{zar}, R^q\psi_*\nu(G)_*\nu(G)^*A\Big). \tag{14}$$

Now
$$R\psi_* \circ \nu(G)_*\nu(G)^* \underset{(*)}{=} R(\psi_*\nu(G)_*) \circ \nu(G)^* \underset{(13)}{=} R(\varphi_*\Gamma_G) \circ \nu(G)^*$$
$$\underset{(**)}{=} \varphi_* \circ (R\Gamma_G \circ \nu(G)^*) \underset{(***)}{=} \varphi_* \circ R\rho;$$

$(*)$ resp. $(**)$ by exactness of $\nu(G)_*$ (cf. (5.9)) resp. φ_* (19.2), and $(***)$ by (6.9.1). Hence (14) is $H^p(X_r, R^q\rho A) = \tilde{E}_2^{pq}$, from which the proposition follows. $\square$

Proposition (19.7) can be used to obtain information on the local-to-global sequence (10). As one application I mention

(19.8) Proposition. — *Let X be a quasi-compact and quasi-separated scheme over $\mathbb{Z}[\frac{1}{2}]$. Assume that there is a non-negative integer d such that X has a Zariski basis of open subschemes U which satisfy $H^n(U'_{et}, \mathbb{Z}/2) = 0$ for $n > d$. Assume moreover that $e := \dim X$ (Krull dimension) is finite. Then in the local-to-global spectral sequence*

$$E_2^{pq} = H^p\big(X_{zar}, \mathcal{H}^q(\mathbb{Z}/2)\big) \implies H^{p+q}(X_{et}, \mathbb{Z}/2) \tag{15}$$

all differentials d_r, $r \geq 2$, vanish whose target (p,q) satisfies $(p,q) \neq (e,d)$ and lies in the half plane $p + q \geq d + e$. For $n > d + e$ the homomorphisms

$$H^n(X_{et}, \mathbb{Z}/2) \longrightarrow H^n_G(X_r, \mathbb{Z}/2) = \bigoplus_{q=0}^{e} H^q(X_r, \mathbb{Z}/2)$$

are isomorphisms.

Proof. By hypothesis the topological space underlying X is spectral, and so the cohomological dimension of X_{zar} is bounded above by $\dim X = e$, by one of the main results of [Sch1]. (The hypothesis that X be quasi-compact and quasi-separated is used only to make sure that X meets the hypothesis of [Sch1, Thm. 4.5]; weaker conditions would suffice, like X quasi-separated and a union of countably many open quasi-compact subschemes, cf. Cor. 4.6 of [loc.cit.].)

Consider the morphism (19.7) of spectral sequences from (15) to

$$\tilde{E}_2^{pq} = H^p(X_r, \mathbb{Z}/2) \quad \Longrightarrow \quad H_G^{p+q}(X_r, \mathbb{Z}/2). \tag{16}$$

The sequence (16) degenerates, i.e. all differentials $\tilde{d}_r$ in (16) are zero, $r \geq 2$ [Gr1, Thm. 4.4.1]. By (19.5), $E_2^{pq} \to \tilde{E}_2^{pq}$ is an isomorphism for $q > d$. Moreover $E_2^{pq} = \tilde{E}_2^{pq} = 0$ for $p > e$. Now one verifies easily by induction on r that the following statement is true for all $r \geq 2$:

$E_r^{pq} \to \tilde{E}_r^{pq}$ is an isomorphism if $q \geq d + r - 1$, or if $p + q \geq d + e$ and $(p, q) \neq (e, d)$.

This clearly implies both assertions of the proposition.   □

For example, the proposition applies with $d = e$ if X is any scheme of finite type over a real closed field of dimension d.

A first instance of degeneration of this spectral sequence was given by Colliot-Thélène and Parimala, namely for $X/\mathbb{R}$ a smooth geometrically connected surface with $X(\mathbb{R})$ compact ([CTP, Prop. 3.1.1]; here $d = e = 2$). Although the proof in [CTP] used quite different methods, this result served as an inspiration for the above proposition.

20. Examples and complements

This final section gives explicit calculations of cohomology groups in various situations, and is meant to illustrate some of the results obtained before. It is divided into four parts: In the first, smooth curves over a real closed field are considered and their étale and b-cohomology is determined. Classical theorems of Witt and Weichold-Geyer are also reproved. The second part makes explicit the results of Sect. 7 for some examples of étale sheaves on a scheme X. Most interesting is the case of the multiplicative sheaf $\mathbb{G}_m$. This part contains also a variety of additional side results and other complements. The third part considers fields again: It relates the b-cohomology of a field to invariants which have been studied elsewhere in the literature, and makes some remarks on the fundamental pro-group of the site $(\operatorname{spec} k)_b$. Finally, the last part comments on some historical aspects of the matters of this book.

20.1 Cohomology of smooth real curves

In the following, smooth (not necessarily projective) curves over a real closed field are considered, and their étale and b-cohomology groups with coefficients $\mu_n^{\otimes i}$ are calculated. It is also shown how various classical theorems about real curves (due to Witt, Weichold, Geyer) can be deduced from results of previous chapters.

(20.1.1) Fix a real closed field R, and let $C = R(\sqrt{-1})$ be its algebraic closure. By a curve X over R I mean in the following a one-dimensional separated scheme of finite type over $\operatorname{spec} R$. Write $X_C := X \otimes_R C = X'$. The sets $X(C)$, $X(R)$ and $X(C)/G$ are always considered as semi-algebraic spaces over R; (co-)homology of these sets is understood to be semi-algebraic (co-)homology, the index "sa" will be omitted. Without comment I use basic concepts from semi-algebraic geometry like semi-algebraic connected components, semi-algebraic completeness, semi-algebraic manifolds etc.; see [DK1] or [BCR], for example. Over $R = \mathbb{R}$ these concepts coincide with the classical ones (completeness corresponds to compactness), and semi-algebraic (co-)homology is the same as Betti (co-)homology.

G is always the group of order 2. The action of G on $\mathbb{Z}$ by inversion is a G-module which will be denoted by $\mathbb{Z}(1)$. More generally put $M(1) := M \otimes \mathbb{Z}(1)$ for any G-module M. Similar notation applies to abelian G-sheaves on a site.

(20.1.2) First I review and discuss some classical results on real curves. Assuming that X/R is a smooth curve, let always s denote the number of (semi-algebraic) connected components of $X(R)$. If X/R is projective then this means that $X(R)$ is a sum of s (semi-algebraic) circles.

Write $\mathrm{Br}(X) := H^2(X_{et}, \mathbb{G}_m)$ for the (cohomological) Brauer group of X; for smooth curves it coincides with the Brauer group defined by means of sheaves of Azumaya algebras [Mi, p. 153].

(20.1.3) Theorem (Witt). — *For any smooth curve X/R there is a canonical isomorphism*

$$\mathrm{Br}(X) \xrightarrow{\sim} H^0\big(X(R), \mathbb{Z}/2\big) = (\mathbb{Z}/2)^s, \qquad (1)$$

given by pointwise evaluation.

That is: Whether or not a class $\alpha \in \mathrm{Br}(X)$ splits over a real point $P \in X(R)$ depends only on the connected component of P; and the so-defined map (1) is an isomorphism.

Essentially this was proved (for $R = \mathbb{R}$ and $X/\mathbb{R}$ projective) by E. Witt in 1934 [Wi, II', p. 5]. Witt's theorem can be obtained as an easy corollary to results of Sect. 19. First recall that canonically $_2\,\mathrm{Br}(X) = H^0(X, \mathcal{H}^2)$ holds, where $\mathcal{H}^i :=$ $\mathcal{H}^i(\mu_2)$ (cf. (19.3.1)). (This is a general fact true for smooth varieties over fields of characteristic $\neq 2$: The first arrow of the Kummer exact sequence

$$0 \longrightarrow \mathrm{Pic}(X)/2 \longrightarrow H^2(X_{et}, \mu_2) \longrightarrow {}_2\,\mathrm{Br}(X) \longrightarrow 0 \qquad (2)$$

is identified with the first arrow of the exact sequence

$$0 \longrightarrow H^1(X, \mathcal{H}^1) \longrightarrow H^2(X_{et}, \mu_2) \longrightarrow H^0(X, \mathcal{H}^2) \longrightarrow 0 \qquad (3)$$

coming from the local-to-global spectral sequence (19.3) for $A = \mu_2$.) Now assume that the curve X/R is geometrically connected. Since $\mathrm{Br}(X)$ injects into $\mathrm{Br}\,R(X)$ by smoothness of X [Mi, IV.2.6] and $\mathrm{Br}\,R(X)(\sqrt{-1}) = 0$ by Tsen's theorem, it follows that $2\,\mathrm{Br}(X) = 0$, and hence $\mathrm{Br}(X) = H^0(X, \mathcal{H}^2)$. By (19.5.1) therefore $\mathrm{Br}(X) \cong (\mathbb{Z}/2)^s$, and the assertion of the theorem follows. The case when X_C is not necessarily connected is immediately reduced to the above. $\qquad \square$

This argument is already covered by the cases of (19.5.1) established in [CTP]. In fact, it is taken from the introduction of that paper.

(20.1.4) Now assume that the smooth curve X/R is projective and connected, and that $X(R) \neq \emptyset$. For brevity write $H^i(X) := H^i(X_{et}, \mathbb{Z}/2)$, and similarly for X_C. The restriction map $\mathrm{Pic}(X)/2 \to \mathrm{Pic}(X_C)/2 = \mathbb{Z}/2$ is surjective since the image of any real point of X generates the second group. From sequences (2) for X and for X_C it follows that also $H^2(X) \to H^2(X_C) = \mathrm{Pic}(X_C)/2$ is surjective. Therefore the exact sequence $0 \to \mathbb{Z}/2 \to \pi_*\mathbb{Z}/2 \to \mathbb{Z}/2 \to 0$ on X_{et} gives rise to an exact sequence

$$0 \to H^0(X) \to H^1(X) \to H^1(X_C) \to H^1(X) \to H^2(X) \to H^2(X_C) \to 0 \quad (4)$$

of finite $\mathbb{Z}/2$-vector spaces, and moreover $H^2(X) \xrightarrow{\sim} H^3(X) \xrightarrow{\sim} H^4(X) \cdots$. Therefore $H^i(X) \cong H^*_G(X(R), \mathbb{Z}/2) \cong (\mathbb{Z}/2)^{2s}$ for $i \geq 2$, by (7.19). Counting dimensions in (4) one also finds

$$H^1(X) \cong (\mathbb{Z}/2)^{g+s}, \tag{5}$$

where $g = $ genus of X_C, since $H^1(X_C) \cong (\mathbb{Z}/2)^{2g}$. (The groups $H^i(X)$ have already been calculated by Cox [Co] by a slightly different argument.) Moreover, $H^2(X) \cong (\mathbb{Z}/2)^{2s}$ together with (2) and Witt's theorem (20.1.3) shows $\text{Pic}(X)/2 \cong (\mathbb{Z}/2)^s$. Now it is easy to see that the map which sends a real closed point of X to the characteristic function of its connected component in $X(R)$ and a complex closed point to 0, descends to a (surjective) map $\text{Pic}(X) \twoheadrightarrow (\mathbb{Z}/2)^s$. (On each oval of $X(R)$, a rational function changes its sign an even number of times.) By what has just been shown, it therefore induces an isomorphism $\text{Pic}(X)/2 \xrightarrow{\sim} (\mathbb{Z}/2)^s$. In different terms, this means:

(20.1.4.1) Theorem (Witt). — *Let X/R be a smooth projective curve which has a real point. Let $P_1, \ldots, P_n$ be finitely many closed points of X such that the number of (real) points among the P_i on any connected component of $X(R)$ is even. Then there is a rational function f on X having odd order at each P_i and even order everywhere else.* □

(The formulation given in Witt's paper [Wi, III, p. 4] is slightly weaker, in that it only considers real closed points.)

(20.1.5) I now turn to Weichold's theorem (1883), later reproved by Geyer in modern terms [Ge1]. Given a smooth projective curve X/R, it determines the number of (semi-algebraic) connected components of $J(R)$, where J is the Jacobian of X. Generally speaking, if A/R is any abelian variety of dimension g, the image of the norm $N \colon A(C) \to A(R)$ is $A(R)_0$, the (semi-algebraic) identity component of $A(R)$, and the extension $0 \to A(R)_0 \to A(R) \to \widehat{H}^0(G, A(C)) \to 0$ splits since $A(C)$ and hence $A(R)_0$ is divisible. Moreover $\widehat{H}^0(G, A(C))$ is isomorphic to $\widehat{H}^1(G, A(C))$ (this follows since A and its dual are isogeneous), and both groups are $\mathbb{Z}/2$-vector spaces of dimension at most g. These general facts are easy to prove, see [Ge2] for example. Weichold's theorem can be rephrased as follows:

(20.1.5.1) Theorem (Weichold, Geyer). — *If X/R is a smooth projective geometrically connected curve and J is its Jacobian, then $J(R)/J(R)_0 \cong \widehat{H}^0(G, J(C)) \cong \widehat{H}^1(G, J(C))$ is isomorphic to $(\mathbb{Z}/2)^a$ where*

$$a = \begin{cases} s-1 & \text{if } X(R) \neq \emptyset, \\ 0 & \text{if } X(R) = \emptyset, \ g \text{ even}, \\ 1 & \text{if } X(R) = \emptyset, \ g \text{ odd}. \end{cases}$$

Here $g = $ genus of X_C and $s = $ number of connected components of $X(R)$.

In the case $X(R) \neq \emptyset$, Weichold's theorem follows immediately from Witt's theorem (20.1.3): The Hochschild-Serre sequence for $X_C \to X$ and coefficients $\mathbb{G}_m$ gives an exact sequence

$$0 \longrightarrow \mathrm{Pic}(X) \longrightarrow \mathrm{Pic}(X_C)^G \longrightarrow \mathrm{Br}\, R \longrightarrow \mathrm{Br}\, X \longrightarrow H^1(G, \mathrm{Pic}\, X_C) \longrightarrow 0. \qquad (6)$$

The existence of a real point implies that $\mathrm{Br}\, R \to \mathrm{Br}\, X$ is non-zero. Since $\mathrm{Br}\, X \cong (\mathbb{Z}/2)^s$ by (20.1.3) one finds $H^1(G, \mathrm{Pic}\, X_C) \cong (\mathbb{Z}/2)^{s-1}$. The exact sequence

$$0 \longrightarrow J(C) \longrightarrow \mathrm{Pic}\, X_C \overset{\deg}{\longrightarrow} \mathbb{Z} \longrightarrow 0 \qquad (7)$$

of G-modules splits since $X(R) \neq \emptyset$, and hence

$$H^1(G, J(C)) \overset{\sim}{\longrightarrow} H^1(G, \mathrm{Pic}\, X_C) \cong (\mathbb{Z}/2)^{s-1}.$$

In the case $X(R) = \emptyset$, however, these techniques only give $H^1(G, J(C)) \hookrightarrow \mathbb{Z}/2$, and some extra argument as in [Ge1, p. 91] is needed to decide whether or not this group is 0.

(20.1.6) After this review of classical results I turn now to a systematic computation of the cohomology of smooth curves X over R. First assume that X/R is smooth, projective and geometrically connected. Let J be the Jacobian of X. Write $Y :=$ $X(C)/G$ for the semi-algebraic quotient space. Then Y is a complete semi-algebraic 2-manifold with boundary $\partial Y = X(R)$. Let g be the genus of X_C and s the number of connected components of $X(R)$. Then $0 \leq s \leq g+1$ (Harnack's inequality, which also follows from Weichold's theorem above and the remark preceding it). Moreover there are two possibilities:

Case 1: $X(C) - X(R)$ has two connected components. Then necessarily $g + 1 - s =: 2p$ is even. Y may be thought of as a closed orientable surface of genus p from which s open disks have been removed. So Y is orientable (meaning that the open surface $Y - \partial Y$ is orientable), and $X(C)$ is the double of Y, i.e. two copies of Y glued together along ∂Y.

Case 2: $X(C) - X(R)$ is connected. Then Y is not orientable. Y is a 2-sphere from which $g+1$ open disks have been deleted, and where for $g+1-s$ of the boundary curves opposite points have been identified.

If $s = g + 1$ one is necessarily in case 1, and $s = 0$ implies case 2. Otherwise all pairs (g, s) with $1 \leq s \leq g$ are possible in both cases, except that $g - s$ must be odd in case 1. (These facts are classical and due to Klein and Weichold. See [Kl, §§ 21-23].)

The (semi-algebraic) homology groups of Y and $(Y, \partial Y)$ with integer coefficients are the following:

Case 1 (Y orientable, $\partial Y \neq \emptyset$): $H_0(Y) = \mathbb{Z}$, $H_1(Y) = \mathbb{Z}^g$, $H_2(Y) = 0$
and $H_0(Y, \partial Y) = 0$, $H_1(Y, \partial Y) = \mathbb{Z}^g$, $H_2(Y, \partial Y) = \mathbb{Z}$.

Case 2a (Y non-orientable, $\partial Y \neq \emptyset$): $H_0(Y) = \mathbb{Z}$, $H_1(Y) = \mathbb{Z}^g$, $H_2(Y) = 0$ and $H_0(Y, \partial Y) = 0$, $H_1(Y, \partial Y) = \mathbb{Z}^{g-1} \oplus \mathbb{Z}/2$, $H_2(Y, \partial Y) = 0$.

Case 2b (Y non-orientable, $\partial Y = \emptyset$): $H_0(Y) = \mathbb{Z}$, $H_1(Y) = \mathbb{Z}^g \oplus \mathbb{Z}/2$, $H_2(Y) = 0$.

So the cohomology groups of Y resp. of $(Y, \partial Y)$ are given by the following table (here M can be any abelian group):

	Y orientable, $\partial Y \neq \emptyset$	Y not orientable, $\partial Y \neq \emptyset$	Y not orientable, $\partial Y = \emptyset$
$H^0(Y, M)$	M	M	M
$H^1(Y, M)$	M^g	M^g	$M^g \oplus {}_2M$
$H^2(Y, M)$	0	0	$M/2M$
$H^0(Y, \partial Y; M)$	0	0	
$H^1(Y, \partial Y; M)$	M^g	$M^{g-1} \oplus {}_2M$	(see above)
$H^2(Y, \partial Y; M)$	M	$M/2M$	

Table 20-1

Note that by the comparison results of Sect. 15, $H^*(Y, M) = H^*(X_b, M)$ and $H^*(Y, \partial Y; M) = H^*(X_b, j_! M)$ if the group M is torsion.

(20.1.7) To calculate étale cohomology of X with constant torsion coefficients M one may suppose that $X(R) \neq \emptyset$. Then the map $H^2(X_b, j_! M) \to H^2(X_{et}, M)$ is zero since it factors through $H^2(X_b, M) = 0$ (see Table 20-1), and so $H^n(X_{et}, M) \xrightarrow{\sim} H^n_G(X(R), M)$ for $n \geq 2$. One finds easily $H^n_G(X(R), M) \cong ({}_2M \oplus M/2M)^s$ for $n \geq 2$, moreover $H^1_G(X(R), M) \cong (M \oplus {}_2M)^s$. It remains to calculate $H^1(X_{et}, M)$. Consider the Hochschild-Serre spectral sequence

$$E_2^{pq} = H^p(G, H^q_{et}(X_C, M)) \implies H^{p+q}(X_{et}, M). \tag{8}$$

The inclusion $x \hookrightarrow X$ of an R-rational point $x \in X(R)$ shows that all differentials d_r ($r \geq 2$) vanish whose target lies on the line $q = 0$. Hence in particular there is a short exact sequence

$$0 \longrightarrow H^1(G, H^0_{et}(X_C, M)) \longrightarrow H^1(X_{et}, M) \longrightarrow H^1_{et}(X_C, M)^G \longrightarrow 0. \tag{9}$$

The first group is ${}_2M$. Again from $X(R) \neq \emptyset$ one concludes that (9) splits. On the other hand, it follows from Weichold's theorem (20.1.5.1) that

$$H^1_{et}(X_C, M) \cong M^{s-1} \oplus M(1)^{s-1} \oplus (\mathbb{Z}G \otimes_{\mathbb{Z}} M)^{g+1-s} \tag{10}$$

as G-modules. (Assuming that M is ℓ-primary torsion, one has $H^1_{et}(X_C, M) = H^1_{et}(J_C, M) = \operatorname{Hom}(T_\ell J_C, M)$ where $T_\ell J_C$ is the Tate module; and for any g-dimensional abelian variety A/R, the G-module $T_\ell A$ is isomorphic to $\mathbb{Z}^a_\ell \oplus \mathbb{Z}_\ell(1)^a \oplus (\mathbb{Z}_\ell G)^{g-a}$ with $a = \dim \widehat{H}^0(G, A(C))$, see [CTS, Appendix].)

The G-invariants of (10) are isomorphic to $M^g \oplus ({}_2M)^{s-1}$. So from (9) one gets $H^1(X_{et}, M) \cong M^g \oplus ({}_2M)^s$. In summary:

(20.1.8) Proposition. — *If X/R is a smooth connected projective curve with $X(R) \neq \emptyset$ and M is a torsion abelian group, then $H^0(X_{et}, M) = M$ and*

$$H^1(X_{et}, M) \cong M^g \oplus ({}_2M)^s, \quad H^n(X_{et}, M) \cong ({}_2M \oplus M/2M)^s \text{ for } n \geq 2.$$

If $X(R) = \emptyset$, see Table 20-1 for the values of $H^*(X_{et}, M) = H^*(Y, M)$. □

(20.1.9) I also calculate the cohomology of the sheaves μ_n of nth roots of unity, $n \geq 1$. Together with the constant sheaf $\mathbb{Z}/n$ this comprises all tensor powers $\mu_n^{\otimes i}$ ($i \in \mathbb{Z}$) since $\mu_n^{\otimes 2} \cong \mathbb{Z}/n$. From the Kummer exact sequence

$$1 \longrightarrow \mu_n \longrightarrow \mathbb{G}_m \xrightarrow{\;n\;} \mathbb{G}_m \longrightarrow 1 \tag{11}$$

one gets the exact sequences

$$1 \longrightarrow R^*/R^{*n} \longrightarrow H^1(X_{et}, \mu_n) \longrightarrow {}_n\operatorname{Pic}(X) \longrightarrow 0 \tag{12}$$

and

$$0 \longrightarrow \operatorname{Pic}(X)/n \longrightarrow H^2(X_{et}, \mu_n) \longrightarrow {}_n\operatorname{Br}(X) \longrightarrow 0. \tag{13}$$

Using $\operatorname{Br} X \xrightarrow{\sim} (\mathbb{Z}/2)^s$ (20.1.3) one sees that (13) splits, by looking at an R-rational point in each connected component of $X(R)$. Also (12) splits: If n is even then the obvious morphism from the Kummer sequence (11) for the power n to the same sequence for the power 2 gives a morphism of exact sequences

$$
\begin{array}{ccccccccc}
1 & \longrightarrow & R^*/R^{*n} & \longrightarrow & H^1(X_{et}, \mu_n) & \longrightarrow & {}_n\operatorname{Pic}(X) & \longrightarrow & 0 \\
 & & \Big\| & & \Big\downarrow & & \Big\downarrow{\scriptstyle n/2} & & \\
1 & \longrightarrow & R^*/R^{*2} & \longrightarrow & H^1(X_{et}, \mu_2) & \longrightarrow & {}_2\operatorname{Pic}(X) & \longrightarrow & 0.
\end{array}
\tag{14}
$$

The lower sequence in (14) splits, hence so does the upper sequence.

It therefore remains to determine torsion and cotorsion of $\operatorname{Pic} X$. First assume $X(R) \neq \emptyset$. From the exact sequence (6) and the argument thereafter one sees $\operatorname{Pic} X = \operatorname{Pic}(X_C)^G$. From the split exact sequence (7) of G-modules one gets therefore a (split) exact sequence

$$0 \longrightarrow J(R) \longrightarrow \operatorname{Pic} X \longrightarrow \mathbb{Z} \longrightarrow 0. \tag{15}$$

Now $J(R) \cong J(R)_0 \oplus (\mathbb{Z}/2)^{s-1}$ by Weichold's theorem ((20.1.5.1) and the discussion before). The torsion subgroup of $J(R)_0$ is isomorphic to $(\mathbb{Q}/\mathbb{Z})^g$ (see [CTS, Appendix] for a proof working over arbitrary real closed fields), and $J(R)_0$ is divisible. Hence, if $n \geq 1$ is even one gets

$$_n\operatorname{Pic}(X) \cong (\mathbb{Z}/n)^g \oplus (\mathbb{Z}/2)^{s-1} \quad \text{and} \quad \operatorname{Pic}(X)/n \cong \mathbb{Z}/n \oplus (\mathbb{Z}/2)^{s-1}; \qquad (16)$$

if n is odd one has to drop the $\mathbb{Z}/2$-summands in (16).

Now assume $X(R) = \emptyset$. Then (6) together with $\operatorname{Br} X = 0$ (20.1.3) gives an exact sequence

$$0 \longrightarrow \operatorname{Pic} X \longrightarrow \operatorname{Pic}(X_C)^G \longrightarrow \mathbb{Z}/2 \longrightarrow 0. \qquad (17)$$

On the other hand, (7) gives the exact sequence

$$0 \longrightarrow J(R) \longrightarrow \operatorname{Pic}(X_C)^G \xrightarrow{\deg} \mathbb{Z}. \qquad (18)$$

So

$$\operatorname{Pic}(X_C)^G \cong J(R) \oplus \mathbb{Z} \cong J(R)_0 \oplus (\mathbb{Z}/2)^a \oplus \mathbb{Z}$$

where $a = \dim H^1(G, J(C))$; and $\operatorname{Pic}(X)$ is a subgroup of index 2 of this group. From Weichold's theorem (20.1.5.1) one concludes: If g is even then $a = 0$, and so $\operatorname{Pic} X \cong J(R)_0 \oplus \mathbb{Z}$. If g is odd then $a = 1$. But in this case the degree of every element in $\operatorname{Pic}(X_C)^G$ is *even* [Gel, p. 91], which implies $\operatorname{Pic}(X) + J(R) = \operatorname{Pic}(X_C)^G$, and hence again $\operatorname{Pic} X \cong J(R)_0 \oplus \mathbb{Z}$. Therefore

$$_n\operatorname{Pic}(X) \cong (\mathbb{Z}/n)^g \quad \text{and} \quad \operatorname{Pic}(X)/n \cong \mathbb{Z}/n \qquad (19)$$

if $X(R) = \emptyset$.

Finally, whether $X(R)$ is empty or not, in the stable range $q > 2$ one has $H^q(X_{et}, \mu_n) \cong (\mathbb{Z}/2)^{2s}$ for n even and $H^q(X_{et}, \mu_n) = 0$ for n odd. This is easily seen either via $H^q(X_{et}, \mu_n) \cong H^q_G(X(R), \mathbb{Z}/n(1))$, or via the spectral sequence $H^p(X_b, R^q j_* \mu_n) \Rightarrow H^{p+q}(X_{et}, \mu_n)$. To sum up the preceding calculations:

(20.1.10) Proposition. — *If X is a smooth geometrically connected projective curve over R and $n \geq 1$, then the cohomology of X with coefficients μ_n is given by*

	n odd	n even, $X(R) \neq \emptyset$	n even, $X(R) = \emptyset$
$H^0(X_{et}, \mu_n)$	0	$\mathbb{Z}/2$	$\mathbb{Z}/2$
$H^1(X_{et}, \mu_n)$	$(\mathbb{Z}/n)^g$	$(\mathbb{Z}/n)^g \oplus (\mathbb{Z}/2)^s$	$(\mathbb{Z}/n)^g \oplus \mathbb{Z}/2$
$H^2(X_{et}, \mu_n)$	$\mathbb{Z}/n$	$\mathbb{Z}/n \oplus (\mathbb{Z}/2)^{2s-1}$	$\mathbb{Z}/n$
$H^q(X_{et}, \mu_n), \, q > 2$	0	$(\mathbb{Z}/2)^{2s}$	0

Table 20-2

$\square$

(20.1.11) Now consider the case where the curve X is (still smooth but) affine. Let $X \hookrightarrow \bar{X}$ be the smooth completion of X, so X is obtained from $\bar{X}$ by taking out finitely many closed points. As before let g be the genus of $\bar{X}_C$ and s the number of connected components of $\bar{X}(R)$. Suppose that $\bar{X}(C) - X(C)$ contains exactly u real and $2w$ non-real points, and that t is the number of circles in $X(R)$. Thus $0 \le t \le s \le t + u$, and $X(R)$ is the disjoint sum of t circles and u open intervals. Note that $u + w \ge 1$ since $X \ne \bar{X}$.

Again let $Y = X(C)/G$, a non-complete 2-manifold with boundary $\partial Y = X(R)$. Hence $H_2(Y, M) = H_2(Y, \partial Y; M) = 0$ for any group of coefficients. Moreover one finds $H_1(Y) \cong Z^{g+w}$, and $H_1(Y, \partial Y) \cong Z^{g+w+u-1}$ in case $\partial Y \ne \emptyset$. Thus also

$$H^0(Y, M) = M, \quad H^1(Y, M) \cong M^{g+w}, \quad H^q(Y, M) = 0 \text{ for } q \ge 2,$$

and in case $\partial Y \ne \emptyset$

$$H^0(Y, \partial Y; M) = 0, \quad H^1(Y, \partial Y; M) \cong M^{g+w+u-1}, \quad H^q(Y, \partial Y; M) = 0 \text{ for } q \ge 2,$$

for any abelian group M. This calculates the b-cohomology if X with constant torsion coefficients. The Leray spectral sequence for j_* gives a short exact sequence

$$0 \longrightarrow H^1(X_b, M) \longrightarrow H^1(X_{et}, M) \longrightarrow H^0(X_r, {}_2M) \longrightarrow H^2(X_b, M) = 0$$

which is split. Hence $H^1(X_{et}, M) \cong M^{g+w} \oplus ({}_2M)^{t+u}$ for any constant sheaf of torsion coefficients M. Since $H^n_G(R, M) = H^n(G, M)$ $(n \ge 0)$ one gets

(20.1.12) **Proposition.** — *Let X be an affine smooth geometrically connected curve over R. For any torsion abelian group M one has, with the notations introduced above,*

$$H^0(X_{et}, M) = M,$$
$$H^1(X_{et}, M) \cong M^{g+w} \oplus ({}_2M)^{t+u},$$
$$H^q(X_{et}, M) \cong \begin{cases} ({}_2M)^t \oplus (M/2M)^{t+u}, & q \ge 2 \text{ even,} \\ ({}_2M)^{t+u} \oplus (M/2M)^t, & q \ge 2 \text{ odd.} \end{cases} \qquad \square$$

(20.1.13) **Remark.** One can use (20.1.12) together with the Hochschild-Serre sequence (8) to derive the Galois module structure of $H^1(X(C), Z)$ (which is a free abelian group of rank $2g + 2w + u - 1$). In this way one finds

$$H^1(X(C), Z) \cong \begin{cases} Z^t \oplus Z(1)^{t+u-1} \oplus (ZG)^{g+w-t} & \text{if } X(R) \ne \emptyset, \\ Z \oplus (ZG)^{g+w-1} & \text{if } X(R) = \emptyset, \end{cases}$$

as G-modules. This in turn can be used to calculate cohomology with coefficients μ_n, using Hochschild-Serre for this sheaf. I state the result without proof: If n is even then

$$H^0(X_{et}, \mu_n) = Z/2,$$
$$H^1(X_{et}, \mu_n) \cong (Z/n)^{g+w+u-1} \oplus (Z/2)^{t+1},$$
$$H^q(X_{et}, \mu_n) \cong (Z/2)^{2t+u} \text{ for } q \ge 2;$$

if n is odd one has to drop all summands $\mathbb{Z}/2$.

20.2 Examples of sheaves

Let always X be a scheme, $X' = X \otimes_{\mathbb{Z}} \mathbb{Z}[\sqrt{-1}]$, and G the group of order 2. In the following some more or less interesting examples of abelian sheaves A on X_{et} will be studied. In particular, I will give some information on the G-sheaves $\nu(G)^* A$ on X_r and their cohomology. This illustrates the main results of Sect. 7, which are saying that the natural homomorphisms

$$h\colon H^n(X_{et}, A) \longrightarrow H^n_G(X_r, \nu(G)^* A) \tag{20}$$

are isomorphisms for $n \gg 0$ and A torsion (or at least "in the limit $n \to \infty$" if $2A = 0$).

(20.2.1) If A is an odd torsion sheaf then $R^q \rho A = 0$ for $q > 0$ (3.12c), and therefore $H^n_G(X_r, \nu(G)^* A) = H^n(X_r, \rho A)$ by (6.9.2). Hence these groups vanish for $n > \dim X_r$. This case is of course the most uninteresting one.

(20.2.2) The most basic interesting example comes from the constant sheaf $A = \mathbb{Z}/2$. By (7.11), $\nu(G)^* A = \mathbb{Z}/2$ has cohomology

$$H^n_G(X_r, \mathbb{Z}/2) = \bigoplus_{i=0}^{n} H^i(X_r, \mathbb{Z}/2), \quad n \geq 0.$$

As in (7.19.1), write (20) in the form

$$\begin{array}{ccc}
h\colon H^n(X_{et}, \mathbb{Z}/2) & \longrightarrow & H^n_G(X_r, \mathbb{Z}/2) = \bigoplus_{i=0}^{n} H^i(X_r, \mathbb{Z}/2) \\
\alpha & \longmapsto & h(\alpha) \quad = h_0(\alpha) + \cdots + h_n(\alpha)
\end{array} \tag{21}$$

with $h_i(\alpha) \in H^i(X_r, \mathbb{Z}/2)$. If $H^n(X'_{et}, \mathbb{Z}/2) = 0$ for $n > d$ then (21) is an isomorphism for $n > d$; and in any case, the map

$$\lim_{n \to \infty} H^n(X_{et}, \mathbb{Z}/2) \longrightarrow H^*(X_r, \mathbb{Z}/2) = \bigoplus_{i=0}^{\infty} H^i(X_r, \mathbb{Z}/2) \tag{22}$$

is a (ring) isomorphism. (Recall that the left hand limit is also the homogeneous localization $H^*(X_{et}, \mathbb{Z}/2)_{(s)}$, where $s := (-1) \in H^1(X_{et}, \mathbb{Z}/2)$.) This was pointed out already in (7.19); all one needs is that X is quasi-compact and quasi-separated over $\mathbb{Z}[\frac{1}{2}]$. In (7.19.1) an explicit description of $h_0(\alpha)$ was given; write again $h_0(\alpha)(\xi) =: \alpha(\xi)$ for $\xi \in X_r$, as done there.

In particular, for a given $\alpha \in H^n(X_{et}, \mathbb{Z}/2)$, the image $h(\alpha)$ vanishes if and only if the cup product $(-1)^N \cdot \alpha$ is zero for large N (in which case we say that α is (-1)-primary torsion).

This fact gives rise to interesting quantitative questions, as follows: Assuming that a specific class α is given for which $h(\alpha) = 0$ is known, which upper bounds can one give for the least integer N with $(-1)^N \cdot \alpha = 0$? There are various situations where such classes α arise naturally. To illustrate this by just one example, consider the most basic case possible, namely when $X_r = \emptyset$. The question is, then, how to bound the degree of nilpotency of (-1) in $H^*(X_{et}, \mathbb{Z}/2)$. If $X = \operatorname{spec} A$ is affine, then $X_r = \emptyset$ if and only if -1 is a sum of squares in A, as one shows by elementary real algebra [KS, p. 104]. One can prove that if -1 is a sum of n squares then $(-1)^n = 0$ in $H^*(X_{et}, \mathbb{Z}/2)$, but $(-1)^{n-1} \neq 0$ in general.

The first to have given bounds of this sort was Burési in his thesis [Bu1]; however his bounds were not best possible. For stronger bounds and other quantitative questions related to the isomorphism (22) see a forthcoming paper by the author.

I state two more corollaries to the isomorphism (22):

(20.2.2.1) Corollary. — *Let X be a scheme with $\frac{1}{2} \in \mathcal{O}(X)$ which is quasi-compact and quasi-separated. For $\alpha \in H^n(X_{et}, \mathbb{Z}/2)$ the following conditions are equivalent:*

(i) *$h_0(\alpha) = 0$, i.e. α vanishes in every point of the real spectrum;*

(ii) *there is $N \geq 0$ such that $(-1)^N \cdot \alpha$ lies in the kernel of the natural map*
$$H^{N+n}(X_{et}, \mathbb{Z}/2) \to H^0(X_{zar}, \mathcal{H}^{N+n}(\mathbb{Z}/2)).$$

Proof. By quasi-compactness of X, (ii) is equivalent to the existence of a Zariski open covering $\{U_\lambda\}$ of X such that $\alpha|_{U_\lambda}$ is (-1)-primary torsion on U_λ for every λ. Therefore (ii) $\Rightarrow$ (i) is obvious. For the converse observe the following fact: Given any class $\beta \in H^i(X_r, \mathbb{Z}/2)$, $i \geq 1$, there is a Zariski open covering $\{V_\mu\}$ of X such that β restricts to 0 in each $(V_\mu)_r$. Indeed, this is just a reformulation of $R^i\varphi_*(\mathbb{Z}/2) = 0$, proved in Theorem (19.2). Hence $h_0(\alpha) = 0$ implies that there is a finite Zariski open covering $\{U_\lambda\}$ with each U_λ quasi-compact such that $h(\alpha|_{U_\lambda}) = 0$ in $H_G^n((U_\lambda)_r, \mathbb{Z}/2)$ for every λ. So $\alpha|_{U_\lambda}$ is (-1)-primary torsion for each λ, by the isomorphism (22). $\square$

If X_r has no sheaf cohomology in positive degrees (e.g. if X is the spectrum of a semilocal ring) then there is a local-global principle for (-1)-primary torsion: The conditions of the corollary are then equivalent to $(-1)^N \cdot \alpha = 0$ for some $N \geq 0$. But as soon as $H^i(X_r, \mathbb{Z}/2) \neq 0$ for some $i > 0$ this becomes definitely false, as is shown by the isomorphism (22).

Another interesting corollary is that connected components of the real spectrum can be separated by étale cohomology classes:

(20.2.2.2) Corollary. — *Let X be a scheme which is quasi-compact and quasi-separated over $\operatorname{spec} \mathbb{Z}[\frac{1}{2}]$. If U, V are disjoint clopen subsets of X_r, there is $\alpha \in H^n(X_{et}, \mathbb{Z}/2)$ (for suitable $n \geq 0$) such that*
$$\alpha(\xi) = 1 \quad \text{for } \xi \in U \quad \text{and} \quad \alpha(\xi) = 0 \quad \text{for } \xi \in V.$$
$\hfill \square$

This is another instance where quantitative questions arise naturally: How small can n be chosen? In general, this seems to be a difficult question. — For X affine, the last corollary has also been proved independently by Burési and by Mahé. See the historical notes at the end of this section.

(20.2.3) More generally, consider now the constant sheaf $A = \mathbb{Z}/m$ for even m. Then $\nu(G)^*A = \mathbb{Z}/m$, the constant G-sheaf with trivial G-operation; and $R^q\rho A = \mathbb{Z}/2$ for $q \geq 1$. The G-cohomology of $\mathbb{Z}/m$ is described by split exact sequences (6.3.1)

$$0 \longrightarrow \bigoplus_{\substack{q < n \\ q \equiv n(2)}} H^q(X_r, \mathbb{Z}/m) \otimes \mathbb{Z}/2 \longrightarrow H^n_G(X_r, \mathbb{Z}/m)$$

$$\longrightarrow H^n(X_r, \mathbb{Z}/m) \oplus \bigoplus_{\substack{q < n \\ q \not\equiv n(2)}} {}_2\big(H^q(X_r, \mathbb{Z}/m)\big) \longrightarrow 0. \tag{23}$$

(20.2.4) The situation is similar for $A = \mu_m$. First observe that the abelian sheaf underlying $\nu(G)^*\mu_m$ (which is $\nu^*\mu_{m,X'}$) is isomorphic to the constant sheaf $\mathbb{Z}/m$. Indeed, consider the embedding $\mu_m \hookrightarrow \mathbb{G}_m$. The sheaf underlying $\nu(G)^*\mathbb{G}_m$ is $\mathcal{N}_c^* = \nu^*\mathbb{G}_{m,X'}$, the sheaf of nowhere vanishing complex Nash functions (5.5.4). Fix $\sqrt{-1} \in H^0(X_r, \mathcal{N}_c^*)$. Then a global section of $\nu^*\mu_{m,X'}$ which generates the sheaf is given by

$\xi \mapsto$ *the element with positive imaginary part and largest real part in* $(\nu^*\mu_{m,X'})_\xi$

($\xi \in X_r$), where real and imaginary part are taken with respect to the fixed section $\sqrt{-1}$.

So $\nu(G)^*\mu_m = (\mathbb{Z}/m)(1)$, the constant sheaf $\mathbb{Z}/m$ on X_r on which G acts by inversion. Clearly $H^*_G(X_r, (\mathbb{Z}/m)(1)) = 0$ for odd m. Below (20.2.7.2) it will be shown that if m is even (and X is quasi-compact and quasi-separated), there are (non-canonical) isomorphisms

$$H^n_G(X_r, (\mathbb{Z}/m)(1)) \cong \bigoplus_{q=0}^{n} H^q(X_r, \mathbb{Z}/2), \quad n \geq 0.$$

(20.2.5) Let F be a Zariski sheaf of $\mathcal{O}_X$-modules on X, and let F_{et} be the induced sheaf of modules on the étale site X_{et} [SGA4 VII.2c]. Then it is easy to see that

$$\nu(G)^*F_{et} = (\rho F_{et}) \oplus (\rho F_{et})(1) = (\rho F_{et}) \otimes_{\mathbb{Z}} \mathbb{Z}G$$

as G-sheaves (where ρF_{et} is a G-sheaf by the trivial G-action). Therefore

$$H^n_G(X_r, \nu(G)^*F_{et}) = H^n(X_r, \rho F_{et}),$$

$n \geq 0$. Note that ρF_{et} is a sheaf of $\mathcal{N}$-modules, which is quasi-coherent if F is so. It is well known that the sheaf $\mathcal{N}$ has badly behaved cohomology in general. For example, on the affine line over $\mathbb{R}$ one has $H^1(\text{sper } \mathbb{R}[t], \mathcal{N}) \neq 0$ [Hb].

(20.2.6) More interesting is the multiplicative sheaf $\mathbf{G}_m$. Let as before $\mathcal{N} = \rho\mathbf{G}_a$ be the sheaf of Nash functions on X_r, $\mathcal{N}_c = \mathcal{N} \otimes_{\mathbb{Z}} \mathbb{Z}[\sqrt{-1}] = \nu^*\mathbf{G}_{a,X'}$ the sheaf of complex Nash functions, and $\mathcal{N}^* = \rho\mathbf{G}_m$ resp. $\mathcal{N}_c^* = \nu^*\mathbf{G}_{m,X'}$ the corresponding sheaves of multiplicative units.

(20.2.6.1) **Lemma.** — *The G-sheaf $\nu(G)^*\mathbf{G}_m$ on X_r decomposes as a direct product*

$$\nu(G)^*\mathbf{G}_m = \mathcal{N}_+^* \times U,$$

where G acts by inversion on U and acts trivially on $\mathcal{N}_+^*$ ($:=$ the sheaf of positive Nash functions).

Proof. The underlying abelian sheaf of $\nu(G)^*\mathbf{G}_m$ is $\mathcal{N}_c^*$. Let σ denote the involution on this sheaf. Let $K := \ker(1+\sigma)$ and $L := \text{im}(1+\sigma)$, considered as subsheaves of $\mathcal{N}_c^*$. One has to show that the obvious inclusion $L \subset \mathcal{N}_+^*$ is an equality, and that the multiplication map $K \times L \to \mathcal{N}_c^*$ is an isomorphism. This can be done stalkwise. Here the claim follows from the fact that in a strictly real local ring every positive unit has a unique positive square root. $\square$

Note that U is the sheaf of complex Nash functions of constant modulus 1. So U is divisible. As far as cohomology is concerned one can replace U by a simpler subsheaf: Let V be the subsheaf of 2-primary torsion elements of U and put $W = U/V$. Then W is uniquely divisible by 2. Since G acts by inversion it follows that $\mathbf{H}_G^n(W) = 0$ for all $n \geq 0$, and hence also $H_G^*(X_r, W) = 0$. Thus $H_G^*(X_r, U) = H_G^*(X_r, V)$. Now clearly V is the ascending union of the G-sheaves $\nu(G)^*\mu_{2^r}$, for $r \to \infty$. Since $\nu(G)^*\mu_{2^r} \cong (\mathbb{Z}/2^r)(1)$ (20.2.4) one has $V \cong \mu_{2^\infty}$, where I'm using the following

(20.2.6.2) **Notation.** If T is any topological space (with trivial G-action) then μ_{2^∞} denotes the G-sheaf on T whose underlying abelian sheaf is constant with stalks $\mathbb{Z}/2^\infty := \varinjlim \mathbb{Z}/2^\nu = \mathbb{Q}_2/\mathbb{Z}_2$, and on which G acts by inversion.

(20.2.6.3) **Proposition.** — *If X is any scheme then*

$$H_G^n(X_r, \nu(G)^*\mathbf{G}_m) = H^n(X_r, \mathcal{N}_+^*) \times H_G^n(X_r, \mu_{2^\infty}), \quad n \geq 0.$$

Proof. By (20.2.6.1) and the above remarks it remains only to show $H_G^n(X_r, \mathcal{N}_+^*) \cong H^n(X_r, \mathcal{N}_+^*)$. But this is clear since $\mathcal{N}_+^*$ is uniquely divisible. $\square$

The cohomology groups $H^n(X_r, \mathcal{N}_+^*)$, $n > 0$, are $\mathbb{Q}$-vector spaces which are usually "large" even in cases like the affine line over $\mathbb{R}$ [Hb]. They do not concern us here. Rather I want to study the groups $H_G^n(X_r, \mu_{2^\infty})$ in more detail.

(20.2.7) For this let more generally T be an arbitrary topological space. Write $H^n(T) := H^n(T, \mathbb{Z}/2)$ in the following. For the G-sheaf μ_{2^∞} on T one has $\mathbf{H}_G^q(\mu_{2^\infty}) = \mathbb{Z}/2$ if $q \geq 0$ is even, and $\mathbf{H}_G^q(\mu_{2^\infty}) = 0$ is q is odd (6.2). Hence in the spectral sequence

$$E_2^{pq} = H^p(T, \mathbf{H}_G^q(\mu_{2^\infty})) \implies H_G^{p+q}(T, \mu_{2^\infty}) \qquad (24)$$

one has $E_2^{pq} = H^p(T)$ if $q \geq 0$ is even, and $E_2^{pq} = 0$ if q is odd.

The spectral sequence (24) has a strong tendency to degenerate. Consider the exact sequence

$$0 \longrightarrow \mathbb{Z}/2 \longrightarrow \mu_{2^\infty} \xrightarrow{2} \mu_{2^\infty} \longrightarrow 0 \qquad (25)$$

of G-sheaves. It gives rise to exact sequences ($n \geq 0$)

$$0 \longrightarrow H_G^{n-1}(T, \mu_{2^\infty})/2 \longrightarrow H_G^n(T, \mathbb{Z}/2) \longrightarrow {}_2\left(H_G^n(T, \mu_{2^\infty})\right) \longrightarrow 0. \qquad (26)$$

Assume for a moment that all groups $H^q(T)$, $q \geq 0$, are finite. Then it follows from (24) and (26), by counting orders of groups, that

(a) the spectral sequence (24) degenerates (i.e. $d_r = 0$, $r \geq 2$), and
(b) the groups $H_G^n(T, \mu_{2^\infty})$ have exponent 2 for all $n \geq 0$;

thus $H_G^n(T, \mu_{2^\infty}) \cong H^n(T) \oplus H^{n-2}(T) \oplus H^{n-4}(T) \cdots$ for all $n \geq 0$.

A moment's reflection shows that (a) and (b) are true more generally if T satisfies the following condition (F):

> (F): *For every $\alpha \in H^*(T)$ there are a continuous map $f: T \to S$ into some space S and an element $\beta \in H^*(S)$ with $\alpha = f^*\beta$, such that $H^q(S)$ is finite for all $q \geq 0$.*

Examples of spaces T which satisfy (F) are:
- Spectral spaces. Indeed, T is a filtering inverse limit of finite spectral spaces T_λ (Proposition (A.2)), and $H^*(T) = \varinjlim H^*(T_\lambda)$.
- CW complexes. If T is a CW complex then sheaf cohomology with constant coefficients coincides with singular cohomology. Hence

$$H^n(T, \mathbb{Z}/2) = [T, K(n, \mathbb{Z}/2)],$$

via a universal element in $H^n(K(n, \mathbb{Z}/2), \mathbb{Z}/2)$, where $K(n, \mathbb{Z}/2)$ is an Eilenberg-MacLane space. Now the cohomology groups $H^q(K(n, \mathbb{Z}/2), \mathbb{Z}/2)$ are explicitly known (they have been determined by Serre [Se1]), and in particular, they are known to be finite.

(20.2.7.1) **Corollary.** — *If T is either a spectral space or a CW complex, the spectral sequence (24) degenerates, and there are (non-canonical) isomorphisms for $n \geq 0$:*

$$H_G^n(T, \mu_{2^\infty}) \cong H^n(T, \mathbb{Z}/2) \oplus H^{n-2}(T, \mathbb{Z}/2) \oplus H^{n-4}(T, \mathbb{Z}/2) \oplus \cdots \qquad (27) \quad \square$$

Observe in passing the following interesting consequence for the cohomology of the G-sheaves $(\mathbb{Z}/2^r)(1)$ (cf. (20.2.4)):

(20.2.7.2) Corollary. — *If T is either a spectral space or a CW complex, and $m > 0$ is an even number, there are (non-canonical) isomorphisms*

$$H_G^n(T, (\mathbb{Z}/m)(1)) \cong \bigoplus_{q=0}^n H^q(T, \mathbb{Z}/2), \quad n \geq 0.$$

Proof. One can assume that $m = 2^r$ is a power of 2 (hence $r \geq 1$). The exact sequence $0 \to (\mathbb{Z}/2^r)(1) \to \mu_{2\infty} \xrightarrow{2^r} \mu_{2\infty} \to 0$ of G-sheaves gives, by (20.2.7.1), exact sequences

$$0 \longrightarrow \bigoplus_{\substack{q \leq n \\ q \not\equiv n(2)}} H^q(T, \mathbb{Z}/2) \longrightarrow H_G^n(T, (\mathbb{Z}/2^r)(1)) \longrightarrow \bigoplus_{\substack{q \leq n \\ q \equiv n(2)}} H^q(T, \mathbb{Z}/2) \longrightarrow 0,$$

(28)

$n \geq 0$. On the other hand, the middle groups in (28) are annihilated by 2, as follows from (6.3.1). Therefore the sequences (28) split. $\qquad\square$

(20.2.7.3) Remark. The direct sum decompositions $H_G^n(T, \mathbb{Z}/2) = \bigoplus_{q=0}^n H^q(T)$ are canonical, they come from the graded ring structure

$$H_G^*(T, \mathbb{Z}/2) = H^*(T)[\gamma], \quad \deg \gamma = 1,$$

with γ the generator of $H^*(G, \mathbb{Z}/2)$. For emphasis write in the following

$$H_G^n(T, \mathbb{Z}/2) = H^n(T) . 1 \oplus H^{n-1}(T) . \gamma \oplus \cdots \oplus H^0(T) . \gamma^n.$$

The decomposition (27), on the other hand, is non-canonical a priori. However it can be made canonical. (This is not used elsewhere.) Consider again the sequences (26), for T a space satisfying (F):

$$0 \longrightarrow H_G^{n-1}(T, \mu_{2\infty}) \xrightarrow{a} H_G^n(T, \mathbb{Z}/2) \xrightarrow{b} H_G^n(T, \mu_{2\infty}) \longrightarrow 0. \qquad (26')$$

Then the following hold:

a) *The composite map $a \circ b : H_G^{n-1}(T, \mathbb{Z}/2) \to H_G^n(T, \mathbb{Z}/2)$ is the map $\xi \mapsto \gamma . \xi + \mathrm{sq}^1(\xi)$, where sq^1 is the first Steenrod square.*

b) *The restriction of b to the subgroup $H^n(T) . 1 \oplus H^{n-2}(T) . \gamma^2 \oplus H^{n-4}(T) . \gamma^4 \cdots$ of $H_G^n(T, \mathbb{Z}/2)$ is bijective.*

Proof. a) follows from the commutative diagram (of G-sheaves) with exact rows

$$
\begin{array}{ccccccccc}
0 & \longrightarrow & \mathbb{Z}/2 & \longrightarrow & \mathbb{Z}/4(1) & \longrightarrow & \mathbb{Z}/2 & \longrightarrow & 0 \\
& & \| & & \uparrow & & \uparrow & & \\
0 & \longrightarrow & \mathbb{Z}/2 & \longrightarrow & \mu_{2\infty} & \xrightarrow{\;2\;} & \mu_{2\infty} & \longrightarrow & 0.
\end{array}
$$

(29)

Indeed, (29) shows that $a \circ b$ is the boundary map corresponding to the upper row in (29), and it is not hard to figure out the explicit formula given in a). (Recall that sq^1 is the boundary map for the exact sequence $0 \to \mathbb{Z}/2 \to \mathbb{Z}/4 \to \mathbb{Z}/2 \to 0$.) To prove b) let K be the subgroup of $H_G^n(T, \mathbb{Z}/2)$ indicated there, and let L be the image of the map $a \circ b$ from a). By (26') one has to show that $K \oplus L = H_G^n(T, \mathbb{Z}/2)$, which is easy using that sq^1 is a derivation and $\mathrm{sq}^1 \circ \mathrm{sq}^1 = 0$. $\qquad\square$

(20.2.7.4) Note that for $T = X_r$ the sequence (26′) is precisely the exact sequence

$$0 \longrightarrow H_G^{n-1}\big(X_r, \nu(G)^*\mathbf{G}_m\big)/2 \longrightarrow H_G^n(X_r, \mathbb{Z}/2) \longrightarrow {}_2\Big(H_G^n\big(X_r, \nu(G)^*\mathbf{G}_m\big)\Big) \longrightarrow 0$$

induced by the Kummer sequence $1 \to \mu_2 \to \mathbf{G}_m \xrightarrow{2} \mathbf{G}_m \to 1$. A particular and very simple application of (20.2.7.3a) is therefore this: Given $u \in \mathcal{O}(X)^*$, let $(u) \in H^1(X_{et}, \mu_2)$ be as usual (cf. (7.10)). Then, cf. (20.2.2), $h_0\big((u)\big) \in H^0(X_r)$ is the characteristic function of $\{u < 0\}$, and $h_1\big((u)\big) = 0$ in $H^1(X_r)$. In particular, $h\big((u)\big) = 0$ if and only if u is everywhere positive on X_r. Another interesting consequence is this:

(20.2.7.5) **Proposition.** — *Under the map h of (21) above, the symbol part of $H^n(X_{et}, \mathbb{Z}/2)$ is mapped into $H^0(X_r, \mathbb{Z}/2)$.* □

Here by the symbol part of $H^*(X_{et}, \mathbb{Z}/2)$ I mean the subring generated by the classes (u), $u \in \mathcal{O}(X)^*$. A curious corollary of the main result is therefore:

(20.2.7.6) **Corollary.** — *If $H^i(X_r, \mathbb{Z}/2) \neq 0$ for some $i > 0$, then $H^*(X_{et}, \mathbb{Z}/2)$ is not generated by symbols.* □

The same conclusion holds if the clopen subsets of X_r are not separated by global units $u \in \mathcal{O}(X)^*$. Also note that the local-global principle for (-1)-torsion discussed after Corollary (20.2.2.1) does always hold for α in the symbol part.

After this detour I will now resume the discussion of the cohomology groups of $\nu(G)^*\mathbf{G}_m$. Applying (20.2.7.1) one gets by (20.2.6.3):

(20.2.8) **Proposition.** — *Let X be a quasi-compact and quasi-separated scheme. Then $R^q\rho\mathbf{G}_m = \mathbb{Z}/2$ or 0 for $q > 0$, depending on whether q is even or odd. Consider the spectral sequence*

$$E_2^{pq} = H^p\big(X_r, R^q\rho\mathbf{G}_m\big) \implies H_G^{p+q}\big(X_r, \nu(G)^*\mathbf{G}_m\big) \tag{30}$$

of (6.9.2). The E_2-terms are

$$E_2^{pq} = \begin{cases} H^p(X_r, \mathcal{N}^*) = H^p(X_r, \mathcal{N}_+^*) \times H^p(X_r, \mathbb{Z}/2) & \text{if } q = 0, \\ H^p(X_r, \mathbb{Z}/2) & \text{if } q > 0 \text{ is even,} \\ 0 & \text{if } q \text{ is odd.} \end{cases}$$

The spectral sequence (30) degenerates. Moreover there are isomorphisms

$$H_G^n\big(X_r, \nu(G)^*\mathbf{G}_m\big) \cong H^n(X_r, \mathcal{N}_+^*) \oplus H^n(X_r, \mathbb{Z}/2) \oplus H^{n-2}(X_r, \mathbb{Z}/2) \oplus \cdots$$

for $n \geq 0$. □

(20.2.9) In particular, for cohomology classes in $H^n(X_{et}, \mathbf{G}_m)$ one gets natural higher invariants with values in groups $H^q(X_r, \mathbf{Z}/2)$, $q \geq 1$. As an example consider the case $n = 2$. Recall that $\mathrm{Br}\, X := H^2(X_{et}, \mathbf{G}_m)$ denotes the cohomological Brauer group. The homomorphism (20) reads here

$$\mathrm{Br}\, X \longrightarrow H^2(X_r, \mathcal{N}_+^*) \times H_G^2(X_r, \mu_{2\infty}). \tag{31}$$

(In many cases $\mathrm{Br}\, X$ is known to be a torsion group, and then the first component of (31) is necessarily zero.) By (20.2.7) there is a natural exact sequence

$$0 \longrightarrow H^2(X_r, \mathbf{Z}/2) \longrightarrow H_G^2(X_r, \mu_{2\infty}) \longrightarrow H^0(X_r, \mathbf{Z}/2) \longrightarrow 0 \tag{32}$$

which splits. The composite map $\theta \colon \mathrm{Br}\, X \to H_G^2(X_r, \mu_{2\infty}) \to H^0(X_r, \mathbf{Z}/2)$ is simply given by evaluation: A class in $\mathrm{Br}\, X$ is pulled back to each point of the real spectrum, where it is either trivial or the class of the quaternions. So $\ker(\theta)$ consists of those classes $\alpha \in \mathrm{Br}\, X$ which are trivial everywhere on X_r. By (32) there is a secondary invariant $\theta_2 \colon \ker(\theta) \to H^2(X_r, \mathbf{Z}/2)$. It can be described as follows. Suppose $\alpha = [A] \in \ker(\theta)$ is the class of an Azumaya algebra A on X. Since $\theta(\alpha) = 0$, A induces an Azumaya algebra $\tilde{A}$ on the locally ringed space $(X_r, \mathcal{N})$. These latter algebras have a natural invariant in $H^2(X_r, \mathcal{N}^*)$, hence in $H^2(X_r, \mathbf{Z}/2)$; and $\theta_2(\alpha)$ is that invariant of $\tilde{A}$.

The invariant θ_2 does not vanish in general. For example, let X be a quasi-projective surface over $\mathbf{R}$. The Kummer exact sequence $1 \to \mathbf{Z}/2 \to \mathbf{G}_m \xrightarrow{2} \mathbf{G}_m \to 1$ shows that there is a surjection

$$\mathrm{coker}\Big(H^2(X_{et}) \to H_G^2(X_r)\Big) \longrightarrow\!\!\!\!\!\to \mathrm{coker}\Big({}_2\mathrm{Br}\, X \to H_G^2(X_r, \mu_{2\infty})\Big)$$

(cohomology groups without coefficients are cohomology groups with coefficients $\mathbf{Z}/2$). Moreover, the first of these cokernels has dimension at most $\dim H^3(X_{\mathbf{C}}) + \dim H^4(X_{\mathbf{C}})$ (over $\mathbf{Z}/2$). This follows from the commutative diagram

$$
\begin{array}{ccccccccc}
H^2(X_{et}) & \xrightarrow{(-1)} & H^3(X_{et}) & \xrightarrow{\mathrm{res}} & H^3(X_{\mathbf{C}}) & \xrightarrow{\mathrm{tr}} & H^3(X_{et}) & \xrightarrow{(-1)} & H^4(X_{et}) & \xrightarrow{\mathrm{res}} & H^4(X_{\mathbf{C}}) \\
\downarrow & & \downarrow & & & & \downarrow & & \downarrow & & \downarrow \\
H_G^2(X_r) & \xrightarrow{\sim} & H_G^3(X_r) & & & & H_G^3(X_r) & \xrightarrow{\sim} & H_G^4(X_r)
\end{array}
$$

with exact top row, in which the rightmost vertical arrow is surjective. Hence if

$$\dim H^2(X_r) > \dim H^3(X_{\mathbf{C}}) + \dim H^4(X_{\mathbf{C}}), \tag{33}$$

then θ_2 cannot be trivial (not even its restriction to the 2-torsion in $\ker \theta$). It is easy to give examples of surfaces which satisfy inequality (33): E.g. take X smooth and affine such that $X(\mathbf{R})$ has at least one compact connected component (here $H^3(X_{\mathbf{C}}) = H^4(X_{\mathbf{C}}) = 0$). As an explicit example take the "algebraic 2-sphere"

$X = \operatorname{spec} \mathbb{R}[u, v, w]/(u^2 + v^2 + w^2 - 1)$: Here $\operatorname{Br} X \cong \mathbb{Z}/2 \oplus \mathbb{Z}/2$, generated by the Hamilton quaternions and the Clifford algebra of a natural positive definite quadratic form on the (algebraic) Hopf bundle (of rank 2) over X [Kn, VIII.6.2]; for the latter generator, θ_2 is defined and is non-zero. Or, to give a projective example, take a smooth projective surface for which $X_{\mathbb{C}}$ is rational and $X(\mathbb{R})$ has at least two connected components (here $H^3(X_{\mathbb{C}}) = 0$ and $H^4(X_{\mathbb{C}}) = \mathbb{Z}/2$).

(20.2.10) If M is an abelian group and ℓ is a prime, write $M_{\ell-\text{tors}}$ for the subgroup of ℓ-primary torsion elements in M.

Let X be a quasi-compact and quasi-separated scheme on which ℓ is invertible. By considering the Kummer exact sequences $1 \to \mu_{\ell^\nu} \to \mathbb{G}_m \overset{\ell^\nu}{\to} \mathbb{G}_m \to 1$ on X_{et} and their associated long cohomology sequences, and passing the latter to the limit $\nu \to \infty$, one obtains short exact sequences

$$0 \longrightarrow H^{q-1}(X_{et}, \mathbb{G}_m) \otimes_{\mathbb{Z}} (\mathbb{Q}_\ell/\mathbb{Z}_\ell) \longrightarrow H^q(X_{et}, \mu_{\ell^\infty}) \longrightarrow H^q(X_{et}, \mathbb{G}_m)_{\ell-\text{tors}} \longrightarrow 0$$
$$(34)$$

for every $q \geq 0$ (see e.g. [Gr3, Thm. II.3.1]). Here $\mu_{\ell^\infty} := \lim_{\nu \to \infty} \mu_{\ell^\nu}$ denotes the subsheaf of $\mathbb{G}_m$ of ℓ-primary roots of unity. From (34) one sees e.g. that $H^q(X_{et}, \mathbb{G}_m)$ is uniquely ℓ-divisible for $q > \operatorname{cd}_\ell(X_{et})$.

Consider in particular the case $\ell = 2$, and the commutative diagram with exact rows

$$0 \to H^{q-1}(X_{et}, \mathbb{G}_m) \otimes (\mathbb{Q}_2/\mathbb{Z}_2) \longrightarrow H^q(X_{et}, \mu_{2^\infty}) \longrightarrow H^q(X_{et}, \mathbb{G}_m)_{2-\text{tors}} \to 0$$
$$\downarrow \qquad\qquad\qquad\qquad \downarrow$$
$$0 \longrightarrow H_G^q(X_r, \mu_{2^\infty}) \longrightarrow H_G^q(X_r, \nu(G)^* \mathbb{G}_m).$$

Its top row is (34), and for the bottom map see (20.2.6.3). If $q > \operatorname{cd}_2(X'_{et})$ then the left vertical arrow is an isomorphism by Corollary (7.20). So one concludes from (20.2.7.1):

(20.2.11) **Theorem.** — *Let X be a quasi-compact and quasi-separated scheme on which 2 is invertible. For every $q > d := \operatorname{cd}_2(X'_{et})$ one has*

$$H^q(X_{et}, \mathbb{G}_m)_{2-\text{tors}} \cong H_G^q(X_r, \mu_{2^\infty})$$
$$\cong H^q(X_r, \mathbb{Z}/2) \oplus H^{q-2}(X_r, \mathbb{Z}/2) \oplus H^{q-4}(X_r, \mathbb{Z}/2) \cdots,$$

and $H^q(X_{et}, \mathbb{G}_m)$ modulo this subgroup is 2-divisible. (The latter is true also for $q = d$.) □

(20.2.12) **Corollary.** — *If X is a quasi-compact and quasi-separated scheme over $\operatorname{spec} \mathbb{Q}$ then for $q > \sup_\ell \{\operatorname{cd}_\ell(X'_{et})\}$*

$$H^q(X_{et}, \mathbb{G}_m) \cong H^q(X_r, \mathbb{Z}/2) \oplus H^{q-2}(X_r, \mathbb{Z}/2) \oplus \cdots \oplus D$$

where D is a uniquely divisible group. If in addition X is noetherian, and either X is regular or $q > \dim X$, then $D = 0$.

Proof. For $q > \dim X$ see (7.23.3a). If X is regular it is known that $H^q(X_{et}, \mathbf{G}_m)$ is a torsion group for $q \geq 2$ [Gr3, Prop. II.1.4]. □

For example, if X is a d-dimensional scheme of finite type over a real closed field, the corollary applies with every $q > 2d$ (or even with $q > d$ if X is affine).

As a final application I will calculate the étale cohomology groups $H^q(X_{et}, \mathbb{Z})$ for $q \gg 0$. The resulting groups turn out to be the same as for the multiplicative sheaf $\mathbf{G}_m$:

(20.2.13) Theorem. — *Let X be a noetherian scheme of finite Krull dimension d. If $q > 1 + \sup\{d, \mathrm{cd}_\ell(X_b) : \ell \text{ prime}\}$, then*

$$H^q(X_{et}, \mathbb{Z}) \cong H^q(X_r, \mathbb{Z}/2) \oplus H^{q-2}(X_r, \mathbb{Z}/2) \oplus H^{q-4}(X_r, \mathbb{Z}/2) \oplus \cdots$$

For example, the theorem applies with $q > 2d + 1$ if X is an algebraic variety over a real closed field, and with $q > d + 1$ if X is in addition affine.

Proof. Let $c := \sup\{d, \mathrm{cd}_\ell(X_b) : \ell \text{ prime}\}$. Since $H^q(X_{et}, \mathbb{Q}) = 0$ for $q > c$ (7.23.3a) it follows that $H^q(X_{et}, \mathbb{Z}) \cong H^{q-1}(X_{et}, \mathbb{Q}/\mathbb{Z})$ for $q > 1 + c$. Moreover

$$H^{q-1}(X_{et}, \mathbb{Q}/\mathbb{Z}) = \bigoplus_\ell H_G^{q-1}(X_r, \mathbb{Q}_\ell/\mathbb{Z}_\ell) = H_G^{q-1}(X_r, \mathbb{Q}_2/\mathbb{Z}_2)$$

for these q, since $q - 1 > \mathrm{cd}_\ell(X_b)$ for all primes ℓ. Now a similar argument as in (20.2.7) proves $H_G^{q-1}(X_r, \mathbb{Q}_2/\mathbb{Z}_2) \cong H^{q-2}(X_r) \oplus H^{q-4}(X_r) \oplus \cdots$, since $q - 1 > \mathrm{cd}(X_r)$. This proves the theorem since $H^q(X_r) = 0$. □

20.3 Fields revisited

In the following, two complementary remarks are made on the b-topology of a (real) field k. First the cohomology $H^*(k_b, \mathbb{Z}/2)$ is related to other invariants of k which have been studied in the literature. The proofs are immediate, apart from the citation of some well-known (but hard) theorems around the Milnor conjectures. Then some comments are made on the homotopy pro-type of $(\mathrm{spec}\, k)_b$, and in particular, on the fundamental pro-group. The latter does not change under a real henselian place. From this one deduces that, in contrast to the étale site, the b-site of a field has in general non-trivial higher homotopy pro-groups.

(20.3.1) Let X be any scheme. From the Leray spectral sequence for j_* one gets an exact sequence

$$0 \to H^1(X_b, j_*A) \to H^1(X_{et}, A) \to H^0(X_r, \mathrm{R}^1\rho A) \to H^2(X_b, j_*A) \to H^2(X_{et}, A)$$
$$(35)$$

for every $A \in \mathrm{Ab}(X_{et})$. Consider $A = \mathbb{Z}/2$. If 2 is invertible on X and if $\mathrm{Pic}(X)$ has no 2-torsion, the Kummer sequence for the exponent 2 together with (35) shows

$$H^1(X_b, \mathbb{Z}/2) = \mathcal{O}(X)_+^* / \mathcal{O}(X)^{*2},$$

where $\mathcal{O}(X)_+^*$ denotes the global units on X which are positive everywhere on X_r. For example X may be the spectrum of a semilocal ring here. In particular, if $X = \mathrm{spec}\, k$ with k a field of characteristic not 2, then $H^1(k_b, \mathbb{Z}/2) = (\Sigma k^{*2})/k^{*2}$. Hence $H^1(k_b, \mathbb{Z}/2) = 0$ if and only if k is pythagorean.

If $H^1(X_r, \mathbb{Z}/2) = 0$ then $H^1(X_b, j_! \mathbb{Z}/2)$ surjects onto $H^1(X_b, \mathbb{Z}/2)$. So if k is a connected semilocal ring with $\mathrm{sper}\, k \neq \emptyset$ in which 2 is a unit, the exact sequence (4) of (2.10) gives an exact sequence

$$0 \longrightarrow \mathbb{Z}/2 \longrightarrow H^0(\mathrm{sper}\, k, \mathbb{Z}/2) \longrightarrow H^1(k_b, j_! \mathbb{Z}/2) \longrightarrow k_+^*/k^{*2} \longrightarrow 1.$$

If k is a formally real field, this shows that $H^1(k_b, j_! \mathbb{Z}/2) = 0$ if and only if k is euclidean.

(20.3.2) To obtain information about higher b-cohomology groups assume now that X is a scheme with $\mathrm{cd}_2(X_r) = 0$, for example the spectrum of a semilocal ring (19.2.1). Then the Leray sequence for j_* becomes a long exact sequence

$$0 \longrightarrow H^1(X_b, j_*A) \longrightarrow H^1(X_{et}, A) \longrightarrow H^0(X_r, \mathrm{R}^1\rho A) \longrightarrow \cdots$$
$$\cdots \longrightarrow H^n(X_b, j_*A) \longrightarrow H^n(X_{et}, A) \longrightarrow H^0(X_r, \mathrm{R}^n\rho A) \longrightarrow \cdots$$
$$(36)$$

which extends (35), for every $A \in \mathrm{Ab}(X_{et})$. Similarly as in (19.5.2) one could use the exact sequence

$$0 \to H^0(X_b, j_!A) \to H^0(X_{et}, A) \to H^0(X_r, \rho A) \to H^1(X_b, j_!A) \to H^1(X_b, j_*A) \to 0$$

and the isomorphisms $H^n(X_b, j_!A) \cong H^n(X_b, j_*A)$, $n \geq 2$, to obtain from (36) a long exact sequence with $j_!A$ instead of j_*A. This resulting sequence is nothing but the fundamental sequence (6.6), in case $\frac{1}{2} \in \mathcal{O}(X)$.

Consider again $A = \mathbb{Z}/2$ and write

$$h^n \colon H^n(X_{et}, \mathbb{Z}/2) \longrightarrow H^0(X_r, \mathbb{Z}/2)$$

for the maps occurring in (36), so that one has short exact sequences

$$0 \longrightarrow \mathrm{coker}(h^{n-1}) \longrightarrow H^n(X_b, \mathbb{Z}/2) \longrightarrow \ker(h^n) \longrightarrow 0, \qquad (37)$$

$n \geq 2$. Assume $X = \operatorname{spec} k$ with k a formally real field. Let $\bar{W}(k)$ be the reduced Witt ring of k and $\bar{I}k$ its fundamental ideal, consisting of the classes of even-dimensional forms. Consider $\bar{W}(k)$ as a subring of $C := H^0(\operatorname{sper} k, \mathbb{Z})$ and put $S(k) := C/\bar{W}(k)$. Then $S(k)$ is a 2-primary torsion group of exponent $2^{\operatorname{st}(k)}$ where $\operatorname{st}(k)$ is the (reduced) *stability index* of k. Now for every $q \geq 0$ there is a natural surjection induced by $2^q C/2^{q+1}C \overset{\sim}{\to} H^0(X_r, \mathbb{Z}/2)$:

$$\frac{2^q C}{\bar{I}^q k + 2^{q+1} C} = (2^q C / \bar{I}^q k) \otimes \mathbb{Z}/2 \longrightarrow \operatorname{coker}(h^q). \tag{38}$$

If $H^q(k_{et}, \mathbb{Z}/2)$ is generated by q-fold cup products from $H^1(k_{et}, \mathbb{Z}/2)$ (hypothesis which is part of the Milnor conjectures) then (38) is an isomorphism. In any case, the left hand side of (38) (and thus $\operatorname{coker}(h^q)$) vanishes for $q > \operatorname{st}(k)$. It is conjectured that $\bar{I}^q k = 2^q C \cap \bar{W}(k)$ holds for all q. This has been verified by Marshall in many cases, see [Ma2], and is in particular known to be true if k has finite chain length. If one assumes this conjecture, it follows that the left hand side of (38) coincides with the group $S_q(k) := (2^q S(k))/(2^{q+1} S(k))$, which may be called the q-th *stability group* of k.

On the other hand there is a conjecture by Marshall [Ma1, Conjecture 1] about the kernel of the K-theoretic analogue of h^n. If one assumes that $H^n(k_{et}, \mathbb{Z}/2)$ is generated by cup products from H^1, this conjecture would imply that $\ker(h^n)$ is the image of the cup product pairing

$$H^1(k_b, \mathbb{Z}/2) \otimes H^{n-1}(k_{et}, \mathbb{Z}/2) \longrightarrow H^n(k_{et}, \mathbb{Z}/2)$$

(and would be equivalent to this assertion if the Milnor conjecture were true). In particular, Marshall's conjecture predicts that h^n should be injective for pythagorean k.

Milnor's conjecture has been proved for $n \leq 4$ (Merkur'ev, Suslin, Rost), and for various special kinds of fields and all n, e.g. by Jacob for pythagorean fields of finite chain length [Ja]. Since Marshall proved his last-mentioned conjecture for all fields of finite chain length [Ma1] one obtains in particular

(20.3.3) Corollary. — *If k is a real pythagorean field of finite chain length then $H^n(k_b, \mathbb{Z}/2)$ is isomorphic to the $(n-1)$st stability group $S_{n-1}(k)$ for $n \geq 2$.* $\square$

The other point I want to make is on the fundamental pro-groups of the sites $(\operatorname{spec} k)_b$, k a field. For the definition of $\pi_1(C)$ for a (suitable) pointed site C see [AM]. Recall only the following (cf. §10 of [loc.cit.]). For any group H one denotes by $\pi^1(C; H)$ the (pointed) set of isomorphism classes of pointed "locally trivial principal H-bundles" on C. By definition $\pi_1(C)$ is the pro-group which pro-represents the functor $H \mapsto \pi^1(C; H)$, i.e.

$$\pi^1(C; H) = \operatorname{Hom}_{\operatorname{pro-gps}}(\pi_1(C), H)$$

holds functorially in H.

Let k be a field with a fixed separable closure $k \subset k_s$, let $\Gamma = \mathrm{Gal}(k_s/k)$ and write T for the subset of involutions in Γ. Then $\pi_1((\mathrm{spec}\, k)_{et})$ is the pro-group of discrete quotient groups of Γ. If $\varphi : \Gamma \to H$ is a continuous homomorphism into a discrete group H then the associated "principal H-bundle" on $(\mathrm{spec}\, k)_{et} \sim (\Gamma\text{-sets})$ is the Γ-set H (on which H acts by right multiplication with the inverse). This object is locally trivial for the b-topology if and only if every involution $t \in T$ acts trivially on H, i.e. iff $T \subset \ker \varphi$. So

(20.3.4) Proposition. — *The fundamental group of $(\mathrm{spec}\, k)_b$ is Γ/N where N is the closed subgroup of Γ generated by T.* □

(This is to mean that $\pi_1((\mathrm{spec}\, k)_b)$ is the pro-group of discrete quotient groups of Γ/N.) I will write $\Gamma/N =: \pi_1(k_b)$ in the following.

A formally real field k is therefore simply connected for the b-topology iff it is the intersection of its real closures (in k_s). Note also that k is hereditarily pythagorean iff $(\mathrm{spec}\, K)_b$ is simply connected for every formally real algebraic extension field K of k [Bel, p. 86].

Unlike the étale site of a field, the site $(\mathrm{spec}\, k)_b$ has non-trivial higher homotopy pro-groups in general (in the sense of [AM]). This is a corollary (see (20.3.6)) to the following invariance of the fundamental group under real henselian places:

(20.3.5) Proposition. — *Let K be a henselian valued field whose residue field k is formally real. Then $\pi_1(k_b) = \pi_1(K_b)$.*

Proof. Let Δ be the value group, and denote the absolute Galois groups of K resp. k by $G(K)$ resp. $G(k)$. It is well known that there is an exact sequence of profinite groups

$$0 \longrightarrow \mathrm{Hom}\Big(\Delta \otimes \mathbb{Q}/\mathbb{Z},\, \mu(k_s)\Big) \longrightarrow G(K) \longrightarrow G(k) \longrightarrow 1,$$

where $\mu(k_s) \cong \mathbb{Q}/\mathbb{Z}$ is the group of roots of unity of k_s. Recall how it is obtained. Let K_{nr} be the maximal unramified extension of K in K_s. The Galois group of K_{nr} over K is naturally isomorphic to $G(k)$. Moreover K_{nr} is a henselian valued field with algebraically closed residue field k_s of characteristic 0. One has the natural pairing

$$\mathrm{Gal}(K_s/K_{nr}) \times K_s^* \longrightarrow \mu(k_s), \quad (\sigma, y) \longmapsto \text{residue class of } \sigma(y)/y, \qquad (39)$$

and (39) induces an isomorphism from $\mathrm{Gal}(K_s/K_{nr})$ to $\mathrm{Hom}(\Delta \otimes \mathbb{Q}/\mathbb{Z}, \mu(k_s))$ [En, §20]. Let T be the set of involutions in $G(K)$, and write M for the kernel of $G(K) \to G(k)$. For $t \in T$ the conjugation action of t on M is by inversion $\sigma \mapsto \sigma^{-1}$: Indeed, if $\sigma \in M$ and $y \in K_s^*$ then

$$\frac{t\sigma t^{-1}(y)}{y} \equiv \frac{t\sigma t^{-1}(ty)}{ty} = t\left(\frac{\sigma y}{y}\right) \equiv \left(\frac{\sigma y}{y}\right)^{-1},$$

so the assertion follows from (39).

Pick any $t \in T$ (note that $T \neq \emptyset$ by hypothesis!). By what has just been shown the coset tM in $G(K)$ is contained in T. Hence the subgroup of $G(K)$ generated by T contains M. This proves the assertion. □

(20.3.6) It is clear from the proposition that there are real fields k with $\pi_1(k_b) = 1$ and arbitrary stability index. If this stability index is > 1 (i.e. if k is not an *SAP*-field) then $(\text{spec } k)_b$ has a non-trivial higher homotopy pro-group. Indeed, the exact sequence (cf. (36))

$$ k^*/k^{*2} = H^1(k_{et}, \mathbb{Z}/2) \longrightarrow H^0(\text{sper } k, \mathbb{Z}/2) \longrightarrow H^2(k_b, \mathbb{Z}/2) $$

shows $H^2(k_b, \mathbb{Z}/2) \neq 0$ for any non-*SAP* field k. So by the profinite Hurewicz theorem [AM, Cor. 4.5] it follows more precisely that $\pi_2((\text{spec } k)_b) \neq 0$, for k as above.

(20.3.7) Without proof I want to mention that one can explicitly determine the Artin-Mazur homotopy pro-type of the site $(\text{spec } k)_b$, at least up to weak ($\natural$-) isomorphism [AM §4]. Let $\Gamma = \text{Gal}(k_s/k)$, and let T be the set of involutions in Γ. Consider the simplicial boolean space

$$ E_b\Gamma := \left(\cdots (\Gamma \amalg T)^3 \mathrel{\substack{\longrightarrow \\ \longrightarrow \\ \longrightarrow}} (\Gamma \amalg T)^2 \mathrel{\substack{\longrightarrow \\ \longrightarrow}} \Gamma \amalg T \right) $$

with the natural face resp. degeneracy maps (deletion resp. repetition). Γ acts naturally on $E_b\Gamma$, so one can form the quotient

$$ (E_b\Gamma)/\Gamma, \tag{40} $$

a simplicial boolean space. The system of discrete simplicial quotient spaces of (40) is a pro-simplicial space. It is $\natural$-isomorphic to the homotopy pro-type of $(\text{spec } k)_b$ in the sense of [AM].

20.4 Some historical remarks

The following lines contain a small review of some previous results from the literature, which are either concerned with the étale cohomology of real varieties or with relations between étale cohomology and orderings on other schemes. It seemed in order to include a few such remarks, since these previous results by other authors — although they can now be identified as particular cases of results of this book — formed the main source of inspiration for the present work, without which its results would never have been found. At the end of this section, I will also make some remarks on recent work by Mahé and Burési which is related to material of this book.

Among the first results which — in today's terminology — are concerned with the étale cohomology of a real algebraic variety are the theorems proved by E. Witt in his fundamental 1934 paper [Wi]. For $X/\mathbb{R}$ a smooth projective curve, Witt showed that the Brauer group of X is canonically isomorphic to $(\mathbb{Z}/2)^s$, where s is the number of ovals of $X(\mathbb{R})$. He also proved: If on any oval a finite even number of points is given, there is a rational function on X which changes its sign in exactly these points. Essentially, this latter theorem calculates the mod 2 Picard group of X. In the language of étale cohomology, Witt calculated therefore the groups $H^1_{et}(X, \mathbb{G}_m)/2$ and $H^2_{et}(X, \mathbb{G}_m)$. See also Sect. 20.1.

In a nutshell, the main result of Part One of this book is already represented in a paper by J.Kr. Arason from 1975 [Ar]. For k a field of characteristic not 2, Arason proves there a cohomological analogue of Pfister's local-global principle for quadratic forms: A cohomology class in $H^*(k, \mathbb{Z}/2)$ restricts to 0 in every real closure of k if and only if its cup-product with a sufficiently high power $(-1)^N$ is zero. In other words, Arason proves the injectivity of the map

$$h: \lim_{n \to \infty} H^n(k, \mathbb{Z}/2) \longrightarrow C(\operatorname{sper} k, \mathbb{Z}/2) \tag{41}$$

of (9.12) (which is also the map (22) of (20.2.2) for $X = \operatorname{spec} A$). Arason also proves that the values of h are continuous, i.e. locally constant, as functions on the real spectrum. As amply explained in Section 9, the fact that (41) is actually bijective is a very particular case of the main result (7.19), to which the proof of the latter is ultimately reduced. Arason's proof of the injectivity of (41) is freely reproduced in (9.13.1). The fact that this map is also surjective seems to have been observed for the first time in [AEJ, 2.4]; in contrast to injectivity, the surjectivity part is very easy (compare again (9.13.1)). I refer to the last pages of Sect. 9 and to Sect. 12 (see also [Sch2]) for a thorough discussion of Arason's theorem and for a different approach to it, which shows that it can be seen as a particular case of profinite Brown type theorems.

In 1979 D.A. Cox, following an idea of M. Artin, identified for arbitrary algebraic varieties $X/\mathbb{R}$ the étale cohomology with finite (locally) constant coefficients as equivariant (bundle) cohomology on the G-space $X(\mathbb{C})$ [Co]. In fact, Cox did more in that he identified the weak étale homotopy type of such X with that of the topologists' free-made quotient $X(\mathbb{C})_G$ (cf. the Introduction). As a corollary he obtained the fundamental long exact sequence which relates étale cohomology of X (with finite constant coefficients) to cohomology of $X(\mathbb{R})$ and of $X(\mathbb{C})/G$. Among the applications given by Cox for his result, there is the calculation of the groups $H^n_{et}(X, \mathbb{Z}/2)$ for $X/\mathbb{R}$ a smooth projective curves (with $X(\mathbb{R}) \neq \emptyset$). Another application is a theorem announced already in 1964 by M. Artin and J.L. Verdier: It says that an algebraic variety $X/\mathbb{R}$ has no real point if and only if its étale cohomological 2-dimension is finite. (This fact is generalized to quite arbitrary schemes in (7.21).)

In their 1990 paper [CTP], J. L. Colliot-Thélène and R. Parimala proved for any smooth algebraic variety X over a real closed field R that canonically $H^0(X, \mathcal{H}^n) \cong (\mathbb{Z}/2)^s$ holds for $n > \dim X$, where $\mathcal{H}^n = \mathcal{H}^n(\mathbb{Z}/2)$ (see (19.3.1)) and s is the number of semi-algebraic connected components of $X(R)$. As explained in the introduction of their paper (see the discussion after (20.1.3) above), this can be seen as a generalization of Witt's theorem on the Brauer group of a smooth real curve. The authors' proof builds upon Arason's above mentioned theorem, on the Bloch-Ogus resolution of $\mathbb{Z}/2 = \mu_2$ and on Mahé's theorem on the separation of connected components by quadratic forms.

At the time of writing their paper, Colliot-Thélène and Parimala were probably unaware of Cox's results [Co]. After I had discussed with Colliot-Thélène some simplifications and generalizations of the material in [CTP] in spring 1990 (the latter arising from the exactness result (19.2)), he pointed out to me Cox's paper, and how to deduce from it the main result of [CTP] without any smoothness assumption, using (19.2). See his (unpublished) talk at Oberwolfach, June 15, 1990, and compare (19.5.1).

The idea that general results like those of Sect. 7 could hold came to me after these discussions with Colliot-Thélène and Mahé. It was supported by the observation that Theorem (7.19) (for $A = \mathbb{Z}/2$) was true both for the spectrum of a field and for an algebraic variety over $\mathbb{R}$, by Arason's resp. Cox's theorem. The main results of Part One of this book were obtained in late 1990 and early 1991, and were announced in some talks in January 1991 and in a preprint circulated in May 1991. My habilitation thesis, of which this book is a slightly revised version, was completed and submitted in November 1992.

Meanwhile, L. Mahé and his student J. Burési had independently pursued a different approach, which was later completed successfully in Burési's 1993 thesis [Bu1]; see also [Bu2] and [Mh]. This approach doesn't use anything like the glued site X_b of this work. Instead, Burési establishes localization techniques for cohomology classes which are inspired by quadratic form theory. Using these techniques he generalizes Arason's theorem to semilocal rings, i.e. shows that the map

$$\lim_{n \to \infty} H^n_{et}(A, \mathbb{Z}/2) \longrightarrow H^0(\operatorname{sper} A, \mathbb{Z}/2) \tag{42}$$

is injective for every semilocal ring A. From Mahé's separation theorem for quadratic forms and from the theory of Stiefel-Whitney classes of such forms he deduces that (42) is surjective for any ring A (always containing $\frac{1}{2}$). For this latter fact, Mahé has recently also given a direct proof [Mh].

One the one hand, Burési's results are under several aspects weaker than those of Sect. 7: They are restricted to affine schemes and to coefficients $\mathbb{Z}/2$, and (more seriously) they do not contain the G-equivariant aspect, i.e. he doesn't construct the higher invariants h_i, $i \geq 1$, from $H^*(X_{et}, \mathbb{Z}/2)$ into $H^i(X_r, \mathbb{Z}/2)$ (20.2.2). As was shown in Sect. 7 of this book, these higher h_i do precisely identify the kernel of (42) for arbitrary rings A (or more general schemes).

On the other hand, he establishes some interesting quantitative results on his way (compare (20.2.2)), which are not covered by the methods of this book. Also, the approach of Mahé and Burési is more direct and less technical than the use of the glued site X_b in this work.

Appendix A:
Some complements on spectral spaces

In this appendix some complementary facts about spectral spaces are gathered. Most of them are concerned with constructible sheaves. The results are elementary, but I am not aware of a reference for them, and so I include the details for the reader's convenience. See (0.6.3) for the notion of spectral resp. locally spectral space.

(A.1) Proposition. — *Let I be a left filtering category and let $\{T_\lambda\}_{\lambda \in I}$ be an inverse system of spectral spaces indexed by I, with all transition maps spectral. Let $T = \varprojlim T_\lambda$ be the inverse limit, formed in the category of topological spaces. Then T is a spectral space, all projections $T \to T_\lambda$ are spectral, and T is also the inverse limit in the category of spectral spaces and spectral maps.*

Proof. Denote the projection $T \to T_\lambda$ by p_λ. Let S_λ be T_λ with the constructible topology, and put $S := \varprojlim S_\lambda$, a boolean topological space. Let $\beta: S \to T$ be the canonical map; β is a continuous bijection.

It is immediate that T is a quasi-compact T_0-space. Let $\lambda_1, \ldots, \lambda_s \in I$, and let K_i be a constructible subspace of T_{λ_i}, $i = 1, \ldots, s$. Then

$$K := \bigcap_{i=1}^{s} p_{\lambda_i}^{-1}(K_i)$$

is a quasi-compact subspace of T, since $\beta^{-1}(K)$ is obviously compact. This shows that T has a basis of open quasi-compact subspaces which is stable under finite intersections. Let $\emptyset \neq Z$ be a closed irreducible subspace of T. Since $p_\lambda(Z)$ is pro-constructible and irreducible in T_λ, there is $z_\lambda \in p_\lambda(Z)$ with $\overline{p_\lambda(Z)} = \overline{\{z_\lambda\}}$. The z_λ are compatible, i.e. there is $z \in T$ with $p_\lambda(z) = z_\lambda$ ($\lambda \in I$); and clearly $z \in Z$. It is checked immediately that $Z = \overline{\{z\}}$.

Hence T is a spectral space, and the maps p_λ are spectral. T is the inverse limit in the category of spectral spaces since every constructible subspace of T has the form $p_\lambda^{-1}(K_\lambda)$, for suitable $\lambda \in I$ and K_λ in T_λ constructible. $\square$

(A.2) Corollary. — *Every spectral space is homeomorphic to a filtering inverse limit of finite spectral spaces.*

Proof. Let T be spectral, and let $\{p_\lambda: T \to T_\lambda\}_{\lambda \in I}$ be the filtering inverse system of all surjective spectral maps from T onto finite spectral spaces (up to isomorphism, so that one has I small). Let $T^* := \varprojlim T_\lambda$, a spectral space. The canonical map $f: T \to T^*$ is spectral and surjective. Let $x, y \in T$ with $x \notin \overline{\{y\}}$. Then there is an open constructible neighborhood U of x with $y \notin U$. Let $E = \{0, 1\}$ be the spectral space with two elements in which $0 \succ 1$. The map $T \to E$ which is 0 on U and 1 on $T - U$ is spectral. So there is $\lambda \in I$ with $p_\lambda(x) \notin \overline{\{p_\lambda(y)\}}$. Hence $f(x) \notin \overline{\{f(y)\}}$. This shows that f is a homeomorphism.   □

(A.3) Definition.

   a) Let T be a spectral space. A sheaf of sets (resp. of abelian groups) A on T is *constructible* if T has a covering $\{K_i\}$ by (finitely many) constructible subspaces K_i for which $A|_{K_i}$ is locally constant with finite stalks (resp. with finitely generated stalks).

   b) Let T be a locally spectral space. A sheaf of sets (resp. of abelian groups) A on T is *constructible* if $A|_U$ is constructible, in the sense of a), for every open spectral subspace U of T.

One may replace "locally constant" by "constant" in a) without affecting the definition. Also one may always assume that the K_i are locally closed in T and pairwise disjoint. In b) it suffices that there exists a basis of open spectral subspaces U for which $A|_U$ is constructible; if T is quasi-separated (i.e. if the constructible topology is Hausdorff) the existence of a covering by such subspaces U is enough. Note that the inverse image of a constructible (abelian) sheaf under a locally spectral map is again constructible (abelian).

   A fact which can be quite useful is that every sheaf on a spectral space is a filtering direct limit of constructible sheaves. This will now be proved, using Corollary (A.2). Note that in general the constructible subsheaves of a sheaf do *not* form a directed system.

(A.4) Proposition. — *Every sheaf (resp. abelian sheaf) on a spectral space T is a filtering direct limit of constructible sheaves (resp. constructible abelian sheaves).*

Proof. Write $T = \varprojlim T_\lambda$ as the inverse limit of a system $\{T_\lambda\}_{\lambda \in I}$ of finite spectral spaces, indexed by a directed set I (A.1). Let $p_\lambda: T \to T_\lambda$ be the projections and $q_{\mu\lambda}: T_\lambda \to T_\mu$ ($\mu \leq \lambda$) the transition maps. Fix a sheaf F on T and put $F_\lambda := p_{\lambda*}F$, a sheaf on T_λ ($\lambda \in I$). Consider the set of pairs

$$J := \Big\{(\lambda, A): \ \lambda \in I, \text{ and } A \text{ is a constructible subsheaf of } F_\lambda \text{ on } T_\lambda\Big\}.$$

For $\mu \leq \lambda$ one has an adjunction morphism

$$\alpha_{\mu\lambda}: \ q_{\mu\lambda}^* F_\mu = q_{\mu\lambda}^* q_{\mu\lambda*} F_\lambda \longrightarrow F_\lambda.$$

The set J is ordered by the relation

$$(\mu, B) \leq (\lambda, A) \quad \text{iff} \quad \mu \leq \lambda \text{ and } \alpha_{\mu\lambda}(q_{\mu\lambda}^* B) \subset A.$$

It is easy to see that J is a directed set. For $(\mu, B) \le (\lambda, A)$ in J, the canonical sheaf map $p_\mu^* F_\mu \to p_\lambda^* F_\lambda$ (on T) restricts to a sheaf map $p_\mu^* B \to p_\lambda^* A$. Therefore one can form the direct limit sheaf

$$F' := \varinjlim_{(\lambda, A) \in J} p_\lambda^* A$$

on T. One has canonical sheaf maps

$$F' = \varinjlim_{(\lambda, A) \in J} p_\lambda^* A \longrightarrow \varinjlim_{\lambda \in I} p_\lambda^* p_{\lambda *} F \longrightarrow F.$$

The first is obviously an isomorphism, and the second is one by the usual limit theorem ([SGA4 VI.8.5], compare (3.4.1a)), which would be easy to prove directly here. Since all sheaves $p_\lambda^* A$ are constructible on T this proves the proposition. $\square$

In the remainder of this appendix some basic facts about constructible sheaves are assembled.

(A.5) Proposition. — *Let T be a locally spectral space.*
 a) *Let A, B be constructible abelian sheaves on T and $\alpha \colon A \to B$ a sheaf homomorphism. Then $\ker(\alpha)$ and $\mathrm{coker}(\alpha)$ are also constructible abelian sheaves.*
 b) *Let $0 \to A' \to A \overset{\alpha}{\to} A'' \to 0$ be an exact sequence in $\mathrm{Ab}(T)$. If two of A', A, A'' are constructible then so is the third.*

Proof. Without loss of generality let T be a spectral space. a) One can assume that A and B are constant sheaves, with stalks M resp. N. Let $s \colon M \to H^0(T, \underline{N})$ be the composite map $M \to H^0(T, \underline{M}) \overset{\alpha}{\to} H^0(T, \underline{N})$. Each $m \in M$ determines a decomposition of T into disjoint clopen pieces, namely the pieces of constant value of the section $s(m) \in H^0(T, \underline{N})$. Since M is finitely generated there are a clopen decomposition $T = U_1 \amalg \cdots \amalg U_n$ and group homomorphisms $\varphi_i \colon M \to N$ such that on U_i, α is induced by φ_i $(i = 1, \dots, n)$. Clearly $\ker(\alpha)$ and $\mathrm{coker}(\alpha)$ are constant sheaves when restricted to any of the U_i. To prove b) it suffices by a) to consider the case where A' and A'' are constructible, and then one can assume that they are constant; say $A' = \underline{M}$ and $A'' = \underline{N}$ with finitely generated abelian groups M and N. Moreover one reduces by induction to the case where N is cyclic. Then locally a generator of N can be lifted to A, and one may assume that this is true globally, i.e. that there is $s \in H^0(T, A)$ such that $\alpha(s)$ generates A'' in every stalk. If N is infinite cyclic then A is constant with stalks $M \oplus \mathbb{Z}$. So assume $N = \mathbb{Z}/n\mathbb{Z}$ with $n \ge 1$, and fix $x \in T$. Let $m := n \cdot s_x \in A'_x$, considered as an element of M. Then $t := n \cdot s - m \in H^0(T, \underline{M})$ is a global section which vanishes in x. Hence there is a clopen neighborhood U of x such that $t|_U \equiv 0$. It follows that $A|_U$ is a constant sheaf, the value in the stalks being the cokernel of the map $\mathbb{Z} \to M \oplus \mathbb{Z}$, $1 \mapsto (-m, n)$. $\square$

If $W \subset T$ is a locally closed subset and M is an abelian group, write M_W for the constant sheaf M on W, extended by zero on $T - W$.

(A.6) Proposition. — *Let T be a spectral space. Let $\mathcal{K} = \mathcal{K}_T$ be the smallest class of abelian sheaves on T with the following properties:*

1) *For every closed constructible subset $Z \subset T$ the sheaf $\mathbb{Z}_Z$ is in $\mathcal{K}$;*
2) *if $0 \longrightarrow F' \longrightarrow F \longrightarrow F'' \longrightarrow 0$ is exact in $\mathrm{Ab}(T)$ and two of F', F, F'' are in $\mathcal{K}$, then so is the third.*

Then $\mathcal{K}$ coincides with the class of all constructible abelian sheaves on T.

Proof. The exact sequence $0 \to \mathbb{Z}_{T-Z} \to \mathbb{Z}_T \to \mathbb{Z}_Z \to 0$ shows that one could replace "closed" by "open" in a). From this remark it is clear that, if Y is an open or closed constructible subspace of T and $F \in \mathcal{K}_Y$, then the extension of F by zero on $T - Y$ is in $\mathcal{K}_T$. By (A.5) every $F \in \mathcal{K}_T$ is constructible. To prove the converse let $A \in \mathrm{Ab}(T)$ be constructible. There are finitely many open quasi-compact subspaces $U_1, \ldots, U_n$ of T such that A is constant on any of the sets $U_J - U_{I-J}$ (where $J \subset \{1, \ldots, n\}$ is any subset and $U_K := \bigcap_{i \in K} U_i$). One shows $A \in \mathcal{K}_T$ by induction over n: If $n = 1$ then there is $U \subset T$, open constructible, such that $A|_U$ and $A|_{T-U}$ are constant. The exact sequence $0 \to A|_U \to A \to A|_{T-U} \to 0$ shows $A \in \mathcal{K}$, by 1) and 2). If in general $n > 1$, then by the inductive hypothesis, $A|_{U_1}$ and $A|_{T-U_1}$ lie in $\mathcal{K}_{U_1}$ and $\mathcal{K}_{T-U_1}$, respectively. The respective extensions by zero of these sheaves are in $\mathcal{K}_T$, and hence also $A \in \mathcal{K}_T$. $\qquad\square$

(A.7) Proposition. — *Let T be a spectral space, and let $\alpha: A \to B$ be a morphism between constructible sheaves of sets on T. Then T has a covering by constructible subsets K_i such that α is isomorphic over K_i to a sheaf map $\underline{M}_i \to \underline{N}_i$ induced by a map $M_i \to N_i$ between two finite sets.*

The proof is similar to the proof of (A.5a). $\qquad\square$

(A.8) Corollary. — *Let T be a locally spectral space. The class of constructible set-valued sheaves on T is closed under taking finite direct and inverse limits in $\widetilde{T}$.* $\qquad\square$

The following fact is needed in Sect. 17:

(A.9) Proposition. — *Let $\{T_\lambda\}_{\lambda \in I}$ be a filtering inverse system of spectral spaces and spectral maps. Let $T := \varprojlim T_\lambda$, and let $p_\lambda: T \to T_\lambda$ $(\lambda \in I)$ be the projections. If A is a constructible (resp. constructible abelian) sheaf on T, there are $\lambda \in I$ and a constructible (resp. constructible abelian) sheaf A_λ on T_λ such that $A \cong p_\lambda^* A_\lambda$.*

The proof of the proposition uses the following two lemmas, the first of which is actually a particular case of it (by (A.2)):

(A.9.1) Lemma. — *Let T be a spectral space and A a constructible sheaf on T. Then there are a spectral map $f\colon T \to S$ to a finite spectral space S and a constructible sheaf B on S with $A \cong f^*B$. Dito for constructible abelian sheaves.*

(A.9.2) Lemma. — *Let $\{T_\lambda\}$ and T be as in (A.9). If $f\colon T \to S$ is a spectral map into a finite spectral space S there are an index $\lambda \in I$ and a commutative diagram*

$$
\begin{array}{ccc}
T & \xrightarrow{\ a\ } X & \xrightarrow{\ g\ } S \\
{\scriptstyle p_\lambda}\downarrow & \ \downarrow{\scriptstyle h} & \\
T_\lambda & \xrightarrow{\ b\ } Y &
\end{array}
\qquad (1)
$$

of spectral spaces and maps, in which X and Y are finite and h is a topological embedding.

Proof of Proposition (A.9). By (A.9.1) one finds a spectral map $f\colon T \to S$ to a finite spectral space S and a constructible sheaf B on S with $A \cong f^*B$. Now choose $\lambda \in I$ and a diagram (1) as in (A.9.2). The sheaf $C := h_*g^*B$ on Y is constructible and satisfies $h^*C \cong g^*B$. Therefore

$$
A \cong f^*B \cong a^*g^*B \cong a^*h^*C \cong p_\lambda^*(b^*C),
$$

and b^*C is a constructible sheaf on T_λ. For constructible abelian sheaves the proof is identical. $\qquad\square$

Proof of Lemma (A.9.1). Write $T = \varprojlim S_\alpha$ as a filtering inverse limit of finite spectral spaces S_α (indexed by a left filtering category J) as in (A.2), and let $q_\alpha\colon T \to S_\alpha$ be the projections. Consider set-valued sheaves in the following. If A is any sheaf on T there is a surjective family of sheaf morphisms $\{q_{\alpha_i}^* A_i \to A\}_i$ with constructible sheaves A_i on S_{α_i} and $\alpha_i \in J$. If A is constructible a finite subfamily suffices to cover A, and hence there are $\alpha \in J$, a constructible sheaf A_α on S_α and a surjective sheaf morphism $q_\alpha^* A_\alpha \twoheadrightarrow A$. Replacing J by J/α one can assume that α is a final object of J.

Form the fibre product of sheaves on T

$$
\begin{array}{ccc}
B & \longrightarrow & q_\alpha^* A_\alpha \\
\downarrow & & \downarrow \\
q_\alpha^* A_\alpha & \longrightarrow & A.
\end{array}
\qquad (2)
$$

Then (2) is also a pushout diagram in $\widetilde{T}$, i.e. A is the cokernel of $B \rightrightarrows q_\alpha^* A_\alpha$. And B is a constructible subsheaf of $q_\alpha^*(A_\alpha) \times q_\alpha^*(A_\alpha) = q_\alpha^*(A_\alpha \times A_\alpha)$. Consider the following assertion:

(*) If F_α is a constructible sheaf on S_α, if F_β denotes its pullback to S_β ($\beta \in J$) and if B is a constructible subsheaf of $F := q_\alpha^* F_\alpha$ on T, there are $\beta \in J$ and a constructible subsheaf B_β of F_β such that $B \cong q_\beta^* B_\beta$.

It suffices to show (*). Indeed, applying (*) with $F_\alpha := A_\alpha \times A_\alpha$ and the sheaf B of (2), it follows that there is a diagram

$$B_\gamma \rightrightarrows A_\gamma \qquad (3)$$

on some S_γ, with a constructible sheaf B_γ on S_γ, such that A is isomorphic to the cokernel of the pullback of (3) by q_γ^*. (The two maps $B = q_\beta^* B_\beta \rightrightarrows q_\alpha^* A_\alpha = q_\beta^* A_\beta$ from (2) "descend" to some S_γ.) By (A.8), and since q_γ^* preserves cokernels, this shows $A \cong q_\gamma^* C_\gamma$ with a constructible sheaf C_γ on S_γ, as desired.

To prove (*) let B_β be the largest subsheaf of F_β for which $q_\beta^* B_\beta$ is contained in B ($\beta \in J$). Note that B_β is a constructible sheaf since S_β is finite. It is easy to see that B is the directed union of the $q_\beta^* B_\beta$, $\beta \in J$. Since B is constructible it follows that $B = q_\beta^* B_\beta$ for some β.

For abelian sheaves the proof is analogous. $\qquad \square$

Proof of Lemma (A.9.2). As a first step, assume that p_λ is surjective for every $\lambda \in I$. Then I show that actually f factorizes throught some p_λ. If $u: X \to Y$ is any spectral map between spectral spaces write

$$G_u := \{(x', x) \in X \times X: u(x') \succ u(x)\},$$

and put $G_X := G_{\mathrm{id}_X}$. The set G_u is pro-constructible in $X \times X$. If $v: X \to Z$ is a second spectral map which is *surjective*, then u factors through v if and only if $G_v \subset G_u$ holds. So if in the situation of (A.9.2) the p_λ are surjective one has to show $G_{p_\lambda} \subset G_f$ for some λ. But this is clear since G_f is constructible in $T \times T$ and the filtering intersection of the G_{p_λ} is G_T which is contained in G_f.

In the general case of (A.9.2) one first finds for some λ a factorization

$$
\begin{array}{ccc}
T & \xrightarrow{\ f\ } & S \\
& {\scriptstyle p_\lambda} \searrow \quad \nearrow {\scriptstyle f'} & \\
& p_\lambda(T) \subset T_\lambda &
\end{array}
\qquad (4)
$$

of f, by the special case just treated. Write $T' := p_\lambda(T)$. Consider now spectral maps $b: T_\lambda \to Y$ from T_λ into finite spectral spaces Y. It suffices to find such b such that the map f' in (4) factors through $b|_{T'}: T' \twoheadrightarrow b(T')$. By the criterion above this means that $G_{b|_{T'}} = G_b \cap (T' \times T')$ must be contained in $G_{f'}$. But $G_{f'}$ is constructible in $T' \times T'$, and the filtering intersection of all sets $G_{b|_{T'}}$ is $G_{T'}$ which is contained in $G_{f'}$. Thus a map b exists as desired. This completes the proof of (A.9.2) and of (A.9). $\qquad \square$

Appendix B:
Cohomology of Artin-Schreier structures

Let k be a field, $k' = k(\sqrt{-1})$, fix a separable closure k_s of k' and put $\Gamma :=$ $\mathrm{Gal}(k_s/k)$, $\Gamma' := \mathrm{Gal}(k_s/k')$. Let $T \subset \Gamma$ be the space of involutions on which Γ acts by conjugation, so that the real spectrum sper k is canonically identified with T/Γ. The triple $\mathfrak{A}(k) := (\Gamma, \Gamma', T)$ is the standard example of an Artin-Schreier structure in the sense of Haran and Jarden. More generally there is an Artin-Schreier structure $\mathfrak{A}(K/k)$ for every Galois extension K/k with $\sqrt{-1} \in K$. It takes into account not only the field automorphisms of K/k but also their action on orderings of intermediate fields.

In [Ha3] Haran introduces a cohomology theory for Artin-Schreier structures. One of its features is a cohomological characterization of real projective groups (see (B.2) below), similar to Serre's characterization of projective groups. The class of real projective groups consists precisely of the absolute Galois groups of pseudo real closed (PRC) fields. The notion of Artin-Schreier structures plays an instrumental role in the proof of this theorem by Haran and Jarden, actually it was introduced for this purpose.

The definition of the cohomology groups of an Artin-Schreier structure $\mathfrak{A} = (\Gamma, \Gamma', X)$, as given in [Ha3], is somewhat mysterious, as the author remarks by himself (see the introduction of his paper). Given a discrete Γ-module A, Haran constructs explicitly a complex of abelian groups which strongly resembles the usual complex of (continuous) cochains by which one can calculate group cohomology $H^*(\Gamma, A)$. But, of course, Haran's complex takes into account the Γ-action on the space X. He then defines $H^*(\mathfrak{A}, A)$ to be cohomology of this complex, and verifies that the so-defined functor $H^n(\mathfrak{A}, -)$ is actually the n-th right derived functor of $H^0(\mathfrak{A}, -)$. The significance of the functor $H^0(\mathfrak{A}, -)$, however, is not at all clear. Haran speculates that there might be a different and more natural functor whose derived functors coincide with the $H^n(\mathfrak{A}, -)$ for $n \geq 1$. He is led to this by the fact that in his cohomology theory there are several instances where the degree 0 case is exceptional [Ha3, 4.2 and 5.2].

Here I want to suggest an alternative approach to Haran's cohomology along quite different lines. When I first came across the paper [Ha3] in early 1991, it struck me that the idea behind this cohomology theory was that it should yield reasonable (finite) dimensions of fields which admit orderings: This was also one of the features of the b-topology of a scheme on which I was working. From this work I knew that it was only to be expected that the Galois action on the space of real closures had

to play a key role. So it was highly plausible that there should be a connection
between both theories. It quickly became clear that Haran's cohomology (of $\mathfrak{A}(k)$,
say) is sheaf cohomology on the site $(\operatorname{spec} k)_b$ (cf. Theorem (B.7) below). This
interpretation also resolved the mysteries about the definition of this cohomology,
see the discussion at the end of this section. Moreover, from the general theorem
on the cohomological dimensions of the sites X_b (Sect. 7) it was an easy (and *very*
special) corollary that the cohomological dimensions of $\mathfrak{A}(k)$ are just those of the
field $k(\sqrt{-1})$. In [Ha3] this had been proved only for algebraic extensions of $\mathbb{Q}$.
Later however I learned from Dan Haran that he had independently determined the
cohomological dimensions of arbitrary Artin-Schreier structures, see [Ha4].

After the necessary definitions have been recalled, it will be proved below that
Haran's cohomology on $\mathfrak{A} = (\Gamma, \Gamma', X)$ has a natural explanation as (sheaf) cohomology on a canonically defined topos. This topos arises by a glueing construction
which parallels the construction of $\widetilde{X}_b$ (Sect. 2); speaking loosely, one glues the quotient space X/Γ to (Γ-sets). In certain cases, e.g. for the absolute Artin-Schreier
structure of a field, this topos is just the *quotient topos* of the G-topos (Γ'-sets)
($G := \Gamma/\Gamma'$), as constructed in Sect. 14 in greater generality. Then the cohomological dimension of arbitrary Artin-Schreier structures is determined. This is an
application of results of Sect. 12. Finally it will be discussed why it is important
to study this cohomology as sheaf cohomology, rather than only by the *ad hoc*
definition of [Ha3].

(B.1) Recall the following definitions from [HJ], [Ha3]. An *Artin-Schreier structure*
is a triple $\mathfrak{A} = (\Gamma, \Gamma', X)$ where Γ is a profinite group, Γ' is an open subgroup of
index ≤ 2 and X is a profinite Γ-space, such that Γ_x has order 2 and $\Gamma_x \cap \Gamma' = \{1\}$
for every $x \in X$. The map $d: X \to \Gamma$ defined by $\Gamma_x = \{1, d(x)\}$ is called the *forgetful
map*. In [HJ] and [Ha3] it is part of the axioms that d be continuous, but this holds
automatically: The composite map $\{(a, x) \in \Gamma \times X: a \cdot x = x\} \subset \Gamma \times X \to G \times X$
($G := \Gamma/\Gamma'$) is a continuous bijection, and hence has a continuous inverse.

A *morphism* $(\alpha, f): (\Gamma, \Gamma', X) \to (\Delta, \Delta', Y)$ of Artin-Schreier structures consists of a group homomorphism $\alpha: \Gamma \to \Delta$ with $\alpha^{-1}(\Delta') = \Gamma'$ and an α-equivariant
map $f: X \to Y$. Following [HJ], (α, f) is called a *cover* if α is surjective and
$\bar{f}: X/\Gamma \to Y/\Delta$ is bijective; or equivalently, if (α, f) is isomorphic to the canonical
morphism $(\Gamma, \Gamma', X) \to (\Gamma/K, \Gamma'/K, X/K)$ for $K := \ker \alpha$.

(B.1.1) The standard example is $\mathfrak{A}(k)$, the *absolute Artin-Schreier structure* of a
field k: This is $\mathfrak{A}(k) = (\Gamma, \Gamma', T)$ where $\Gamma = \operatorname{Gal}(k_s/k)$, Γ' is the subgroup which
fixes $\sqrt{-1}$ and $T \subset \Gamma$ is the space of involutions in Γ, i.e. the space of real closures
of k in k_s. More generally, if $k(\sqrt{-1}) \subset K \subset k_s$ is an intermediate field which
is Galois over k, and $N \subset \Gamma$ is the closed normal subgroup which fixes K, then
$\mathfrak{A}(K/k) := (\Gamma/N, \Gamma'/N, T/N)$ is an Artin-Schreier structure. The space T/N can
be interpreted as the space of maximal ordered intermediate fields of K/k.

(B.2) The notion of Artin-Schreier structures was introduced in [HJ] as an instrument for the study of absolute Galois groups of pseudo real closed (PRC) fields. To formulate the main result of this paper, recall that Haran and Jarden call a profinite group Γ *real projective* iff Γ is virtually 2-torsion free (i.e. the involutions form a closed subset of Γ) and if for every real lifting problem (solid arrows)

$$
\begin{array}{ccc}
 & & \Gamma \\
 & \diagup & \downarrow{\scriptstyle \varphi} \\
B & \xrightarrow[\ \alpha\ ]{\ \kappa\ } A & \longrightarrow 1
\end{array}
\tag{1}
$$

with finite groups A, B, a lift (dotted arrow) exists. Here the lifting problem (1) (with α surjective) is called *real* if the involutions do not lead to an obstruction, i.e. if for every involution $t \in \Gamma$ with $\varphi(t) \neq 1$ there is an involution $b \in B$ with $\alpha(b) = \varphi(t)$.

The main result of [HJ] states that real projective groups are related to PRC fields in the same way as projective (profinite) groups are to pseudo-algebraically closed (PAC) fields: The absolute Galois group of every PRC field is real projective; and conversely, every real projective group is isomorphic to the absolute Galois group of such a field. For the proof of this theorem in [HJ] the notion of Artin-Schreier structures is instrumental. The corresponding results for PAC fields are due to Ax and Lubotzky – Van den Dries.

(B.3) I now reproduce the definition of cohomology of Artin-Schreier structures from [Ha3]. Let $\mathfrak{A} = (\Gamma, \Gamma', X)$ be an Artin-Schreier structure and A a Γ-module. Haran puts $C^0(\mathfrak{A}, A) := A \oplus \mathrm{Hom}_{\Gamma'}(X, A)$ and

$$
C^n(\mathfrak{A}, A) := \left\{ f: \Gamma^{n-1} \times (\Gamma \amalg X) \to A: \begin{array}{l} f \text{ is continuous and } f(a_1, \ldots, a_n) = 0 \\ \text{whenever } a_i = 1 \text{ for some } i,\ 1 \leq i \leq n \end{array} \right\}
$$

for $n \geq 1$. He defines differentials $\partial: C^n(\mathfrak{A}, A) \to C^{n+1}(\mathfrak{A}, A)$ for $n \geq 1$ by a formula which is identical to the usual formula in the cochain definition of group cohomology (which arises from the bar resolution). (Although the last position in the argument of $f \in C^n(\mathfrak{A}, A)$ may not be an element of Γ, the bar formula makes perfect sense!) The first differential $\partial: C^0(\mathfrak{A}, A) \longrightarrow C^1(\mathfrak{A}, A) \subset \mathrm{Hom}(\Gamma \amalg X, A)$ is defined by

$$
\partial(a, f): \quad \sigma \longmapsto \sigma a - a \ (\sigma \in \Gamma) \quad \text{and} \quad x \longmapsto (1 + d(x)) . f(x) - a \ (x \in X)
$$

for $(a, f) \in C^0(\mathfrak{A}, A)$. These definitions make $C^{\bullet}(\mathfrak{A}, A)$ into a complex, and Haran puts

$$
H^n(\mathfrak{A}, A) := H^n C^{\bullet}(\mathfrak{A}, A).
$$

A posteriori he verifies that the $H^n(\mathfrak{A}, -)$ form a ∂-functor which is effaçeable, i.e. that $H^n(\mathfrak{A}, -)$ is the n-th right derived functor of (the left exact functor) $H^0(\mathfrak{A}, -)$. Note that this H^0 is the strange functor

$$
H^0(\mathfrak{A}, A) = \left\{ (a, f) \in A^\Gamma \times \mathrm{Hom}_{\Gamma'}(X, A): (1 + d(x)) . f(x) = a \text{ for every } x \in X \right\}.
\tag{2}
$$

(B.4) Let $\mathfrak{A} = (\Gamma, \Gamma', X)$ be an Artin-Schreier structure with forgetful map $d: X \to \Gamma$. If A is a Γ-set let

$$X(A) := \{(x, a) \in X \times A: d(x) . a = a\}.$$

As a subset of $X \times A$ this set is clopen and stable under the action of Γ. Hence the espace étalé $X(A) \to X$ is a Γ-sheaf on X (Sect. 8). Let $\rho: (\Gamma\text{-sets}) \to \widetilde{X}(\Gamma')$ be the functor $A \mapsto X(A)$, i.e. ρA is $X(A)$ considered as a Γ'-sheaf on X. The functor ρ is left exact.

(B.4.1) Definition. Given an Artin-Schreier structure $\mathfrak{A}$ as above, define the topos $E(\mathfrak{A})$ as

$$E(\mathfrak{A}) := \Big(\widetilde{X}(\Gamma'), (\Gamma\text{-sets}), \rho\Big).$$

Thus $E(\mathfrak{A})$ is the category of all triples (B, A, α) with B a Γ'-sheaf on X, A a Γ-set and $\alpha: B \to \rho A$ a morphism of Γ'-sheaves. (Cf. [SGA4 IV.9.5] or Sect. 2 for the glueing construction.) As usual denote by $i: \widetilde{X}(\Gamma') \to E(\mathfrak{A})$ and $j: (\Gamma\text{-sets}) \to E(\mathfrak{A})$ the canonical topos embeddings.

Note that the topos $\widetilde{X}(\Gamma')$ is equivalent to the category of sheaves on the quotient space $X/\Gamma' = X/\Gamma$ (8.7.2), since Γ' acts freely on X.

It is straightforward how the topos $E(\mathfrak{A})$ depends functorially on $\mathfrak{A}$, but I shall not dwell on this point.

(B.4.2) Examples. 1. Let k be a field and $\mathfrak{A}(k)$ the absolute Artin-Schreier structure of k. Then $E\big(\mathfrak{A}(k)\big)$ is (equivalent to) the topos of sheaves on $(\operatorname{spec} k)_b$, cf. (9.4c).

2. Let $1 \to \Gamma' \to \Gamma \to G \to 1$ be an extension of profinite groups, with G of order 2. If Γ' acts freely on $X := \{t \in \Gamma: t^2 = 1, t \notin \Gamma'\}$ then $\mathfrak{A} := (\Gamma, \Gamma', X)$ is an Artin-Schreier structure. The associated glued topos $E(\mathfrak{A})$ is just the quotient topos $(\Gamma'\text{-sets})/G$ as defined in Sect. 14. This follows from Sect. 11.2 where it is shown that $\widetilde{X}(\Gamma')$ is the fixtopos of the G-topos $(\Gamma'\text{-sets})$.

(B.5.1) Let $\mathfrak{A} = (\Gamma, \Gamma', X)$ be an Artin-Schreier structure. Always put $G := \Gamma/\Gamma'$ in the following. Write π^* resp. π_* for the two functors $\operatorname{res}_{\Gamma'}^{\Gamma}$ resp. $\operatorname{coind}_{\Gamma'}^{\Gamma}$: $(\Gamma\text{-mod}) \leftrightarrows (\Gamma'\text{-mod})$, cf. (0.5.3). Consider now Γ as a subgroup of $\Gamma \times G$ via the *diagonal* embedding $\Gamma \hookrightarrow \Gamma \times G$, $x \longmapsto (x, x \bmod \Gamma')$. The restriction resp. coinduction functors with respect to *this* embedding will be denoted by $\pi(G)^*$ resp. $\pi(G)_*$: $(\Gamma\text{-sets}) \leftrightarrows (\Gamma \times G\text{-sets})$. This is in accordance with the general notations for G-toposes, cf. (10.8). Since $(\Gamma \times G\text{-sets})$ is canonically identified with the category of G-objects in $(\Gamma\text{-sets})$ one has three functors $(\Gamma \times G\text{-sets}) \rightrightarrows (E(\mathfrak{A}))(G)$ induced by $j_!$, j_*, $i_*\rho$: $(\Gamma\text{-sets}) \rightrightarrows E(\mathfrak{A})$, respectively. As usual I will not make a notational distinction between the former and the latter. Similarly ρ may also denote the functor $(\Gamma \times G\text{-sets}) \to (\widetilde{X}(\Gamma'))(G)$ induced by the usual ρ of (B.4). By (8.8.3) the category $(\widetilde{X}(\Gamma'))(G)$ is canonically equivalent to the category $\widetilde{X}(\Gamma)$ of Γ-sheaves on X.

If B is any Γ'-set then $\rho\pi_* B$ is canonically isomorphic to the Γ'-sheaf $\mathrm{pr}_1 : X \times B \to X$ on X. In particular $\rho\pi_* : (\Gamma'\text{-mod}) \to \mathrm{Ab}_{\Gamma'}(X)$ is an exact additive functor. Similarly, if A is any Γ-set then $\rho\pi(G)_* A$ is the Γ-sheaf $\mathrm{pr}_1 : X \times A \to X$ on X. The proofs are the same as for d) resp. a) of (9.4).

On the topos $E(\mathfrak{A})$ one has long exact sequences similar to those on quotient toposes (14.6): Every Γ-module A gives rise to an exact sequence $0 \to j_! \pi(G)_* A \to j_* \pi(G)_* A \to i_* \rho\pi(G)_* A \to 0$ in $\mathrm{Ab}_G(E(\mathfrak{A}))$. Exactly the same arguments as in the proof of (14.6) show that the associated long exact sequence is the following sequence (3):

(B.5.2) Proposition. — *For every Γ-module A there is a natural long exact sequence*

$$\cdots \longrightarrow H^n\big(E(\mathfrak{A}), j_! A\big) \longrightarrow H^n(\Gamma, A) \longrightarrow H^n_\Gamma(X, A) \longrightarrow \cdots. \qquad (3) \quad \square$$

Note that $H^n_{\Gamma'}(X, A) = 0$ for $n \geq 1$ and $H^n_\Gamma(X, A) = H^n(G, \mathrm{Hom}_{\Gamma'}(X, A))$ ($n \geq 0$), both since Γ' acts freely on X (8.9).

Consider now the following functor φ:

(B.5.3) Definition. For every Γ-module A one has the trace map $\mathrm{tr} : \pi_* \pi^* A \to A$. Let

$$\varphi(A) := \Big(\rho\pi_* \pi^* A, \; A, \; \rho(\mathrm{tr}) : \rho\pi_* \pi^* A \to \rho A\Big),$$

an object in $\mathrm{Ab}(E(\mathfrak{A}))$. This defines an additive functor $\varphi : (\Gamma\text{-mod}) \to \mathrm{Ab}(E(\mathfrak{A}))$ which is clearly exact.

(B.5.4) Let us calculate $H^0\big(E(\mathfrak{A}), \varphi A\big)$ for a Γ-module A. If $C = (B, A, \alpha : B \to \rho A)$ is an object of $\mathrm{Ab}(E(\mathfrak{A}))$ then $H^0(E(\mathfrak{A}), C)$ is the group of all pairs $(b, a) \in H^0_{\Gamma'}(X, B) \times A^\Gamma$ for which $\alpha(b) = \rho(a)$. (Here ρ denotes also the canonical map $A^{\Gamma'} \to H^0_{\Gamma'}(X, \rho A)$.) For any Γ-module M, $H^0_{\Gamma'}(X, \rho M) = \mathrm{Hom}_{\Gamma'}(X, \rho M)$ is the group of Γ'-maps $f : X \to M$ such that $d(x) . f(x) = f(x)$ holds for $x \in X$. Such maps f correspond bijectively to Γ-equivariant maps, and so $\mathrm{Hom}_{\Gamma'}(X, \rho M) \approx \mathrm{Hom}_\Gamma(X, M)$ canonically. In particular, $\mathrm{Hom}_{\Gamma'}(X, \rho\pi_* \pi^* A)$ becomes identified with $\mathrm{Hom}_\Gamma(X, \pi_* \pi^* A) = \mathrm{Hom}_{\Gamma'}(X, A)$. Moreover, if $f \in \mathrm{Hom}_{\Gamma'}(X, A)$ then $\rho(\mathrm{tr})(f)$ is the map $X \to A$, $x \mapsto (1 + d(x)) . f(x)$. Hence $H^0(E(\mathfrak{A}), \varphi A)$ is

$$\Big\{(f, a) \in \mathrm{Hom}_{\Gamma'}(X, A) \times A^\Gamma : \; (1 + d(x)) . f(x) = a \text{ for every } x \in X\Big\}.$$

This is exactly $H^0(\mathfrak{A}, A)$ as defined by Haran (see (2))! Now the following proposition allows the identification of Haran's cohomology with sheaf cohomology on $E(\mathfrak{A})$:

(B.6) Proposition. — *The exact additive functor* $\varphi: (\Gamma\text{-mod}) \longrightarrow \mathrm{Ab}(E(\mathfrak{A}))$ *sends injective* Γ-*modules to objects which are acyclic for* $H^0(E(\mathfrak{A}), -)$.

(B.6.1) Lemma. — *For any* Γ-*module* A *and* $x \in X$ *the stalk of the* Γ'-*sheaf* $R^n\rho A$ *in* x *is* $H^n(\Gamma_x, A)$ ($n \geq 0$).

To prove the proposition let A be a Γ-module, and let B be the kernel of the trace map $\pi_*\pi^*A \to A$. There is an exact sequence in $\mathrm{Ab}(E(\mathfrak{A}))$

$$0 \longrightarrow i_*\rho B \longrightarrow \varphi A \longrightarrow j_*A \longrightarrow i_*R^1\rho B \longrightarrow 0. \tag{4}$$

Indeed, writing out both components, (4) can be displayed symbolically as

$$
\begin{array}{ccccccccc}
0 & \longrightarrow & \rho B & \longrightarrow & \rho\pi_*\pi^*A & \xrightarrow{\rho(\mathrm{tr})} & \rho A & \longrightarrow & R^1\rho B & \longrightarrow & 0 \\
 & & \downarrow & & \rho(\mathrm{tr})\downarrow & & \mathrm{id}\downarrow & & \downarrow & & \\
0 & \longrightarrow & 0 & \longrightarrow & A & \xrightarrow{\mathrm{id}} & A & \longrightarrow & 0 & \longrightarrow & 0.
\end{array}
$$

The top line is the beginning of the long exact sequence for ρ derived from $0 \to B \to \pi_*\pi^*A \to A \to 0$. It is exact since $\rho\pi_*$ is an exact functor (see (B.5.1)). So (4) is clearly exact. Let now A be injective as a Γ-module. By exactness of $\rho\pi_*$ one has isomorphisms $R^n\rho A \xrightarrow{\sim} R^{n+1}\rho B$ for $n \geq 1$, and since A is injective it follows that $R^n\rho B = 0$ for $n \geq 2$. From (B.6.1) one concludes that also $R^1\rho B = 0$, since cohomology $H^*(\Gamma_x, B)$ is periodic. Hence if A is injective, (4) becomes an exact sequence

$$0 \longrightarrow i_*\rho B \longrightarrow \varphi A \longrightarrow j_*A \longrightarrow 0. \tag{5}$$

Since j_*A is injective in $\mathrm{Ab}(E(\mathfrak{A}))$ and the first term has no cohomology in positive degrees anyway it follows from (5) that $H^n(E(\mathfrak{A}), \varphi A) = 0$ for $n \geq 1$. □

Proof of the lemma. Let $0 \to A \to I^0 \to I^1 \cdots$ be an injective resolution of A, so that $R^n\rho A$ is the n-th cohomology object of the complex $\rho(I^\bullet)$ in $\mathrm{Ab}_{\Gamma'}(X)$. Since the stalk of ρI^n in x is $H^0(\Gamma_x, I^n)$ it follows that the stalk of $R^n\rho A$ in x is the n-th cohomology group of the complex $H^0(\Gamma_x, I^\bullet)$. For every closed subgroup H of Γ the restriction functor res^Γ_H maps injective Γ-modules to acyclic H-modules: This is clear if $[\Gamma : H] < \infty$ (res^Γ_H has an exact left adjoint) and follows in general by passage to the limit. From this the assertion of the lemma follows. □

(B.7) Theorem. — *Let* $\mathfrak{A}$ *be any Artin-Schreier structure. For every* Γ-*module* A *there are canonical isomorphisms*

$$H^n(\mathfrak{A}, A) \cong H^n(E(\mathfrak{A}), \varphi A), \quad n \geq 0.$$

Proof. This has been shown explicitly for $n = 0$ (B.5.4). Since the $H^n(\mathfrak{A}, -)$ are the right derived functors of $H^0(\mathfrak{A}, -)$, the claim follows for all n from exactness of φ and from (B.6). □

(B.8) Thus Haran's cohomology of $\mathfrak{A}$ is identified as cohomology on the glued topos $E(\mathfrak{A})$. From the view point of this topos the functor φ is somewhat unnatural; more canonical choices would be $j_!$ or j_*. The next proposition shows that all three choices give the same cohomology in degrees ≥ 2. For an additional discussion of this point see (B.12) below.

(B.8.1) Proposition. — *Let $\mathfrak{A}$ be an Artin-Schreier structure and A any Γ-module. Then there are natural morphisms $j_!A \to \varphi A \to j_*A$ in $\mathrm{Ab}\big(E(\mathfrak{A})\big)$. The induced maps*

$$H^n\big(E(\mathfrak{A}), j_!A\big) \longrightarrow H^n\big(E(\mathfrak{A}), \varphi A\big) = H^n(\mathfrak{A}, A) \longrightarrow H^n\big(E(\mathfrak{A}), j_*A\big)$$

are surjective for $n = 1$ and are isomorphisms for $n \geq 2$.

Proof. Since $j^*\varphi A = A$ the maps $j_!A \to \varphi A \to j_*A$ arise by adjunction. Both maps have kernel and cokernel in the closed subtopos $\widetilde{X}(\Gamma')$ of $E(\mathfrak{A})$; i.e. there is an exact sequence

$$0 \longrightarrow i_*F \longrightarrow j_!A \longrightarrow \varphi A \longrightarrow i_*F' \longrightarrow 0$$

with $F, F' \in \mathrm{Ab}_{\Gamma'}(X)$, and a similar one for $\varphi A \to j_*A$. The assertion follows from an inspection of the hypercohomology spectral sequences of these complexes. $\square$

(B.9) Now it will be shown how the identification of Artin-Schreier cohomology in Theorem (B.7) leads to the determination of the cohomological dimensions of arbitrary Artin-Schreier structures. Essentially the proof is a corollary to the main result of Sect. 12. As pointed out in the Introduction, the following theorem was proved independently by Dan Haran ([Ha4], see also [Ha3 §6]).

Let $\mathfrak{A} = (\Gamma, \Gamma', X)$ be an Artin-Schreier structure. If ℓ is any prime then $\mathrm{cd}_\ell(\mathfrak{A})$ denotes the largest integer n for which there is an ℓ-primary Γ-module A with $H^n(\mathfrak{A}, A) \neq 0$. If no largest n exists one puts $\mathrm{cd}_\ell(\mathfrak{A}) = \infty$.

(B.9.1) Theorem. — *Let $\mathfrak{A} = (\Gamma, \Gamma', X)$ be an Artin-Schreier structure.*
a) *If ℓ is an odd prime then $\mathrm{cd}_\ell(\mathfrak{A}) = \mathrm{cd}_\ell(\Gamma) = \mathrm{cd}_\ell(\Gamma')$.*
b) *If the forgetful map $d: X \to \Gamma$ is a bijection from X to the space T of all involutions of Γ then $\mathrm{cd}_2(\mathfrak{A}) = \mathrm{cd}_2(\Gamma')$. In all other cases $\mathrm{cd}_2(\mathfrak{A}) = \infty$.*

Proof. Let B be any Γ'-module. By Proposition (B.8.1) and Theorem (B.7) there are natural surjective maps

$$H^n(\mathfrak{A}, \pi_*B) = H^n\big(E(\mathfrak{A}), \varphi\pi_*B\big) \longrightarrow H^n\big(E(\mathfrak{A}), j_*\pi_*B\big) = H^n(\Gamma', B)$$

for $n \geq 1$. The last equality holds since $j_*\pi_*$ is exact. Hence it is clear that $\mathrm{cd}_\ell(\mathfrak{A}) \geq \mathrm{cd}_\ell(\Gamma')$ for all primes ℓ. On the other hand, if A is an *odd* torsion Γ-module,

$R^n j_* A = 0$ for $n \geq 1$ by (B.6.1). Hence for such A one has $H^*\big(E(\mathfrak{A}), j_* A\big) = H^*(\Gamma, A)$, and (B.8.1) gives isomorphisms $H^n(\mathfrak{A}, A) \cong H^n(\Gamma, A)$ for $n \geq 2$. Thus if ℓ is odd, $\mathrm{cd}_\ell(\mathfrak{A}) = \mathrm{cd}_\ell(\Gamma)$ if $\mathrm{cd}_\ell(\Gamma) \geq 1$, and at least $\mathrm{cd}_\ell(\mathfrak{A}) \leq 1$ if $\mathrm{cd}_\ell(\Gamma) = 0$. Actually $\mathrm{cd}_\ell(\mathfrak{A}) = 0$ holds in the latter case, see below.

Now let $\ell = 2$, and assume first that $X = T$ is the Γ-space of all involutions of Γ. Then the hypotheses of (12.13) are satisfied, so this theorem says that

$$ H^n(\Gamma, A) \longrightarrow H^n_\Gamma(T, A) $$

is an isomorphism for $n > d := \mathrm{cd}_2(\Gamma')$ and every 2-primary Γ-module A. The long exact sequence (3) for $\mathfrak{A}$ implies $H^n\big(E(\mathfrak{A}), j_! A\big) = 0$ for $n \geq d+2$. From (B.8.1) it follows that $\mathrm{cd}_2(\mathfrak{A}) \leq d+1$.

Now actually $\mathrm{cd}_2(\mathfrak{A}) \leq d$ (and hence equality) holds. To see this let A be any Γ-module and put $B := \ker\big(\mathrm{tr}\colon \pi_* \pi^* A \to A\big)$. There is a short exact sequence

$$ 0 \longrightarrow j_! B \longrightarrow j_* \pi_* \pi^* A \longrightarrow \varphi A \longrightarrow 0 \tag{6} $$

in $\mathrm{Ab}\big(E(\mathfrak{A})\big)$: Written in components this is

$$
\begin{array}{ccccccccc}
0 & \longrightarrow & 0 & \longrightarrow & \rho \pi_* \pi^* A & \xrightarrow{\ \mathrm{id}\ } & \rho \pi_* \pi^* A & \longrightarrow & 0 \\
 & & \downarrow & & \ \downarrow{\scriptstyle \mathrm{id}} & & \ \downarrow{\scriptstyle \rho(\mathrm{tr})} & & \\
0 & \longrightarrow & B & \longrightarrow & \pi_* \pi^* A & \xrightarrow{\ \mathrm{tr}\ } & A & \longrightarrow & 0.
\end{array}
$$

Now (6) shows for any prime ℓ that $\mathrm{cd}_\ell(\mathfrak{A}) \leq 1 + \mathrm{cd}_\ell(\Gamma')$ actually implies $\mathrm{cd}_\ell(\mathfrak{A}) = \mathrm{cd}_\ell(\Gamma')$. To see why assume that $d := \mathrm{cd}_\ell(\Gamma') < \infty$. Then $\mathrm{cd}_\ell(\mathfrak{A}) \leq d+1$ implies $H^{d+2}\big(E(\mathfrak{A}), j_! A\big) = 0$ using (B.8.1), and so (6) gives a surjection (A any ℓ-primary Γ-module)

$$ H^{d+1}\Big(E(\mathfrak{A}), j_* \pi_* \pi^* A\Big) \longrightarrow\!\!\!\!\!\rightarrow H^{d+1}(\mathfrak{A}, A). \tag{7} $$

But since $j_* \pi_*$ is exact the first group in (7) is $H^{d+1}(\Gamma', \pi^* A)$, which is zero.

To prove the theorem it only remains to show $\mathrm{cd}_2(\mathfrak{A}) = \infty$ if $d\colon X \to T$ is not a bijection. A short proof for this is given in [Ha3 6.7] which uses the restrictions of $\mathfrak{A}$ from Γ to its subgroups of order 2. Since this requires some more definitions I give another proof instead. The idea is that $\mathrm{cd}_2(\mathfrak{A}) < \infty$ would imply $H^n(\Gamma, A) \xrightarrow{\sim} H^n_\Gamma(X, A)$ for $n \gg 0$ and 2-primary Γ-modules A. On the other hand one knows $H^n(\Gamma, A) \xrightarrow{\sim} H^n_\Gamma(T, A)$ for $n \gg 0$, by Sect. 12. The assertion will be proved by showing that d cannot induce isomorphisms in equivariant cohomology for general A and $n \gg 0$, unless it is bijective.

One can assume $\mathrm{cd}_2(\Gamma') < \infty$. Since the map d is Γ-invariant it induces pullback maps

$$ H^n_\Gamma(T, A) \longrightarrow H^n_\Gamma(X, A) \tag{8} $$

$(n \geq 0)$ for every Γ-module A. Consider the commutative square

$$
\begin{array}{ccc}
X & \xrightarrow{\ d\ } & T \\
{\scriptstyle q}\downarrow & & \downarrow{\scriptstyle p} \\
Y & \xrightarrow{\ f\ } & U
\end{array}
$$

in which the vertical maps are the quotients by the Γ-actions and f is induced by d. There are isomorphisms

$$
H_\Gamma^n(T, A) \cong H^0(U, R^n p_*^\Gamma A) \quad \text{and} \quad H_\Gamma^n(X, A) \cong H^0(Y, R^n q_*^\Gamma A),
$$

compare (8.9). Under these isomorphisms (8) is the map induced by f and the canonical sheaf map

$$
f^* R^n p_*^\Gamma A \longrightarrow R^n q_*^\Gamma A \tag{9}
$$

on Y. For $x \in X$ and $t := d(x) \in T$ the stalk of (9) in $q(x) \in Y$ is the restriction map

$$
H^n(C_\Gamma(t), A) \longrightarrow H^n(\langle t \rangle, A) \tag{10}
$$

($C_\Gamma(t)$ is the centralizer of t in Γ). I will show that if $d: X \to T$ is not bijective, (8) fails to be an isomorphism for suitable $(\mathbb{Z}/2)\Gamma$-modules A and arbitrary n. By the long exact sequence (3) this will imply $\mathrm{cd}_2(\mathfrak{A}) = \infty$, since there are factorizations

$$
\begin{array}{ccccccc}
 & & & & H_\Gamma^n(T, A) & & \\
 & & & {\scriptstyle \nu(G)^*}\nearrow & \downarrow{\scriptstyle (8)} & & \\
\cdots \ H^n(\mathfrak{A}, A) & \longrightarrow & H^n(\Gamma, A) & \longrightarrow & H_\Gamma^n(X, A) & \cdots
\end{array}
$$

in which the maps labelled $\nu(G)^*$ are isomorphisms for $n \gg 0$, again by Theorem (12.13). (This theorem is applicable since Γ' has no 2-torsion.)

First suppose that $f: Y \to U$ is not surjective. Since the stalks of $R^n p_*^\Gamma(\mathbb{Z}/2)$ are non-zero everywhere and since U is a boolean space, it is clear that

$$
f^*: H^0(U, R^n p_*^\Gamma(\mathbb{Z}/2)) \longrightarrow H^0(Y, f^* R^n p_*^\Gamma(\mathbb{Z}/2))
$$

cannot be injective. Hence neither can (8) be injective for $A = \mathbb{Z}/2$ and any n. Next suppose that f is not injective. Pick $y \neq y'$ in Y with $f(y) = f(y')$. If (8) were surjective for $A = \mathbb{Z}/2$ there would be $a \in H^0(U, R^n p_*^\Gamma(\mathbb{Z}/2))$ such that the image b of a in $H^0(Y, R^n q_*^\Gamma(\mathbb{Z}/2)) = H^0(Y, \mathbb{Z}/2)$ has $b(y) = 1$ and $b(y') = 0$. But by (10) always $b(y) = b(y')$ holds.

So assume that f is a homeomorphism. Then in particular d is surjective. Now observe the following elementary

Fact: Let X be a boolean space and $\alpha: F \to G$ a homomorphism between sheaves of abelian groups on X. Suppose that $\alpha: H^0(X, F) \to H^0(X, G)$ is bijective. Then α is an isomorphism. $\qquad\square$

Assume that d is not injective. Then $C := C_\Gamma(t) \neq \langle t \rangle$ for some $t \in T$. I claim that one can find a $(\mathbb{Z}/2)\Gamma$-module A such that (10) is not surjective for any $n \geq 0$. This will prove the theorem: For such A the sheaf map (9) will fail to be an isomorphism for any n, and so also (8) will fail to be one, by the fact just mentioned.

To see why the claim is true it suffices to consider the case where Γ is finite. Let M be the permutation $(\mathbb{Z}/2)C$-module over $C/\langle t \rangle$. The image of the restriction map $H^n(C, M) \to H^n(\langle t \rangle, M)$ is contained in the C-invariants of $H^n(\langle t \rangle, M)$, and so this map is never surjective. Let A be the Γ-module induced by M from C. Then M is a direct summand of $\mathrm{res}_C^\Gamma(A)$. So (10) is not surjective, and the theorem is proved. $\square$

(B.9.2) Corollary. — *If k is any field and ℓ is any prime then* $\mathrm{cd}_\ell \mathfrak{A}(k) = \mathrm{cd}_\ell k(\sqrt{-1})$. $\square$

(B.9.3) Observe that this is (almost) exactly the main result of Sect. 7 in the very special case where the scheme X is the spectrum of a field. (Only the small argument has to be added which shows $\mathrm{cd}_\ell \mathfrak{A}(k) \neq 1 + \mathrm{cd}_\ell k(\sqrt{-1}) < \infty$, see the last proof.) So essentially the proof of the corollary can be reduced to the Arason's theorem (9.12). This is not true for Theorem (B.9.1) in general, since the group Γ may lack the multiplicative cohomological properties of an absolute Galois group (see (9.13.2)). To prove (B.9.1) in general one needs projective resolutions by profinite Γ-modules as in Sect. 12. They are also used in Haran's proof [Ha4], which otherwise proceeds within the setup of (generalized) Artin-Schreier structures.

(B.10) Let Γ be a profinite group and T its space of involutions. In [HJ, Prop. 7.7] it is shown that Γ is real projective (cf. (B.2)) if and only if there is an open subgroup Γ' of index ≤ 2 such that $\mathfrak{A} := (\Gamma, \Gamma', T)$ is an Artin-Schreier structure which is projective in the sense of Artin-Schreier structures. As shown in [Ha3, Cor. 3.4], an Artin-Schreier structure $\mathfrak{A}$ is projective iff $\mathrm{cd}_\ell(\mathfrak{A}) \leq 1$ for all primes ℓ. (The reason is that $H^2(\mathfrak{A}, A)$ classifies extensions of $\mathfrak{A}$ by the Γ-module A.) By Theorem (B.9.1) it is equivalent that $\mathrm{cd}_\ell(\Gamma') \leq 1$ for all primes ℓ, which in turn means that Γ' is a projective group [Se4, pp. I-23 and I-74]. In summary this proves (cf. [Ha4, Prop. 2.2])

(B.10.1) Corollary. — *A profinite group Γ is real projective if and only if $C_\Gamma(t) = \{1, t\}$ for every involution $t \in \Gamma$ and Γ contains an open subgroup of index ≤ 2 which is projective.* $\square$

In particular, the absolute Galois group of any field k is real projective if and only if the absolute Galois group of $k(\sqrt{-1})$ is projective.

(B.11) Remark on H^1. Let $\mathfrak{A} = (\Gamma, \Gamma', X)$ be an Artin-Schreier structure and A a Γ-module. An *extension* of $\mathfrak{A}$ by A is a cover $(\alpha, f) \colon \mathfrak{A}' \to \mathfrak{A}$ of Artin-Schreier structures (B.1) with $\ker \alpha = A$ (as a Γ-module). In particular there is the *split* extension of $\mathfrak{A}$ by A, namely $A_0\mathfrak{A} := (\Delta, \Delta', Y)$ with $\Delta = A \rtimes \Gamma$ (semi-direct product), $\Delta' = A \rtimes \Gamma'$ and $Y = A \times X$, with Δ acting on Y by $(a, \sigma) \cdot (b, x) = (a + \sigma.b, \sigma.x)$. There is a natural notion of isomorphism between extensions. Haran proves [Ha3 §3] that the group $H^2(\mathfrak{A}, A)$ stands in natural bijection with the set of isomorphism classes of extensions of $\mathfrak{A}$ by A. The proof uses 2-cocycles to describe such extensions, as in the usual group-theoretic case. So by analogy with group cohomology one would hope that $H^1(\mathfrak{A}, A)$ classifies the splittings in the split extension $A_0\mathfrak{A}$, up to inner conjugacy. But this turns out not to be true. However if one uses the approach via the glued topos $E(\mathfrak{A})$ as sketched in this section, one is well off since the cohomology group $H^1(E(\mathfrak{A}), j_!A)$ turns out to do the job! Note that $H^2(\mathfrak{A}, A)$ is isomorphic to $H^2(E(\mathfrak{A}), j_!A)$ (B.8.1), so one gets a unified treatment of these "classical" interpretations of low-dimensional cohomology by using $j_!A$ instead of φA.

For completeness here is a brief sketch of proof for the above assertion. If $X = \emptyset$ it reduces to the classical description of $H^1(\Gamma, A)$, so assume $X \neq \emptyset$. Let $Z^1(\mathfrak{A}, A)$ denote the group of 1-cocycles in Haran's complex. So $Z^1(\mathfrak{A}, A)$ consists of all pairs (f, s) where $f: \Gamma \to A$ is a (usual) 1-cocycle and $s: X \to A$ is a continuous map, such that $\sigma s(x) - s(\sigma x) + f(\sigma) = 0$ for all $(\sigma, x) \in \Gamma \times X$. One checks that via $(f, s) \leftrightarrow s$ this group is isomorphic to

$$Z := \left\{ s \in C(X, A): \ \forall \sigma \in \Gamma \text{ the map } X \to A, \ x \mapsto s(\sigma x) - \sigma s(x) \text{ is constant} \right\}.$$

The splittings of the extension $A_0\mathfrak{A} \to \mathfrak{A}$ correspond bijectively to the elements $(f, s) \in Z^1(\mathfrak{A}, A)$ [Ha3, p. 148]. Two such splittings are conjugate under $A \rtimes \Gamma$ if and only if their difference has the form $\left(\sigma \mapsto (\sigma - 1)a, \ x \mapsto a \right) \in Z^1(\mathfrak{A}, A)$ for some $a \in A$. So the group $K^1(\mathfrak{A}, A)$ of splittings mod conjugation fits into a short exact sequence

$$0 \longrightarrow A \overset{i}{\longrightarrow} Z \longrightarrow K^1(\mathfrak{A}, A) \longrightarrow 0 \qquad (11)$$

where i is the map which sends $a \in A$ to the constant map $a: X \to A$.

On the other hand let $P(A) = C(X, A)/A$ be the Γ-module which is the cokernel of the injective map $A \to C(X, A)$. Then also $P(A)^\Gamma$ is canonically isomorphic to the cokernel of i in (11). As in Remark (6.7) one shows that there are natural isomorphisms $H^n(\Gamma, P(A)) \cong H^{n+1}(E(\mathfrak{A}), j_!A)$ for all n. So in particular $K^1(\mathfrak{A}, A) \cong P(A)^\Gamma \cong H^1(E(\mathfrak{A}), j_!A)$.

(B.12) Remarks. Let $\mathfrak{A} = (\Gamma, \Gamma', X)$ be an Artin-Schreier structure. As already remarked, the definition of $H^0(\mathfrak{A}, A)$ is little enlightening. And looking at the definition of $H^0(\mathfrak{A}, A) = H^0(E(\mathfrak{A}), \varphi A)$ from the view point of the topos $E(\mathfrak{A})$, also the choice of φA has an artificial flavor. More natural choices would be j_*A or $j_!A$. But each of them has both its advantages and its shortcomings. Observe that

in general neither $H^n(E(\mathfrak{A}), j_*A)$ nor $H^n(E(\mathfrak{A}), j_!A)$ is the n-th derived functor
of the corresponding H^0, as functors in A. The sheaf j_*A has the advantage that
it leads to well-behaved corestriction mappings in cohomology. This is neither true
for Haran's definition (see the discussion in [Ha3 §4]) nor for a definition which uses
$j_!A$. On the other hand the sheaf $j_!A$ is the only one which yields the "correct"
H^1: Indeed, $H^1(E(\mathfrak{A}), j_!A)$ is canonically isomorphic to the group of splittings
mod conjugation in the split extension $A_0\mathfrak{A}$ (Remark (B.11)), while in general
$H^1(E(\mathfrak{A}), \varphi A) = H^1(\mathfrak{A}, A)$ and $H^1(E(\mathfrak{A}), j_*A)$ are not.

But in any case, the question which choice to make affects only cohomology in
degrees 0 and 1. Indeed, if F is any abelian group object in $E(\mathfrak{A})$ with $j^*F = A$,
an analogue of (B.8.1) shows that canonically

$$H^n(E(\mathfrak{A}), j_!A) \cong H^n(E(\mathfrak{A}), F) \cong H^n(E(\mathfrak{A}), j_*A), \quad n \geq 2,$$

and that moreover the two maps $H^1(E(\mathfrak{A}), j_!A) \to H^1(E(\mathfrak{A}), F) \to H^1(E(\mathfrak{A}), j_*A)$
are both surjective. In particular, $j_!A$ (resp. j_*A) yields the maximal (resp. the
minimal) H^1 among those F.

Perhaps the conclusion from all this could be that one should not try to single
out one of the possible choices for the definition of a cohomology $H^*(\mathfrak{A}, A)$ and
forget about the others. This would mean deliberately depriving oneself of the
technical advantages of these other functors, and giving up much flexibility without
getting anything in return. Rather, each of φA, $j_!A$, j_*A may be the proper
choice, depending on the situation one wants to handle. Hence one should work
with the glued topos $E(\mathfrak{A})$ as a whole, and with the various direct and inverse
image functors between these toposes (associated with morphisms of Artin-Schreier
structures), bearing in mind the above remarks.

References

[AR]   M. E. Alonso, M.F. Roy: Real strict localizations. *Math. Z.* **194**, 429-441 (1987).

[ABR]   C. Andradas, L. Bröcker, J. M. Ruiz: Minimal generation of basic open semianalytic sets. *Invent. math.* **92**, 409-430 (1988).

[Ar]   J. Kr. Arason: Primideale im graduierten Wittring und im mod 2 Cohomologiering. *Math. Z.* **145**, 139-143 (1975).

[AEJ]   J. Kr. Arason, R. Elman, B. Jacob: The graded Witt ring and Galois cohomology, I. In: Quadratic and Hermitian Forms, Conf. Hamilton, Ontario, 1983, *Canad. Math. Soc. Conf. Proc.* **4** (1984), pp. 17-50.

[AM]   M. Artin, B. Mazur: Etale Homotopy. *Lect. Notes Math.* **100**, Springer, Berlin Heidelberg New York, 1969.

[Be1]   E. Becker: Hereditarily-pythagorean fields and orderings of higher level. *IMPA Monografias de Matemática* **29**, Rio de Janeiro, 1978.

[Be2]   E. Becker: Summen *n*-ter Potenzen in Körpern. *J. reine angew. Math.* **307/ 308**, 8-30 (1979).

[BnI]   D. J. Benson: Representations and Cohomology. I: Basic Representation Theory of Finite Groups and Associative Algebras. Cambridge University Press, Cambridge, 1991.

[BnII] D. J. Benson: Representations and Cohomology. II: Cohomology of Groups and Modules. Cambridge University Press, Cambridge, 1991.

[BO]   S. Bloch, A. Ogus: Gersten's conjecture and the homology of schemes. *Ann. scient. Éc. Norm. Sup. (4)* **7**, 181-202 (1974).

[BLR]   S. Bosch, W. Lütkebohmert, M. Raynaud: Néron Models. Ergebnisse der Mathematik und ihrer Grenzgebiete, 3. Folge, Band 21, Springer, Berlin Heidelberg New York, 1990.

[BCR]   J. Bochnak, M. Coste, M.-F. Roy: Géométrie Algébrique Réelle. Ergebnisse der Mathematik und ihrer Grenzgebiete, 3. Folge, Band 12, Springer, Berlin Heidelberg New York, 1987.

[Bo]   N. Bourbaki: Éléments de Mathématique. Algèbre: Chap. 4 à 7, Masson, Paris 1981; Chap. 10, Masson, Paris, 1980.

[Bd1]   G. E. Bredon: Sheaf Theory. McGraw-Hill, New York St. Louis etc., 1967.

[Bd2]   G. E. Bredon: Introduction to Compact Transformation Groups. Academic Press, New York London, 1972.

[Br]   K. S. Brown: Cohomology of Groups. Graduate Texts in Mathematics **87**, Springer, New York Heidelberg Berlin, 1982.

[Bm]   A. Brumer: Pseudocompact algebras, profinite groups and class formations. *J. Algebra* **4**, 442-470 (1966).

[Bu1]   J. Burési: Cohomologie étale et algèbre réelle. Thèse, Université de Rennes I, 1993.

[Bu2]   J. Burési: Local-global principle for étale cohomology. Preprint.

[CaC]   M. Carral, M. Coste: Normal spectral spaces and their dimensions.
           J. Pure Applied Algebra **30**, 227-235 (1983).

[CE]   H. Cartan, S. Eilenberg: Homological Algebra. Princeton University
           Press, Princeton, N.J., 1956.

[Ch]   L. G. Chouinard: Projectivity and relative projectivity over group rings.
           J. Pure Applied Algebra **7**, 287-302 (1976).

[CTP]   J.-L. Colliot-Thélène, R. Parimala: Real components of algebraic vari-
           eties and étale cohomology. *Invent. math.* **101**, 81-99 (1990).

[CTS]   J.-L. Colliot-Thélène, C. Scheiderer: Zero cycles and cohomology on real
           algebraic varieties. Preprint 1993.

[CC]   M.-F. Coste, M. Coste: Topologies for real algebraic geometry. In: Topos
           Theoretic Methods in Geometry, ed. A. Kock, Various Publications Se-
           ries no. 30, Aarhus University 1979, pp. 37-100.

[CR]   M. Coste, M.-F. Roy: La topologie du spectre réel. In: Proc. Special Ses-
           sion on Ordered Fields and Real Algebraic Geometry, ed. D. W. Dubois,
           T. Recio, *Contemp. Math.* **8** (1982), pp. 27-59.

[CRC]   M.-F. Coste-Roy, M. Coste: Le spectre étale réel d'un anneau est spatial.
           C. R. Acad. Sci. Paris **290**, ser. A, 91-94 (1980).

[Co]   D. A. Cox: The étale homotopy type of varieties over $\mathbb{R}$. *Proc. Am.
           Math. Soc.* **76**, 17-22 (1979).

[Cu]   F. Cucker: Sur les anneaux de sections globales du faisceau structural
           sur le spectre réel. *Comm. Algebra* **16**, 307-323 (1988).

[Df]   H. Delfs: Homology of Locally Semialgebraic Spaces. *Lect. Notes Math.*
           1484, Springer, Berlin Heidelberg New York, 1991.

[DK1]   H. Delfs, M. Knebusch: Semialgebraic topology over a real closed field
           II: Basic theory of semialgebraic spaces. *Math. Z.* **178**, 175-213 (1981).

[DK2]   H. Delfs, M. Knebusch: On the homology of algebraic varieties over real
           closed fields. *J. reine angew. Math.* **335**, 122-163 (1981).

[tD]   T. tom Dieck: Transformation Groups. De Gruyter, Berlin New York,
           1987.

[En]   O. Endler: Valuation Theory. Springer, Berlin Heidelberg New York,
           1972.

[Er]   Yu. L. Ershov: Galois groups of maximal 2-extensions. (English trans-
           lation.) *Math. Notes* **36**, 956-961 (1985).

[Ga]   J. M. Gamboa: Un exemple d'ensemble constructible à adhérence non
           constructible. *C. R. Acad. Sci. Paris* **306**, sér. I, 617-619 (1988).

[Ge1]   W.-D. Geyer: Ein algebraischer Beweis des Satzes von Weichold über
           reelle algebraische Funktionenkörper. In: Algebraische Zahlentheorie,
           Bericht einer Tagung des Math. Forschungsinstitutes Oberwolfach, hrsg.
           H. Hasse und P. Roquette, Mannheim 1966, pp. 83-98.

[Ge2]   W.-D. Geyer: Dualität bei abelschen Varietäten über reell abgeschlosse-
           nen Körpern. *J. reine angew. Math.* **293/294**, 62-66 (1977).

[Go]   R. Godement: Topologie Algébrique et Théorie des Faisceaux. Her-
           mann, Paris, 1964.

[Gr1]   A. Grothendieck: Sur quelques points d'algèbre homologique. *Tôhoku
           Math. J.* **9**, 119-221 (1957).

[Gr2]   A. Grothendieck: Local Cohomology. *Lect. Notes Math.* **41**, Springer,
           Berlin Heidelberg New York, 1967.

[Gr3]   A. Grothendieck: Le groupe de Brauer I–III. In: Dix Exposés sur la Cohomologie des Schémas, ed. A. Grothendieck, N. H. Kuiper, North Holland, Amsterdam and Masson, Paris, 1968.

[Ha1]   D. Haran: Closed subgroups of $G(\mathbb{Q})$ with involutions. *J. Algebra* **129**, 393-411 (1990).

[Ha2]   D. Haran: A proof of Serre's theorem. *J. Indian Math. Soc.* **55**, 213-234 (1990).

[Ha3]   D. Haran: Cohomology theory of Artin-Schreier structures. *J. Pure Applied Algebra* **69**, 141-160 (1990).

[Ha4]   D. Haran: On the cohomological dimension of Artin-Schreier structures. *J. Algebra* **156**, 219-236 (1993).

[HJ]   D. Haran, M. Jarden: The absolute Galois group of a pseudo real closed field. *Ann. Sc. Norm. Sup. Pisa (4)* **12**, 449-489 (1985).

[Ht]   R. Hartshorne: Residues and Duality. *Lect. Notes Math.* **20**, Springer, Berlin Heidelberg New York, 1966.

[HiSt] P. J. Hilton, U. Stammbach: A Course in Homological Algebra. Springer, New York Heidelberg Berlin, 1971.

[Hb]   J. H. Hubbard: On the cohomology of Nash sheaves. *Topology* **11**, 265-270 (1972).

[Hu1]   R. Huber: Isoalgebraische Räume. Dissertation, Universität Regensburg, 1984.

[Hu2]   R. Huber: Spezialisierungen im reellen Spektrum zu einer komplexen Varietät. *Geom. Dedicata* **38**, 309-327 (1991).

[HS]   R. Huber, C. Scheiderer: A relative notion of local completeness in semialgebraic geometry. *Arch. Math.* **53**, 571-584 (1989).

[Ja]   B. Jacob: On the structure of pythagorean fields. *J. Algebra* **68**, 247-267 (1981).

[Kl]   F. Klein: Über Riemanns Theorie der algebraischen Funktionen und ihrer Integrale. Eine Ergänzung der gewöhnlichen Darstellungen. Teubner, Leipzig, 1882. In: Gesammelte Abhandlungen III, pp. 499-573.

[KS]   M. Knebusch, C. Scheiderer: Einführung in die reelle Algebra. Vieweg, Wiesbaden, 1989.

[Kn]   M.-A. Knus: Quadratic and Hermitian Forms over Rings. Grundlehren der mathematischen Wissenschaften in Einzeldarstellungen, Band 294, Springer, Berlin Heidelberg New York, 1991.

[LW]   A. T. Lundell, S. Weingram: The Topology of CW Complexes. Van Nostrand Reinhold, New York Cincinnati etc., 1969.

[Mh]   L. Mahé: On the separation of connected components by étale cohomology. Preprint.

[Ma1]   M. Marshall: Some local-global principles for formally real fields. *Canad. J. Math.* **29**, 606-614 (1977).

[Ma2]   M. Marshall: Quotients and inverse limits of spaces of orderings. *Canad. J. Math.* **31**, 604-616 (1979).

[Mi]   J. S. Milne: Étale Cohomology. Princeton University Press, Princeton, N.J., 1980.

[Pa]   D. S. Passman: The Algebraic Structure of Group Rings. Wiley, New York, 1977.

[Pe]   A. R. Pears: Dimension Theory of General Spaces. Cambridge University Press, Cambridge, 1975.

[Pu]   M. J. de la Puente: Riemann surfaces of a ring and compactifications of semi-algebraic sets. Dissertation, Stanford University, 1988.

[R]   M.-F. Roy: Faisceau structural sur le spectre réel et fonctions de Nash. In: Géométrie Algébrique Réelle et Formes Quadratiques. Proc. Rennes 1981, ed. J.-L. Colliot-Thélène, M. Coste, L. Mahé, M.-F. Roy, *Lect. Notes Math.* **959**, Springer, Berlin Heidelberg New York 1982, pp. 406-432.

[Sch1]   C. Scheiderer: Quasi-augmented simplicial spaces, with an application to cohomological dimension. *J. Pure Applied Algebra* **81**, 293-311 (1992).

[Sch2]   C. Scheiderer: Farrell cohomology and Brown theorems for profinite groups. In preparation.

[Schw]   N. Schwartz: The basic theory of real closed spaces. *Mem. Am. Math. Soc.* **397**, Providence, R. I., 1989.

[Sg]   G. Segal: Classifying spaces and spectral sequences. *Inst. Hautes Études Sci. Publ. Math.* **34**, 105-112 (1968).

[Se1]   J.-P. Serre: Cohomologie modulo 2 des complexes d'Eilenberg-MacLane. *Comm. Math. Helv.* **27**, 198-232 (1953).

[Se2]   J.-P. Serre: Sur la dimension cohomologique des groupes profinis. *Topology* **3**, 413-420 (1965).

[Se3]   J.-P. Serre: Cohomologie des groupes discrets. In: *Ann. Math. Studies* **70**, Princeton University Press, Princeton, N.J., 1971, pp. 77-169.

[Se4]   J.-P. Serre: Cohomologie Galoisienne. *Lect. Notes Math.* **5**, 4ème ed., Springer, Berlin Heidelberg New York, 1973.

[Sw]   R. G. Swan: Groups of cohomological dimension one. *J. Algebra* **12**, 585-601 (1969).

[Wi]   E. Witt: Zerlegung reeller algebraischer Funktionen in Quadrate, Schiefkörper über reellem Funktionenkörper. *J. reine angew. Math.* **171**, 4-11 (1934).

The following standard references of algebraic geometry are cited in the traditional way:

[EGA I*] A. Grothendieck, J. A. Dieudonné: Eléments de Géométrie Algébrique I. Grundlehren der mathematischen Wissenschaften in Einzeldarstellungen, Band 166, Springer, Berlin Heidelberg New York, 1971.

[EGA IV] A. Grothendieck: Éléments de Géométrie Algébrique. *Inst. Hautes Études Sci. Publ. Math.* **20**, **24**, **28**, **32**, 1964–1967.

[FGA]   A. Grothendieck: Fondements de la Géométrie Algébrique. Extraits du Séminaire Bourbaki 1957-1962. Sécretariat mathématique, Paris, 1962.

[SGA1]   A. Grothendieck: Revêtements Étales et Groupe Fondamental (SGA 1). *Lect. Notes Math.* **224**, Springer, Berlin Heidelberg New York, 1971.

[SGA4]   M. Artin, A. Grothendieck, J. L. Verdier: Théorie des Topos et Cohomologie Étale des Schemas (SGA 4). *Lect. Notes Math.* **269**, **270**, **305**, Springer, Berlin Heidelberg New York, 1972–1973.

Symbol Index

(Sch), (Sch/X)	xxi	category of schemes, resp. of schemes over X
Et/X	xxi	category of étale X-schemes
X_{et}, $\widetilde{X}_{et}$	xxi	étale site, étale topos of X
$\mathbb{Z}_\ell$, $\mathbb{Q}_\ell$	xxi	ring of ℓ-adic integers, field of ℓ-adic numbers
sper A	xxi	real spectrum of A
$k(\xi)$	xxi	real closed field at ξ
supp, φ	xxi	support map
X_r	xxi	real spectrum of X
f_r	xxii	map between real spectra induced by f
T^G	xxii	fixpoints of G-action on T
T/G	xxii	quotient space of T mod G
G_x	xxii	isotropy group of x in G
$C_G(H)$	xxii	centralizer of H in G
$N_G(H)$	xxii	normalizer of H in G
(G-sets)	xxii	category of discrete (left) G-spaces
(G-mod)	xxii	category of discrete (left) G-modules
$H^n(G, -)$	xxii	group cohomology of G
res_H^G	xxii	restriction from G to H
coind_H^G	xxii	coinduction from H to G
ind_H^G	xxiii	induction from H to G
$\succ$	xxiii	specialization
$C(T, S)$	xxiii	set of continuous maps $T \to S$
$\widetilde{T}$	xxiii	category of set-valued sheaves on T
Ab(T)	xxiii	category of abelian sheaves on T
A_x	xxiii	stalk of A at x
s_x	xxiii	value of section s at x
ret	1	real étale topology
X_{ret}, $\widetilde{X}_{ret}$	1	real étale site, real étale topos
$G^\sharp$, $F^\flat$	6	($\sharp$, $\flat$ are equivalences between $\widetilde{X}_{ret}$ and $\widetilde{X}_r$)
$P^\dagger$	7	(presheaf on X_r associated to presheaf P on Et/X)
$f_{ret} = (f_{ret}^*, f_{ret*})$	8	topos morphism induced by f
b	11	intersection of the topologies et and ret on Et/X
X_b	11	the site (Et/X, b)
$j = (j^*, j_*)$	11	open embedding $\widetilde{X}_{et} \to \widetilde{X}_b$
$i = (i^*, i_*)$	11	closed embedding $\widetilde{X}_{ret} \to \widetilde{X}_b$
ρ	12	glueing functor $\widetilde{X}_{et} \to \widetilde{X}_{ret}$
$j_!$	12	empty extension, resp. extension by zero
$i^!$	12	right adjoint of i_*
a_{et}, a_b, a_{ret}	14	functor "associated t-sheaf", for $t = et$, b, ret
ϵ_{et}, ϵ_b, ϵ_{ret}	14	functor Et/$X \to \widetilde{X}_t$, for $t = et$, b, ret
$\underline{M}_{et}$, $\underline{M}_b$, $\underline{M}_{ret}$	15	constant sheaf M on X_{et}, X_b, X_{ret}

Subject Index

Springer-Verlag
and the Environment

We at Springer-Verlag firmly believe that an international science publisher has a special obligation to the environment, and our corporate policies consistently reflect this conviction.

We also expect our business partners – paper mills, printers, packaging manufacturers, etc. – to commit themselves to using environmentally friendly materials and production processes.

The paper in this book is made from low- or no-chlorine pulp and is acid free, in conformance with international standards for paper permanency.

Lecture Notes in Mathematics

For information about Vols. 1–1411
please contact your bookseller or Springer-Verlag

Montecatini Terme 1988. Seminar. Editors: M. Francaviglia, F. Gherardelli. IX, 197 pages. 1990.

Vol. 1452: E. Hlawka, R.F. Tichy (Eds.), Number-Theoretic Analysis. Seminar, 1988–89. V, 220 pages. 1990.

Vol. 1453: Yu.G. Borisovich, Yu.E. Gliklikh (Eds.), Global Analysis – Studies and Applications IV. V, 320 pages. 1990.

Vol. 1454: F. Baldassari, S. Bosch, B. Dwork (Eds.), p-adic Analysis. Proceedings, 1989. V, 382 pages. 1990.

Vol. 1455: J.-P. Françoise, R. Roussarie (Eds.), Bifurcations of Planar Vector Fields. Proceedings, 1989. VI, 396 pages. 1990.

Vol. 1456: L.G. Kovács (Ed.), Groups – Canberra 1989. Proceedings. XII, 198 pages. 1990.

Vol. 1457: O. Axelsson, L.Yu. Kolotilina (Eds.), Preconditioned Conjugate Gradient Methods. Proceedings, 1989. V, 196 pages. 1990.

Vol. 1458: R. Schaaf, Global Solution Branches of Two Point Boundary Value Problems. XIX, 141 pages. 1990.

Vol. 1459: D. Tiba, Optimal Control of Nonsmooth Distributed Parameter Systems. VII, 159 pages. 1990.

Vol. 1460: G. Toscani, V. Boffi, S. Rionero (Eds.), Mathematical Aspects of Fluid Plasma Dynamics. Proceedings, 1988. V, 221 pages. 1991.

Vol. 1461: R. Gorenflo, S. Vessella, Abel Integral Equations. VII, 215 pages. 1991.

Vol. 1462: D. Mond, J. Montaldi (Eds.), Singularity Theory and its Applications. Warwick 1989, Part I. VIII, 405 pages. 1991.

Vol. 1463: R. Roberts, I. Stewart (Eds.), Singularity Theory and its Applications. Warwick 1989, Part II. VIII, 322 pages. 1991.

Vol. 1464: D. L. Burkholder, E. Pardoux, A. Sznitman, Ecole d'Eté de Probabilités de Saint- Flour XIX-1989. Editor: P. L. Hennequin. VI, 256 pages. 1991.

Vol. 1465: G. David, Wavelets and Singular Integrals on Curves and Surfaces. X, 107 pages. 1991.

Vol. 1466: W. Banaszczyk, Additive Subgroups of Topological Vector Spaces. VII, 178 pages. 1991.

Vol. 1467: W. M. Schmidt, Diophantine Approximations and Diophantine Equations. VIII, 217 pages. 1991.

Vol. 1468: J. Noguchi, T. Ohsawa (Eds.), Prospects in Complex Geometry. Proceedings, 1989. VII, 421 pages. 1991.

Vol. 1469: J. Lindenstrauss, V. D. Milman (Eds.), Geometric Aspects of Functional Analysis. Seminar 1989-90. XI, 191 pages. 1991.

Vol. 1470: E. Odell, H. Rosenthal (Eds.), Functional Analysis. Proceedings, 1987-89. VII, 199 pages. 1991.

Vol. 1471: A. A. Panchishkin, Non-Archimedean L-Functions of Siegel and Hilbert Modular Forms. VII, 157 pages. 1991.

Vol. 1472: T. T. Nielsen, Bose Algebras: The Complex and Real Wave Representations. V, 132 pages. 1991.

Vol. 1473: Y. Hino, S. Murakami, T. Naito, Functional Differential Equations with Infinite Delay. X, 317 pages. 1991.

Vol. 1474: S. Jackowski, B. Oliver, K. Pawałowski (Eds.), Algebraic Topology, Poznań 1989. Proceedings. VIII, 397 pages. 1991.

Vol. 1475: S. Busenberg, M. Martelli (Eds.), Delay Differential Equations and Dynamical Systems. Proceedings, 1990. VIII, 249 pages. 1991.

Vol. 1476: M. Bekkali, Topics in Set Theory. VII, 120 pages. 1991.

Vol. 1477: R. Jajte, Strong Limit Theorems in Noncommutative L_2-Spaces. X, 113 pages. 1991.

Vol. 1478: M.-P. Malliavin (Ed.), Topics in Invariant Theory. Seminar 1989-1990. VI, 272 pages. 1991.

Vol. 1479: S. Bloch, I. Dolgachev, W. Fulton (Eds.), Algebraic Geometry. Proceedings, 1989. VII, 300 pages. 1991.

Vol. 1480: F. Dumortier, R. Roussarie, J. Sotomayor, H. Żołądek, Bifurcations of Planar Vector Fields: Nilpotent Singularities and Abelian Integrals. VIII, 226 pages. 1991.

Vol. 1481: D. Ferus, U. Pinkall, U. Simon, B. Wegner (Eds.), Global Differential Geometry and Global Analysis. Proceedings, 1991. VIII, 283 pages. 1991.

Vol. 1482: J. Chabrowski, The Dirichlet Problem with L^2-Boundary Data for Elliptic Linear Equations. VI, 173 pages. 1991.

Vol. 1483: E. Reithmeier, Periodic Solutions of Nonlinear Dynamical Systems. VI, 171 pages. 1991.

Vol. 1484: H. Delfs, Homology of Locally Semialgebraic Spaces. IX, 136 pages. 1991.

Vol. 1485: J. Azéma, P. A. Meyer, M. Yor (Eds.), Séminaire de Probabilités XXV. VIII, 440 pages. 1991.

Vol. 1486: L. Arnold, H. Crauel, J.-P. Eckmann (Eds.), Lyapunov Exponents. Proceedings, 1990. VIII, 365 pages. 1991.

Vol. 1487: E. Freitag, Singular Modular Forms and Theta Relations. VI, 172 pages. 1991.

Vol. 1488: A. Carboni, M. C. Pedicchio, G. Rosolini (Eds.), Category Theory. Proceedings, 1990. VII, 494 pages. 1991.

Vol. 1489: A. Mielke, Hamiltonian and Lagrangian Flows on Center Manifolds. X, 140 pages. 1991.

Vol. 1490: K. Metsch, Linear Spaces with Few Lines. XIII, 196 pages. 1991.

Vol. 1491: E. Lluis-Puebla, J.-L. Loday, H. Gillet, C. Soulé, V. Snaith, Higher Algebraic K-Theory: an overview. IX, 164 pages. 1992.

Vol. 1492: K. R. Wicks, Fractals and Hyperspaces. VIII, 168 pages. 1991.

Vol. 1493: E. Benoît (Ed.), Dynamic Bifurcations. Proceedings, Luminy 1990. VII, 219 pages. 1991.

Vol. 1494: M.-T. Cheng, X.-W. Zhou, D.-G. Deng (Eds.), Harmonic Analysis. Proceedings, 1988. IX, 226 pages. 1991.

Vol. 1495: J. M. Bony, G. Grubb, L. Hörmander, H. Komatsu, J. Sjöstrand, Microlocal Analysis and Applications. Montecatini Terme, 1989. Editors: L. Cattabriga, L. Rodino. VII, 349 pages. 1991.

Vol. 1496: C. Foias, B. Francis, J. W. Helton, H. Kwakernaak, J. B. Pearson, H_∞-Control Theory. Como, 1990. Editors: E. Mosca, L. Pandolfi. VII, 336 pages. 1991.

Vol. 1497: G. T. Herman, A. K. Louis, F. Natterer (Eds.), Mathematical Methods in Tomography. Proceedings 1990. X, 268 pages. 1991.

Vol. 1498: R. Lang, Spectral Theory of Random Schrödinger Operators. X, 125 pages. 1991.

Vol. 1499: K. Taira, Boundary Value Problems and Markov Processes. IX, 132 pages. 1991.

Vol. 1500: J.-P. Serre, Lie Algebras and Lie Groups. VII, 168 pages. 1992.

Vol. 1501: A. De Masi, E. Presutti, Mathematical Methods for Hydrodynamic Limits. IX, 196 pages. 1991.

Vol. 1502: C. Simpson, Asymptotic Behavior of Monodromy. V, 139 pages. 1991.

Vol. 1503: S. Shokranian, The Selberg-Arthur Trace Formula (Lectures by J. Arthur). VII, 97 pages. 1991.

Vol. 1504: J. Cheeger, M. Gromov, C. Okonek, P. Pansu, Geometric Topology: Recent Developments. Editors: P. de Bartolomeis, F. Tricerri. VII, 197 pages. 1991.

Vol. 1505: K. Kajitani, T. Nishitani, The Hyperbolic Cauchy Problem. VII, 168 pages. 1991.

Vol. 1506: A. Buium, Differential Algebraic Groups of Finite Dimension. XV, 145 pages. 1992.

Vol. 1507: K. Hulek, T. Peternell, M. Schneider, F.-O. Schreyer (Eds.), Complex Algebraic Varieties. Proceedings, 1990. VII, 179 pages. 1992.

Vol. 1508: M. Vuorinen (Ed.), Quasiconformal Space Mappings. A Collection of Surveys 1960-1990. IX, 148 pages. 1992.

Vol. 1509: J. Aguadé, M. Castellet, F. R. Cohen (Eds.), Algebraic Topology - Homotopy and Group Cohomology. Proceedings, 1990. X, 330 pages. 1992.

Vol. 1510: P. P. Kulish (Ed.), Quantum Groups. Proceedings, 1990. XII, 398 pages. 1992.

Vol. 1511: B. S. Yadav, D. Singh (Eds.), Functional Analysis and Operator Theory. Proceedings, 1990. VIII, 223 pages. 1992.

Vol. 1512: L. M. Adleman, M.-D. A. Huang, Primality Testing and Abelian Varieties Over Finite Fields. VII, 142 pages. 1992.

Vol. 1513: L. S. Block, W. A. Coppel, Dynamics in One Dimension. VIII, 249 pages. 1992.

Vol. 1514: U. Krengel, K. Richter, V. Warstat (Eds.), Ergodic Theory and Related Topics III, Proceedings, 1990. VIII, 236 pages. 1992.

Vol. 1515: E. Ballico, F. Catanese, C. Ciliberto (Eds.), Classification of Irregular Varieties. Proceedings, 1990. VII, 149 pages. 1992.

Vol. 1516: R. A. Lorentz, Multivariate Birkhoff Interpolation. IX, 192 pages. 1992.

Vol. 1517: K. Keimel, W. Roth, Ordered Cones and Approximation. VI, 134 pages. 1992.

Vol. 1518: H. Stichtenoth, M. A. Tsfasman (Eds.), Coding Theory and Algebraic Geometry. Proceedings, 1991. VIII, 223 pages. 1992.

Vol. 1519: M. W. Short, The Primitive Soluble Permutation Groups of Degree less than 256. IX, 145 pages. 1992.

Vol. 1520: Yu. G. Borisovich, Yu. E. Gliklikh (Eds.), Global Analysis - Studies and Applications V. VII, 284 pages. 1992.

Vol. 1521: S. Busenberg, B. Forte, H. K. Kuiken, Mathematical Modelling of Industrial Process. Bari, 1990. Editors: V. Capasso, A. Fasano. VII, 162 pages. 1992.

Vol. 1522: J.-M. Delort, F. B. I. Transformation. VII, 101 pages. 1992.

Vol. 1523: W. Xue, Rings with Morita Duality. X, 168 pages. 1992.

Vol. 1524: M. Coste, L. Mahé, M.-F. Roy (Eds.), Real Algebraic Geometry. Proceedings, 1991. VIII, 418 pages. 1992.

Vol. 1525: C. Casacuberta, M. Castellet (Eds.), Mathematical Research Today and Tomorrow. VII, 112 pages. 1992.

Vol. 1526: J. Azéma, P. A. Meyer, M. Yor (Eds.), Séminaire de Probabilités XXVI. X, 633 pages. 1992.

Vol. 1527: M. I. Freidlin, J.-F. Le Gall, Ecole d'Eté de Probabilités de Saint-Flour XX - 1990. Editor: P. L. Hennequin. VIII, 244 pages. 1992.

Vol. 1528: G. Isac, Complementarity Problems. VI, 297 pages. 1992.

Vol. 1529: J. van Neerven, The Adjoint of a Semigroup of Linear Operators. X, 195 pages. 1992.

Vol. 1530: J. G. Heywood, K. Masuda, R. Rautmann, S. A. Solonnikov (Eds.), The Navier-Stokes Equations II - Theory and Numerical Methods. IX, 322 pages. 1992.

Vol. 1531: M. Stoer, Design of Survivable Networks. IV, 206 pages. 1992.

Vol. 1532: J. F. Colombeau, Multiplication of Distributions. X, 184 pages. 1992.

Vol. 1533: P. Jipsen, H. Rose, Varieties of Lattices. X, 162 pages. 1992.

Vol. 1534: C. Greither, Cyclic Galois Extensions of Commutative Rings. X, 145 pages. 1992.

Vol. 1535: A. B. Evans, Orthomorphism Graphs of Groups. VIII, 114 pages. 1992.

Vol. 1536: M. K. Kwong, A. Zettl, Norm Inequalities for Derivatives and Differences. VII, 150 pages. 1992.

Vol. 1537: P. Fitzpatrick, M. Martelli, J. Mawhin, R. Nussbaum, Topological Methods for Ordinary Differential Equations. Montecatini Terme, 1991. Editors: M. Furi, P. Zecca. VII, 218 pages. 1993.

Vol. 1538: P.-A. Meyer, Quantum Probability for Probabilists. X, 287 pages. 1993.

Vol. 1539: M. Coornaert, A. Papadopoulos, Symbolic Dynamics and Hyperbolic Groups. VIII, 138 pages. 1993.

Vol. 1540: H. Komatsu (Ed.), Functional Analysis and Related Topics, 1991. Proceedings. XXI, 413 pages. 1993.

Vol. 1541: D. A. Dawson, B. Maisonneuve, J. Spencer, Ecole d'Eté de Probabilités de Saint-Flour XXI - 1991. Editor: P. L. Hennequin. VIII, 356 pages. 1993.

Vol. 1542: J. Fröhlich, Th. Kerler, Quantum Groups, Quantum Categories and Quantum Field Theory. VII, 431 pages. 1993.

Vol. 1543: A. L. Dontchev, T. Zolezzi, Well-Posed Optimization Problems. XII, 421 pages. 1993.

Vol. 1544: M. Schürmann, White Noise on Bialgebras. VII, 146 pages. 1993.

Vol. 1545: J. Morgan, K. O'Grady, Differential Topology of Complex Surfaces. VIII, 224 pages. 1993.

Vol. 1546: V. V. Kalashnikov, V. M. Zolotarev (Eds.), Stability Problems for Stochastic Models. Proceedings, 1991. VIII, 229 pages. 1993.